Introductory Combinatorics

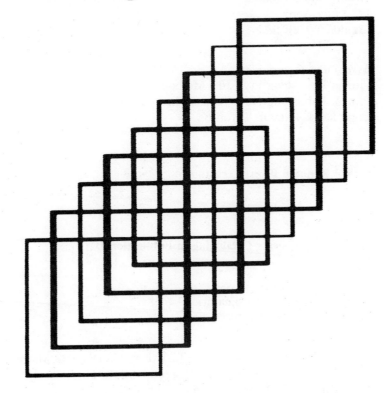

Kenneth P. Bogart

Dartmouth College

Department of
Mathematics and
Computer Science

Pitman

Boston · London · Melbourne · Toronto

Pitman Publishing Inc.
1020 Plain Street
Marshfield, Massachusetts 02050

Pitman Books Limited
128 Long Acre
London WC2E 9AN

Associated Companies
Pitman Publishing Pty Ltd., Melbourne
Pitman Publishing New Zealand Ltd., Wellington
Copp Clark Pitman, Toronto

Library of Congress Cataloging in Publication Data

Bogart, Kenneth P.
    Introductory combinatorics.

    Includes index.
    1. Combinatory analysis. I. Title. II. Discrete Mathematics.
III. Computer Science. IV. Discrete structure
QA164.B63   1983   511′.6   82-22392
ISBN 0-273-01923-6

Manufactured in the United States of America

10 9 8 7 6 5 4 3 2 1

Contents

4 Graph Theory   117

Preface

Discrete mathematics in general and combinatorial mathematics in particular have become an increasingly important part of the university mathematics curriculum. A clear reason for this can be found in the wealth of applications of combinatorial mathematics in computer science, operations research, statistics and some of the social and physical sciences. The rapid growth in the breadth and depth of the field of combinatorics in the last several decades, first in graph theory and designs, and more recently in enumeration and ordered sets, has led to a recognition of combinatorics as a field with which the aspiring mathematician should become familiar.

The purpose of this book is to present a broad and comprehensive survey of modern combinatorics at an introductory level. The book begins with an introduction of concepts fundamental to all branches of combinatorics in the context of combinatorial enumeration. The second chapter is devoted to enumeration problems that involve counting the number of equivalence classes of an equivalence relation, while the third chapter discusses somewhat less direct methods of enumeration, the principle of inclusion and exclusion and generating functions. The remainder of the book is devoted to a study of combinatorial structures. Chapters 4 and 5 provide a solid introduction to graph theory and some of its applications in computer science, operations research and other areas. Chapter 6 is an introduction to experimental design, covering Latin squares and block designs, and concluding with a discussion of projective planes intended to unify these topics. Chapter 7 introduces partially ordered sets from the points of view of various applications and concludes with Möbius inversion as an example of the interplay between combinatorial structures and the theory of enumeration.

In choosing material for this book I have aimed for core material of value to students in a wide variety of fields: mathematics, computer science, statistics, operations research and the physical and behavioral sciences. A course from this book might be taught in departments of mathematics, computer science and statistics, so I have tried to write a book that can be taught by a mathematician or statistician with a traditional background to a varied audience as well as by a computer scientist to a group consisting primarily of computer science students. A course in discrete mathematics could easily include material that is covered in courses in algebra, analyses of algorithms, data structures, digital circuit design, logic, probability and the statistical design of experiments. Including a great deal of material from these areas would interfere with these well established courses and lead away from combinatorics. Thus, I have consciously attempted only to introduce ideas from these areas rather than devote significant effort to them. Prior study of this book should enrich the student's experience in any of these courses. Because the book includes only introductory study of subjects in computer science (such as the analysis of algorithms) I have been able to write it with the style and flavor of a traditional mathematics textbook. On the other hand, I have also tried to present the tantalizing algorithmic "aroma" that makes combinatorics so important in computer science.

As any author who has written a book in a developing field of mathematics will attest, it is all too tempting to introduce a subject by giving the fundamental definitions, discussing a few examples and then stopping. To the reader the result will appear as a disjointed list of ideas with no real substance. To overcome this problem, an author may try to connect the definitions up with relevant (and useful) theorems. The result may then appear to a student as a futile exercise in abstract mathematics. Because of these possible pitfalls, I have decided to introduce a subject only if it has a major application or has inherent mathematical interest, and only if this application or interest can be explained in elementary terms. Virtually every section begins with a problem to be solved, develops tools to solve it and give examples of how these tools are used. At times these motivating problems and examples are intentionally unsophisticated (e.g., distributing candy to children), so that the student does not carry the double burden of learning a sophisticated application and a new mathematical idea at the same time. As much as possible, definitions are introduced so as to name an idea that has come up in a motivating problem or example. Similarly, theorems typically arise as answers to questions suggested by motivating problems or examples. Proofs of theorems are written as informal explanations. Unless there is a clear reference to it from elsewhere, each proof is written so that the reader may skip it. Thus, the instructor or reader who chooses to de–emphasize proofs should have no difficulty. On the other hand, proofs—especially those using induction—have been written carefully. The proofs can serve as models in courses intended in part to develop the student's mathematical sophistication. Proofs or topics which are either overly complex or highly technical have been indicated with an asterisk; there is no harm in passing over such material. Exercises which depend on such material or are especially difficult are also marked with asterisks. The end of a proof or example is marked by a solid square (■).

This book should be useful in a variety of courses. The Committee on the Undergraduate Program in Mathematics of the Mathematical Association of America has recently recommended a course in discrete mathematics as an option to follow the calculus sequence. It is straightforward to design a course covering their outline from this book; in fact most of the possible supplementary topics they recommend are also covered in this book. By covering Chapters 1–3 and the Appendix on matrix algebra in one term, leaving Chapters 4–7 for the second term, it is possible to teach a two–term discrete mathematics course from this book as well. Such a course could be profitably followed by courses in probability and linear algebra. The standard course in discrete structures in computer science typically covers the material in Chapter 1, the first two sections of Chapter 2, Chapter 4 with the possible exception of planarity and coloring, and the first four sections of Chapter 7. Both Chapter 3 and Chapter 5 could be used in such a course as a supplement in order to provide a background for future work in the analysis of algorithms. At Dartmouth, we have used earlier versions of the book for an upper-division course. The prerequisites for this course are freshman calculus and either finite mathematics or a term of multivariable mathematics that includes matrix algebra. Our audience has consisted of roughly equal numbers of students majoring in mathematics, computer science and other subjects. Although virtually all the material in the book has been taught at one time or another, there is more material here than can be covered in one semester or quarter at this level. Typically, we have covered a selection of material from at least six of the seven chapters. We have found it possible to assign Chapter 6 along with an appropriate selection of exercises chosen by the student as a term project. Students are pleased to discover that they have learned how to teach themselves mathematics. Since Chapter 6 depends only on Chapter 1 and the first two sections of Chapter 2, it provides a basis for a project students can begin early in the term.

Although there is a natural progression in sophistication as one moves through the book, I have designed it with a minimum of dependence between chapters. Chapter 1 and the first two sections of Chapter 2 are fundamental to the rest of the book. Generally, each section in a chapter assumes familiarity with the preceding sections, though the material on planarity, coloring and orientability is not used in subsequent sections of Chapter 4, and the material on Boolean algebras is not used in subsequent sections of Chapter 7. An outline of the other dependencies of sections on earlier sections and prerequisite material is shown in Table 0.1. Some exercises do not follow the pattern of the table; however both the statement of the exercise and the solutions given in the answer key should make such dependencies clear. The answer key is available to instructors from Pitman Publishing Inc.

Many people have had an impact on this book as it has developed over a number of years. Several of my classes in combinatorics have cheerfully served as guinea pigs and offered valuable comments. Among the students in these classes, Drew Golfin deserves special thanks. Fred McMorris used an early version of the book in an experimental course at Bowling Green State University; his advice and experience have had a significant impact on the book. Robert Norman taught from the next to last draft of the book at Dartmouth; the enthusiastic advice and encouragement I received from him and his students has also shaped the final

SECTION	DEPENDS DIRECTLY ON	COULD BENEFIT FROM
3.1		2.3
3.2–3.5	power series	2.4, derivatives
3.5	exponential function	2.3
4.2		3.2–3.3
4.7	matrix products matrix sums	*determinants
5.1	4.1	4.2, 4.3
5.2	4.2	
5.3–5.4	4.1, 4.5	4.2
6.2	matrix products transposes	determinants
6.4		vector subspaces
7.1	4.1, 4.2, 4.5	
7.2		5.3, 5.4
7.5–7	matrix inverses and systems of equations	

Table 0.1

* only asterisked subsection uses determinants

product significantly. The reviewers of the manuscript in various stages of completion, some of them anonymous, have also influenced the book in many ways. Agnes Chan, Jay Goldman, Eugene Lawler, Steven Maurer and Anthony Ralston made special contributions to the final draft.

Almost every author stands on the shoulders of those who have written before, and in this I am no exception. All the books in the suggested reading lists are books I have found valuable for my courses in combinatorics. In particular, without what I have learned from the books and articles of, or personal associations with, Claude Berge, Robert Dilworth, Shimon Even, D. Ray Fulkerson, Curtis Greene, Jay Goldman, Marshall Hall Jr., Frank Harary, László Lovász, Robert Norman, Gian–Carlo Rota, Herbert Ryser, J. Laurie Snell, and Herbert Wilf, this book would not be what it is, and probably would not exist at all.

Finally, one person has invested a great deal of time and effort in this book over the last few years. Marie Slack typed the manuscript into our computer and has used the computer to make seemingly endless revisions to prepare class notes and the final draft. Her sixth sense for what is right and wrong, her cheerful and careful work on revisions and her dedication to the project have made it possible for me to experiment with new ideas and make changes to a much greater extent than I otherwise could have. In thanks for her efforts, this book is dedicated to the staff of the mathematics department at Dartmouth College.

1 An Introduction to Enumeration

SECTION 1   ELEMENTARY COUNTING PRINCIPLES

What is Combinatorics?

In combinatorial mathematics, we study arrangements of objects. The objects might be minicomputers; the arrangement might be a network connecting them. The objects might be headache relief tablets. They might be arranged into groups for testing on patients. Usually, the set of objects we wish to arrange is finite or the objects themselves are arranged among a finite number of sets. The problems studied are quite varied. Often, more than one kind of problem arises from a given application. One may ask if there are *any* arrangements at all satisfying certain conditions, or one may ask *how many* arrangements are there that satisfy these conditions. One may ask if all arrangements satisfying a given set of conditions must also satisfy certain additional conditions. Finally, one might ask for an efficient way to generate one arrangement or all arrangements satisfying certain conditions. Frequently, in combinatorial mathematics we attempt to represent an arrangement in concrete, but mathematical, terms in which we may analyze our problems more easily.

The Sum Principle

We begin our study of combinatorial mathematics with elementary principles that show us how to count the number of arrangements that satisfy certain conditions, or lie in a certain set of arrangements built up from other sets. Suppose we have two sets A and B of objects (e.g., arrangements). Their *union*, $A \cup B$, is the set of objects in A or B, or in both. Their intersection, $A \cap B$, is the set of objects in both A and B. The union of a collection of sets is the set of objects *in at least one of the original sets;* the intersection of a collection is the set of objects *in each set of the collection.* Probably the most elementary "counting principle" deals with the union of a collection of sets no two of which have any elements in common. (In other words, the intersection of every pair of sets has no elements. We shall

1

say the sets in such a collection are mutually disjoint.) The *sum principle* tells us
that *the number of elements in the union of a finite collection of mutually disjoint
sets is the sum of the numbers of elements in each of the sets.* The illustrations in
Figure 1.1 are called Venn diagrams and show where the principle does and
doesn't apply.

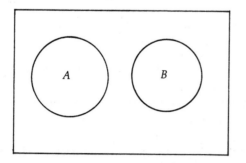 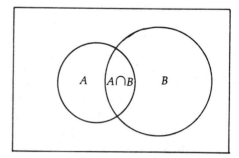

Figure 1.1 (a) The circles represent sets A and B chosen from some bigger set represented by the rectangle. Since the circles don't overlap, the sum principle applies.
(b) Since the circles representing A and B overlap nontrivially, the size of $A \cup B$
is not the sum of the sizes of A and B.

Example 1.1.   The symbols that can be typed on a standard typewriter include the eight digits (2–9), 26 lower-case letters, 26 upper-case letters and 24
punctuation and special symbols. What is the total number of symbols available?
It is obvious that the total number is $8 + 26 + 26 + 24 = 84$. ∎

By itself, the principle is so obvious as to be trivial. However, sometimes
in the solution of a major problem, you can get started by asking yourself if the
sum principle applies. In other words, you divide the problem into cases, count
the possibilities in each case and add the results. Exercise 6 at the end of this section is an example where breaking a problem into cases is helpful.

At times, it will be useful to have a standarized notation when we use the
sum principle. If we have two sets A and B, we use the symbol $A \cup B$ to stand for
their union. For any set S, we use $|S|$ to stand for the size of S. Thus for two sets
A and B, the sum principle says that if A and B are disjoint,

$$|A \cup B| = |A| + |B|.$$

To deal with an arbitrary collection $A_1, A_2, \ldots, A_n$ of sets, we use either the
notation

$$\bigcup_{i=1}^{n} A_i \text{ or } A_1 \cup A_2 \cup \cdots \cup A_n$$

to stand for the union of the sets. Thus, for n sets $A_1, A_2, \ldots, A_n$ which are mutually disjoint, we can write either

$$|A_1 \cup A_2 \cup \cdots \cup A_n| = |A_1| + |A_2| + \cdots + |A_n|$$

or

$$\left| \bigcup_{i=1}^{n} A_i \right| = \sum_{i=1}^{n} |A_i|.$$

We read the second equation as "the size of the union from $i = 1$ to n of A_i is the sum from $i = 1$ to n of the size of A_i."

     Example 1.2. Suppose a club has 15 members and must choose a president and secretary–treasurer from among its members. In how many ways may it choose its officers?

     We may regard a list of two names of club members as a choice first of a president and then of a secretary–treasurer. Since these lists are the arrangements we are interested in, we now ask how many such lists of two different club members there are. It may be obvious to you that there are $15 \cdot 14 = 210$ such lists. If so, you are right. This result also follows from an application of the sum principle. Suppose we divide the lists into piles (sets) such that each pile has all the lists with a given person as president and no other lists. Then we have 15 piles, one for each presidential candidate. Each pile has 14 lists, because once we choose a president, there are 14 people left for the second place on the list. Now, no two piles have any lists in common, so the number of elements in the union of the piles is the sum of the number of elements in each pile. Thus we add 14 to itself 15 times to get a total of $15 \cdot 14 = 210$ lists in the union of all the piles. ∎

The Multiplication Principle

The example suggests a second "counting principle," called the *multiplication principle*. One abstract statement of the multiplication principle is that a union of m disjoint sets each of size n has $m \cdot n$ elements. Clearly, we can *prove* this principle by applying the sum principle as above. Another more concrete statement of the multiplication principle suggested by our example says that *if the lists in a set of two element lists have m possible first elements, and if for each first ele-*

ment, n different second elements are paired with it in lists in the set, then the set has m·n lists.

This concrete version of the principle takes more words than the abstract version, but it suggests a question we should ask: what if the lists have three or four elements? Does some kind of multiplication principle apply?

Example 1.3. If the club has to choose a president, a secretary, and a treasurer, then we would want to count three–element lists. If the club needs a vice–president too, then the lists we would count are four–element lists. Our intuition suggests that the number of lists of four different people chosen from among the 15 club members is $15 \cdot 14 \cdot 13 \cdot 12$. ∎

This example suggests that a more general form of the multiplication principle could deal with lists of length k. Then the principle we already know would be the special case for $k = 2$. The general form of the multiplication principle can be stated as follows:

If the lists in a set S of lists of length k have the properties that

(1) there are m_1 different first elements of lists in S;
(2) for each way of specifying the first $i - 1$ entries of a list, there are m_i ways to   specify the i–th entry in the list;

then S contains $m_1 \cdot m_2 \cdot \ldots \cdot m_k$ lists. (Later on we shall use the symbol $\prod\limits_{i=1}^{k} m_i$ for the product from $i = 1$ to k of m_i.)

Since, in Example 1.3, there are 15 different choices for president; and given a choice of president, there are 14 choices of vice–president; and once we have chosen a president and vice–president, there are 13 choices of secretary; and once we have chosen a president, vice–president and secretary, there are 12 choices for treasurer; then the multiplication principle tells us that there are $15 \cdot 14 \cdot 13 \cdot 12$ ways to choose the four officers in the club.

Lists with Distinct Elements

There are some useful formulas we can derive from the multiplication principle that relieve us of the need to appeal to it to justify trivial computations. The symbol $n!$ is called n factorial and stands for the product of the first n positive integers; that is

$$n! = n(n - 1) \ldots 2 \cdot 1.$$

The symbol $0!$ is taken to mean the number 1.

Theorem 1.1. The number of ways of listing all the elements of an n-element set (without repeating any elements) is $n!$

Theorem 1.1 does not require a formal proof because it is an immediate result of the multiplication principle.

A list of all the elements of a set (without repeats) is sometimes called a *permutation* of the set. A list of k distinct elements chosen from an n-element set S is sometimes referred to as a permutation of k elements chosen from S. The number of such lists is then referred to as "the number of permutations of n things taken k at a time." The symbol $(n)_k$, called a "falling factorial" or "factorial power," is sometimes used to stand for $\dfrac{n!}{(n-k)!}$. $P(n, k)$ is another symbol for $\dfrac{n!}{(n-k)!}$, because this quantity is the number of permutations of n things taken k at a time. This is the meaning of the next theorem.

Theorem 1.2. The number of lists of k distinct elements of an n-element set is $n!/(n-k)! = (n)_k = n(n-1) \ldots (n-k+1)$ (for $k \le n$).

Proof. The proof is a natural application of the multiplication principle. If $k = 1$, the number of lists in question is simply the number of elements in the set, which is $n = \dfrac{n!}{(n-1)!}$. For an arbitrary number k, there are n choices for the first element in the list, $n-1$ choices for the second, $n-2$ choices for the third, and so on down to $n-k+1$ choices for the k-th position. Thus, by the multiplication principle, there are

$$n(n-1) \cdot (n-2) \ldots (n-k+1) = n!/(n-k)!$$

lists. ∎

The proof given above is somewhat distasteful because of the "and so on," especially since it comes at the point which requires the most thought, namely the reference to $n-k+1$. By learning the principle of mathematical induction and becoming comfortable with it, we may replace such vague "and so on"'s with clearer explanations. A later section of this chapter will be devoted to this principle, and will derive the most general form of the multiplication principle from the simpler version. For the remainder of this section, we will concentrate on applications of the principles we already have.

Lists with Repeats Allowed

Sometimes we will be interested in lists which are allowed to have repeat entries.

Example 1.4.   The central processing unit of a computer must retrieve four files from disk storage units. Each file may be stored on any of the five different disk storage units connected to the central processing unit. In how many ways may the files be located on the disk storage units?

A description of where the files are located is a list which tells us which of the units has file 1, which has file 2, which has file 3 and which has file 4. Since there are five possibilities for each entry in the list, there are $5 \cdot 5 \cdot 5 \cdot 5 = 625$ possible arrangements of the files on the disk drives. ■

Theorem 1.3.   The number of lists of length k, each of whose entries is chosen from a set with m elements (with repeats allowed) is m^k.

Proof.   For each position in the list, there are m choices, so the multiplication principle tells us there are m^k lists. ■

EXERCISES

1. A city council with seven members must elect a mayor and vice–mayor from among its members. In how many ways may these officers be chosen?

2. A local restaurant offers a meat and cheese sandwich. You can choose one of three kinds of bread, one of four kinds of meat, and one of three kinds of cheese. How many sandwiches are possible?

3. A 16-bit computer word is a list of 16 symbols, each a 0 or 1. How many such 16–bit computer words are there?

4. A hamburger shop offers you a "personalized hamburger." You may choose to have any of the following on your hamburger: lettuce, tomato, cheese, pickles, special sauce, catsup, mustard, onions. How many different kinds of personalized hamburgers can you choose? (Hint: Try to use a list of 0's and 1's to describe the toppings on a sandwich.)

5. A coffee machine allows you to choose your coffee either plain or with single–or double–portions of sugar and/or cream. In how many ways may you choose your coffee?

6. In order to hike along trails from Bing Mountain to South Mountain, you must either cross Joe's Peak or go through Narrow's Pass and cross Greene Mountain. There are two trails each from Bing Mountain to Joe's Peak and Narrow's Pass. There are three trails from Narrow's Pass to Greene Mountain, two trails from Greene Mountain to South Mountain and two trails from Joe's Peak to South Mountain. In how many ways may you plan a hike from Bing Mountain to South Mountain?

7. A local ice cream shop offers the following banana split. You get three scoops of ice cream, each having any of eight flavors. You get your choice of four toppings, whipped cream if you want it and your choice of shredded coconut, chopped peanuts or no nuts. All this sits on a split banana. How many different banana splits are possible?

8. A computer–aided instruction program asks a student each question on a list of 20 different questions. For each question, there are ten choices for the answer. If the student answers a question correctly, the program says so. If the student gets the answer wrong, the program gives the student a second chance. If the answer is wrong the second time, the computer types the correct answer with a brief explanation and, in any case, goes on to the next question.

(a) If we watch the process, observing only the computer's responses, how many different patterns could we conceivably see?

(b) If we watch the process observing both the computer's responses *and* the student's responses, how many outcomes could we conceivably see? (Note that this is considerably more complex than (a) unless you first divide the set of possible responses into those with a correct answer on the first try, those with a correct answer on the second try, and those with two incorrect answers. This is the point of the earlier remark in the text that if you *search* for a way to apply the sum principle, you can make an apparently complex problem manageable.)

9. A professor has six test questions. Three of them form a unit and must be kept together in the order the professor has chosen. In how many ways may the professor arrange the test questions?

10. Answer the question of Exercise 9 if the three questions kept together may be answered in any order.

11. Three reporters from the school newspaper and three reporters from the school radio station form the panel for a discussion. In how many ways they be seated in a row behind a table if

(a) Each group must sit together?

(b) No two members of a group may sit together?

12. Use $Y - X$ to denote the set of elements of Y not in X. If every element of the set X is an element of the set Y, use the sum principle to explain why

$$|Y - X| = |Y| - |X|.$$

13. The intersection of two sets A and B, denoted by $A \cap B$, is the set of elements they have in common, and their union $A \cup B$ is the set of all elements in one or the other set or in both of them. Explain why

$$|A \cup B| = |A| + |B| - |A \cap B|.$$

(Hint: You can find three disjoint sets whose union is $A \cup B$ by looking at part b of Figure 1.1. Problem 12 may be applied to compute the size of two of these sets.)

14. Stirling showed that $n!$ is approximately $\sqrt{2\pi n}\, n^n e^{-n}$. Using a computer or scientific calculator, determine the ratio of $n!$ to this approximation for values of n equal to multiples of 10 up to 50.

SECTION 2   ORDERED PAIRS, RELATIONS AND FUNCTIONS

Ordered Pairs

A list with two elements is normally called an *ordered pair;* if the first element of the list is a and the second element is b, we write (a, b) as the symbol for the ordered pair.   A list of n objects is sometimes called an n–tuple.   Just as we can represent selections of club officers or disk storage units as sets of lists, we can represent many practical or mathematical ideas as sets of lists, frequently as sets of ordered pairs.

Example 2.1.   A delicatessen offers a "simple sandwich" consisting of your choice of whole wheat, rye or white bread filled with corned beef, roast beef, ham or turkey.   How many different simple sandwiches are possible?

Each sandwich may be represented as an ordered pair (bread, meat) with three choices for the first entry and four choices for the second.   By the multiplication principle, there are 12 sandwiches possible. ■

With the same computation, we can prove another formula.

Theorem 2.1.   If the set M has m elements and the set N has n elements, then there are mn ordered pairs whose first entry is in M and whose second entry is in N.

Cartesian Product of Sets

The set of all ordered pairs whose first entry is in M and whose second entry is in N is sometimes called the *cartesian product* of M and N and is denoted by $M \times N$.   This notation is suggestive, since the size of $M \times N$ is mn.   Once again we use $|S|$ to stand for the number of elements of a set S.   In this notation, Theorem 2.1 says that $|M \times N| = |M| \cdot |N|$.

Functions

We can also use ordered pairs to give combinatorial descriptions of functions. Typically, you will see the phrase "f is a function from A to B" defined by the statement, "f is a rule which associates with each element x of A a unique element $f(x)$ of B."   For most purposes, this definition is entirely adequate.   However, if we ask ourselves, "How do we tell whether two functions are really different?", we see a small problem.

The rule $f(x) = x + 2$ clearly describes a function from $\{-1, 0, 1, 2\}$ to $\{1, 2, 3, 4\}$.   The rule given by $g(x) = x^4 - 2x^3 - x^2 + 3x + 2$ is also a rule for a function from $\{-1, 0, 1, 2\}$ to $\{1, 2, 3, 4\}$.   In fact, $f(x) = g(x)$ for each x in the domain set $\{-1, 0, 1, 2\}$.   Thus we have two apparently different rules which describe the same function, and not two different functions.   If we agreed upon a standardized way of stating the relationship the rule specifies, then we could say that two functions are the same if they describe the same relationship.   One way to describe the relationship f specifies is to write down the set $\{(-1, f(-1)), (0, f(0)), (1, f(1)), (2, f(2))\}$ of all ordered pairs whose first entry is a domain element and whose second entry is the related range element which the function rule assigns to the first entry.   For f we get

$$f = \{(-1, 1), (0, 2), (1, 3), (2, 4)\}$$

and for g we get

$$g = \{(-1, 1), (0, 2), (1, 3), (2, 4)\}.$$

In other words, by using ordered pairs we have found an unambiguous way to describe functions.   Some mathematicians call the set of ordered pairs associated with a function the *graph* of this function.   In combinatorial mathematics, we have another concept of graph, and so to avoid confusion we will call the set of ordered pairs associated with some function the *relation* (short for relationship) of the function.   A way to visualize this relation is shown in Figure 1.2.

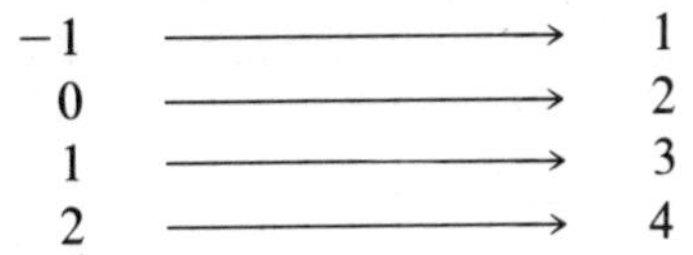

```
−1  ─────────────────▶  1
 0  ─────────────────▶  2
 1  ─────────────────▶  3
 2  ─────────────────▶  4
```

Figure 1.2 A way of visualizing the relation of the function given by $f(x) = x + 2$.

The idea of using a set of ordered pairs to describe a relationship is useful in many situations, so we will try to make this idea more specific.

Relations

We define a *relation from a set A to a set B* to be a set of ordered pairs whose first entries are in A and whose second entries are in B. A is called the *domain* of the relation and B is called the *range*.   If (a, b) is an ordered pair in a relation, we say that the relation *associates* b with a.

Example 2.2.   The set $S = \{(1, 2), (1, 3), (1, 4), (2, 3), (2, 4), (3, 4)\}$ is a familiar relation.   What is the usual description of the relation and what are its domain and range?

An ordered pair (a, b) is in S if and only if $a < b$.   Thus we would ordinarily say S is the "less than" relation.   There is more than one answer to the question about domain and range.   For example, we could agree to let the domain A and range B both be the set $\{1, 2, 3, 4\}$.   However, we could use $\{1, 2, 3\}$ for A and

{2, 3, 4} for B without changing our informal description that S is the "less than" relation. Figure 1.3 shows how to visualize this relation.

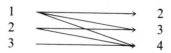

Figure 1.3 A way of visualizing the relation of Example 2.

Definition of Function

Now we *define* a *function from A to B* to be a relation from A to B that associates with each element a of A a unique element b of B. Can you find reasons why the relation of Example 2.2 is not a function? This definition of a function has the advantage that if we use just our intuitive understanding that the word "relation" (or "relationship") means about the same as the word "rule," then our definition of function is the same as the definition we gave using the word "rule." If, on the other hand, we use the definition of a relation as a set of ordered pairs, we have a very precise and unambiguous definition of a function. With this definition, it makes perfectly good sense to ask how many functions there are from a finite set A to a finite set B.

The Number of Functions

    Theorem 2.2. If A has n elements and B has m elements, then there are m^n functions from A to B.

    Proof. List the elements of A in some order, say $a_1, a_2, \ldots, a_n$. Then each function f gives a list of n elements chosen from B, namely $f(a_1)$, $f(a_2), \ldots, f(a_n)$. Also each list of n elements chosen from B, say $b_1, b_2, \ldots, b_n$ gives a function, namely the one whose relation is

$$\{(a_1, b_1), (a_2, b_2), \ldots, (a_n, b_n)\}.$$

Clearly, different lists give different functions, so the number of functions from A to B is the same as the number of lists of n elements chosen from B (with repeats allowed). By Theorem 1.3, there are m^n such lists, thus there are m^n functions from A to B. ∎

    Example 2.3. You have five different chairs to paint and four different colors of paint. In how many ways can you paint the chairs?
    Assigning a color of paint to each chair gives a function from the set of

chairs to the set of paints. Different functions yield different paint jobs. Thus, by Theorem 2.2, there are $4^5 = 1024$ possible paint jobs. ∎

One–to–One Functions

A function f is called a *one–to–one function* if it associates *different* values in the range with different values in the domain. Figure 2.1 may help you visualize the concept of a one–to–one function. Note how each arrow points to a different number. None of the paint jobs in Example 2.3 corresponds to a one–to–one function because with five chairs and four colors of paint, two chairs will have to have the same color. This is one example of the *pigeon–hole principle*. Informally, the principle says that if more than n pigeons are to be placed in n pigeon holes, one hole will end up with more than one pigeon. Another way to say the same thing is that there are no one–to–one functions from an m–element set to an n–element set if $m > n$.

> **Theorem 2.3.** If $n \leq m$, there are $(m)_n = \dfrac{m!}{(m-n)!}$ one–to–one functions from an n–element set A to an m–element set B.
>
> *Proof.* As in Theorem 2.2, we may associate lists of m elements chosen from B with functions from A to B. To say that some function is one–to–one is to say that the associated list is a list of distinct elements of B. By Theorem 1.2, the number of lists is $\dfrac{m!}{(m-n)!} = (m)_n$. ∎

Example 2.4. If five different chairs are to be painted different colors and seven colors are available, how many ways are there to paint the chairs?

Since the paint jobs correspond to one–to–one functions, there are $7!/(7-5)! = 7!/2! = 7 \cdot 6 \cdot 5 \cdot 4 \cdot 3 = 2520$ possible paint jobs. ∎

Other Kinds of Functions and Relations

A function f from A *onto* B is defined as function from A to B such that each member of B is associated with at least one member of A. As Exercise 14 illustrates, the computation of the number of functions from A onto B is more involved than the kinds of computations we have made so far. A one–to–one function from a set onto itself is usually called a *permutation* of that set. In Section 1 we defined a permutation differently. Given a set listed in one way, say (a, b, c, d), a second list such as (b, c, a, d) determines a one–to–one function from the set onto itself; in this case $\{(a, b), (b, c), (c, a), (d, d)\}$. Thus, the two uses of the word permutation are closely related. It is necessary to determine from the context whether a particular use of the word ''permutation'' refers to a listing of a set or a one–to–one function from a set onto itself.

Just as two objects may or may not be in some relationship with each other, we might have a mutual relationship involving three or more objects. For example, we might have a relationship among students, courses and professors given by "student X takes course Y from professor Z." A natural way to represent this relation is as a set of lists (X, Y, Z), that is a set of 3–tuples, where (X, Y, Z) is in the set if and only if X takes Y from Z. Here there is no natural idea of going from one set to another, so we can't distinguish between domain and range. It has become traditional, especially in computer science, to define an *n–ary relation with domains* $A_1, A_2, \ldots, A_n$ to be a set of n–tuples in which the first entry of a tuple lies in A_1, the second lies in A_2, and so on. In this context, the relations we have defined in terms of ordered pairs are called *binary relations*.

EXERCISES

1. Write down all the sets of ordered pairs that correspond to possible functions from $\{0, 1\}$ to $\{1, 2, 3\}$.

2. Write down all the sets of ordered pairs that correspond to possible functions from $\{1, 2, 3\}$ to $\{0, 1\}$.

3. Write down all the sets of ordered pairs that correspond to possible relations from the set $\{0, 1\}$ to the set $\{a, b\}$.

4. Write down the set of ordered pairs that corresponds to the "greater than" relation from the set $\{1, 2, 3, 4\}$ to the set $\{1, 2, 3, 4\}$.

5. Let $f(x) = x^2 - 1$. Write down the set of ordered pairs that is the relation f if the domain of f is $\{1, 2, 3, 4, 5\}$.

6. Do Exercise 5 with domain $\{-3, -2, -1, 0, 1, 2, 3\}$.

7. Draw pictures like those in Figures 2.1 and 2.2 to represent the functions of the last two problems. Which of these functions is one–to–one?

8. Which of the following relations are functions and why?
   (a) $\{(a, 1), (b, 2), (c, 1)\}$
   (b) $\{(1, a), (2, b), (3, b), (1, c)\}$
   (c) $\{(-1, 1), (0, 0), (1, 1), (2, 4), (-2, 4)\}$
   (d) $\{(1, -1), (0, 0), (1, 1), (4, 2), (4, -2)\}$
   (e) $\{(0, 0), (1, 1), (4, 2)\}$

9. On the set of integers from 1 to 12 inclusive, there is a relation containing (a, b) if a is a factor of b (but not b itself). Write down the set of ordered pairs corresponding to this relation.

10. How many functions from a four–element set to a five–element set are not one–to–one? How many functions from a five–element set to a four–element set are not one–to–one?

11. A computer must assign each of four jobs to one of ten different slave computers.
   (a) In how many ways may it do so?
   (b) In how many ways may it make the assignments if no slave is to get more than one job?

12. A group organizing a faculty–student tennis match must match five faculty vol-

unteers with five of the 12 students who volunteered to be in the match. In how many ways may they do this?

13. The chairperson of a department has to assign advisors to three seniors from among ten faculty members in the department. How many different assignments are there? How many give no faculty member three advisees? How many give no faculty member two or three advisees?

14. (a) In how many ways may you pass out ten different pieces of candy to three children?

(b) What if each child must get one piece? (Hint: To answer this question, ask yourself in how many distributions does one child get candy? In how many distributions do exactly two children get candy?)

15. A department has three freshmen courses, course A (algebra–trigonometry), course C (calculus), and course F (finite mathematics). For each course there are three possible textbooks that have departmental approval. Five professors are willing to be in charge of A, six professors are willing to be in charge of C, and three professors are willing to be in charge of course F. Assuming only one professor can be in charge of a course, how many triples (3–tuples) are there in the relation: professor X is in charge of course Y and uses textbook Z? How many such relations are possible?

16. Using the pigeon–hole principle, show that in a list of $n^2 + 1$ numbers, there are either $n + 1$ numbers (not necessarily consecutive) in increasing order or $n + 1$ numbers in decreasing order. (For example, in the list 1, 5, 3, 4, 2 we have both the increasing list 1, 3, 4 and the decreasing lists 5, 4, 2 and 5, 3, 2.)

SECTION 3   SUBSETS

The Number of Subsets of a Set

We say a set A is a subset of a set B (written $A \subseteq B$) if each element of A is an element of B. A typical way to visualize the subset relation is shown in Figure 1.4. It is convenient to imagine a set called the *empty set,* denoted by $\emptyset$, which has no elements whatsoever. We say that $\emptyset \subseteq A$ for every set A. To help you visualize

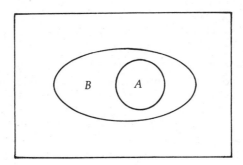

Figure 1.4 A Venn Diagram showing a subset A of a set B.

the empty set, suppose that the law requires the energy committee of the U.S. Senate to have an oil shale subcommittee, but the head of the energy committee refuses to appoint anyone to the subcommittee. Then the set of members of the subcommittee is the empty set.

Theorem 3.1.   The number of subsets of an n–element set is 2^n.
Proof.   Let the n–element set be given by $N = \{1, 2, \ldots, n\}$. To decide on a subset A, for each element of N we decide whether or not it is in A. This gives a list of n choices, each choice being yes or no. By the multiplication principle, there are 2^n such lists. ∎

Subsets and functions are intimately related. For each subset A of $N = \{1, 2, \ldots, n\}$, we define the function f_A by

$$f_A(x) = \begin{cases} 1 \text{ if } x \in A \\ 0 \text{ if } x \notin A. \end{cases}$$

Note that two different subsets A and B will have different functions f_A and f_B. (These functions are called the characteristic functions of A and B.) Thus each subset A of N determines a function from N into $\{0, 1\}$.

Now suppose f is a function from N to $\{0, 1\}$. Let A be the set of all x in N such that $f(x) = 1$. Then $f = f_A$. Thus, every function from N to $\{0, 1\}$ is an f_A for some subset A of N. This gives us a second proof that the number of subsets of N is 2^n because this is the number of functions from N to $\{0, 1\}$.

The Correspondence Principle

We just showed that each set in the collection of all subsets of N corresponds to exactly one function in the collection of all functions from N into $\{0, 1\}$. We say there is a "one–to–one correspondence" between the collections S and T if there is a one–to–one function from S onto T. (Such a function is sometimes called a *bijection*.) In Theorem 3.1 we used (without saying so) the "*correspondence principle*," which states that *if there is a one–to–one correspondence between two sets, then they have the same number of elements*. This is our third basic counting principle. It captures our intuitive understanding of what it means for two collections to have the same size. The first proof we gave for Theorem 3.1 used a one–to–one correspondence between subsets and lists of yesses and nos. We have used this principle earlier in this book and will use it in the future without explicitly referring to it.

Example 3.1.   A psychology professor asks for volunteers from a nine person class to participate in a perception experiment. How many different groups could the professor get?

Since any subset of the class (including the empty set!) might be the set of volunteers, there are $2^9 = 512$ possible experimental groups. ■

Frequently we will be interested in subsets of a set of a certain size.

Example 3.2. The professor in Example 3.1, realizing that there may be no volunteers, decides instead to choose three students and assign the experiment to them as a special project. How many different experimental groups can the professor choose?

First, the number of distinct lists of three different names that may be chosen from the names of all the nine students is $9!/6! = 9 \cdot 8 \cdot 7 = 504$. However, each three–element subset occurs in $3! = 6$ different lists. Thus if s is the number of sets of three names that may be chosen, $6 \cdot s$ is, by the multiplication principle, the number of lists. Therefore,

$$6 \cdot s = 504$$

so that $s = 504/6 = 84$ is the number of ways of choosing three names from among all nine possible names. ■

Binomial Coefficients

We use the symbol $\binom{n}{k}$ or $C(n; k)$ to stand for the number of k–element subsets of an n–element set. (It is standard to read either symbol as "n choose k".) The usual name for these quantities is *binomial coefficient;* the reason for the name will become apparent later. $C(n; k)$ is also called the "number of combinations of n things taken k at a time."

k–Element Subsets

Theorem 3.2. The number of k–element subsets of an n–element set is given by the formula

$$\binom{n}{k} = \frac{n!}{k!(n - k)!}$$

Proof. The set of all lists of k elements chosen from the n–element set contains $\dfrac{n!}{(n-k)!}$ lists. However, each k–element subset may be listed in $k!$ ways. The number of ways to first choose a k–element subset and to then list the elements of that subset is, by the multiplication principle,

$$\binom{n}{k} \cdot k!$$

However, each of these lists is a different list of k elements chosen from among all the n elements. Thus

$$\binom{n}{k} \cdot k! = \frac{n!}{(n-k)!}$$

which yields

$$\binom{n}{k} = \frac{n!}{k! \cdot (n-k)!} \quad \blacksquare$$

Labellings with Two Labels

Just as we found a useful generalization of the multiplication principle by re-interpreting ordered pairs as lists, we can find another interpretation of $\binom{n}{k}$ that will lead eventually to a useful generalization. Suppose we have a set of n elements and two kinds of labels; k labels of one kind and $(n-k)$ labels of the other kind. In how many ways may we assign these labels to the elements of the set so that each element is labelled?

Example 3.3. We have seven different chairs and we are supposed to paint two of them blue and five of them red. In how many different ways may we paint the chairs?

Once we select the two chairs to paint blue, we have determined the color of the rest of the chairs also, so the number of different paint jobs is the number of ways of choosing two objects out of seven, namely $C(7; 2) = 7!/(5! \cdot 2!) = 21$. Of course, we might have chosen the red chairs first, but we should end up with the same number of paint jobs anyhow. In fact, our formula says $C(7; 5) = 7!/(2! \cdot 5!) = 21$, just as it must. ∎

Two theorems are immediately suggested by Example 3.3; their proofs consist in picking out the ideas displayed in the example and are left as exercises.

Theorem 3.3.   The number of ways to label n objects with k labels of one kind and $n - k$ labels of a second kind is $\binom{n}{k}$.

A labelling of a set N with the labels 0 and 1 (or the labels "out" and "in") assigns the number 1 to k elements of N, and the number 0 to the other $n - k$ elements of N. Thus, the labelling gives us a function f from N to $\{0, 1\}$. This is the same as the "characteristic function" description given for subsets following the proof of Theorem 3.1. We shall examine labelling with more than two labels in a future section.

The other fact Example 3.3 suggests is that the number of labellings with k labels of type 1 and $n - k$ of type 2 must be the same as the number of labellings with k labels of type 2 and $n - k$ of type 1.

Theorem 3.4. $\binom{n}{k} = \binom{n}{n - k}$.

Pascal's Triangle

If we wanted to compute the number of 15–element subsets of a 25–element set by means of our formula, we would have to compute

$$\frac{25!}{15! \cdot 10!} = \frac{25 \cdot 24 \cdot \ldots \cdot 16}{10 \cdot 9 \cdot 8 \cdot 7 \cdot \ldots \cdot 1}.$$

The numerator in this fraction will have 10 factors whose average value is 20, so it would be "close" in size to

$$20^{10} = 2^{10} \cdot 10^{10} = 1024 \cdot 10^{10} \doteq 10^{13}.$$

(The symbol "$\doteq$" stands for "approximately equal".) The calculation of such a large product involves considerable effort, with considerable chance for error. In fact, a typical hand calculator would be of no use to us in such a computation, and storing a 13–place integer in some digital computers might require considerable ingenuity on our part. On the other hand, a rough estimate of $C(25; 15)$ can be obtained by replacing each term in both the numerator *and* the denominator by their average values. We get

$$C(25; 15) \doteq \frac{20^{10}}{5^{10}} = 4^{10} = 2^{20} = (1024)^2 \doteq 10^6.$$

Thus the number $C(25; 15)$, though large, is not itself out of the range of a hand calculator or a computer. This suggests that we should search for a method of computing the numbers $C(n; k)$ which could be used to build them up from smaller numbers. An array of numbers called *Pascal's triangle* gives us a method of computation. Table 1.1 was created by listing all nonzero values of $\binom{0}{k}$ in one row, next listing all nonzero values of $\binom{1}{k}$ in the next row, then all values of $\binom{2}{k}$ and so on, so each row starts right below where the last one started. Thus in row n and column k, we find the number $\binom{n}{k}$.

Row Number	Column Number 0	1	2	3	4	5	6
0	1						
1	1	1					
2	1	2	1				
3	1	3	3	1			
4	1	4	6	4	1		
5	1	5	10	10	5	1	
6	1	6	15	20	15	6	1

Table 1.1

Table 1.2, usually called Pascal's triangle, is formed from Table 1.1 by removing the row and column labels and arranging the rows so that their centers line up instead of their left–most entries. Table 1.2 has two important features.

```
              1
            1   1
          1   2   1
        1  3     3  1
      1 4    6     4 1
    1 5  10    10   5 1
  1 6 15   20   15 6 1
```

Table 1.2. Pascal's
Triangle

First, the left and right sides of the table consists entirely of 1's. This corresponds to the fact that $C(n; 0) = C(n; n) = 1$. Not quite so obvious is the fact that each entry not on either the left or right border is the sum of the two elements above and to the left and right of it. This suggests an important theorem about binomial coefficients. To understand what the theorem says, we first note that the corresponding observation about Table 1.1 is that each entry is the sum of a pair of entries in the row directly above it, with the first entry of the pair in the same column and the second in the preceding column. The k–th entry in row i is thus apparently the sum of entry $k - 1$ and entry k in the row above, which is a list of the binomial coefficients $C(i - 1, h)$. This suggests the relationship in Theorem 3.5.

Theorem 3.5. $\dbinom{i}{k} = \dbinom{i - 1}{k - 1} + \dbinom{i - 1}{k}$ whenever $0 < k < i$.

Proof.   The formula says that the the number of k–element subsets of an i–element set is the sum of the number of $(k - 1)$–element subsets *and* the number of k–element subsets of an $(i - 1)$–element set. This is what we now prove. Divide the k–element subsets of $I = \{1, 2, \ldots, i\}$ into two collections of sets; collection 1 consists of those subsets containing i; collection 2 consists of those subsets not containing i. By the sum principle, the number of k–element subsets is the sum of the sizes of collection 1 and collection 2.

Let $I' = \{1, 2, \ldots, i - 1\}$. Each set S in collection 1 yields a $(k - 1)$–element subset of I', the set $S - \{i\}$ consisting of the elements in S other than i. Similarly, each $(k - 1)$–element subset S' of I' yields a set in collection 1, namely $S' \cup \{i\}$. Thus the number of sets in collection 1 is the same as the number of $(k - 1)$–element subsets of I', an $(i - 1)$–element set.

Each set in collection 2 is a k–element subset of I', and each k–element subset of I' is in collection 2. Thus the number of sets in collection 2 is the number of k–element subsets of I', an $(i - 1)$–element set. Thus $C(i; k) = C(i - 1; k - 1) + C(i - 1; k)$. ∎

Example 3.4.   From Table 1.2, $C(6; 1) = 6$, $C(6; 2) = 15$, and $C(6; 3) = 20$. Thus $C(7; 2) = 6 + 15 = 21$, $C(7; 3) = 15 + 20 = 35$ and therefore, $C(8; 3) = 21 + 35 = 56$. ∎

Recursion and Iteration

The formula for computing the numbers $C(i; k)$ given in Theorem 3.5 is called a *recursion formula*, because we make use of the numbers $C(j; k)$ with j smaller than i in the computation. The regularity of the computation lets us develop an *iterative method* or *iterative algorithm* for computing row i of the Pascal triangle. First write down $C(1; 0) = C(0; 1) = 1$. Then apply Theorem 3.5 to get row 2. Repeat (or "iterate") this process until you reach row i. We shall use the phrase "iterative method" to describe a computation of a function that repeats a process again and again for increasing values of one or more parameters.

A Recursive Algorithm

Another way to make use of the recursion formula is by means of a recursive algorithm. The idea is to state directly that to compute $C(n; k)$ we first compute $C(n - 1, k - 1)$ and $C(n - 1, k)$ and add the results. Briefly speaking, to create such an algorithm we first decide on a name for it and then write a sequence of instructions such as the following:

```
PROCEDURE BINOM(N,K) /* ACCEPTS TWO INPUTS  N  AND  K */
IF  K=0  or  K=N  LET  B=1
OTHERWISE
LET  C = RESULT OF BINOM WITH  N-1  AND  K-1
LET  D = RESULT OF BINOM WITH  N-1  AND  K
LET  B = C+D
FINALLY MAKE  B  THE RESULT
```

Recursive methods like this look complicated at first view, but often can be used when iterative methods would be difficult to describe. On the other hand, recursive methods often are less efficient than iterative methods. The example below shows the steps we would go through if asked to execute the algorithm named BINOM with inputs of 3 and 2. Note how the symbols B, C and D stand for different things at different times.

```
EXECUTE BINOM(3,2)
2 ≠ 0,  2 ≠ 3
C = RESULT OF BINOM(2,1)
    EXECUTE BINOM(2,1)
    1 ≠ 0,  1 ≠ 2
    C = RESULT OF BINOM(1,0)
        EXECUTE BINOM(1,0)
        B = 1
    C = 1
    D = RESULT OF BINOM(1,1)
        EXECUTE BINOM(1,1)
        B = 1
    D = 1
    B = 2
C = 2
D = RESULT OF BINOM(2,2)
        EXECUTE BINOM(2,2)
        B = 1
    D = 1
    B = 3
```

EXERCISES

1. Suppose you have a final exam with eight questions. The instructions state "Choose six of the eight questions and answer them in the order given." In how many ways may you choose the questions to complete the exam?

2. Write out all subsets of the set $\{a, b, c\}$.

3. How many two–element subsets does the four–element set $\{A, B, C, D\}$ have? List them.

4. A candy store stocks 10 different kinds of chocolate candy. In how many ways may someone select a bag of six pieces of candy, each piece of a different kind?

5. A test has two sections of five questions each. The instructions say to answer three questions in each section. In how many ways may a student choose the questions to complete the exam?

6. Write down row 7 of Pascal's triangle (this means the row corresponding to subsets of a seven–element set).

7. Some rows of Pascal's triangle consist of even numbers with the exception of the two entries at the ends. Show that row 8 has this property. Do any rows beyond row 8 (i.e. with $n > 8$) have this property? Explain why not or give an example.

8. In how many ways may a three–person executive committee and a four–person administrative committee be chosen from a 20–member club if the committees may overlap? What if the membership of the committees may not overlap?

9. A test has three sections of four questions each. According to the instructions, the student must answer two questions from each of any two sections and one question from the other section. In how many ways may the student choose questions to answer?

10. A gift basket consists of two of four different kinds of cheeses, two of five different kinds of tinned meats, five of eight different kinds of fancy fruits, two of three different kinds of crackers and three of six different kinds of cookies. In how many ways may the gift basket be completed?

11. In how many ways may a two–, a three–, and a four–element set containing all the elements of a nine–element set be chosen?

12. In how many ways may the committees of Exercise 8 be chosen if they may have at most one member in common?

13. Prove Theorem 3.3.

14. Prove Theorem 3.4.

15. Locate several uses of the correspondence principle in Sections 1–3.

16. Write a brief paragraph outlining which values of $C(j; k)$ you *must* compute if you want to find $C(m; n)$ by the iterative method of Pascal's triangle. (For example, to compute $C(5; 2)$, you need not compute $C(4; 3)$ or $C(4; 4)$. The values of $C(j; k)$ you *do* have to compute include $C(4; 1)$, $C(4; 2)$, $C(3; 0)$, $C(3; 1)$, $C(3; 2)$ and some others. Try to make a general statement rather than just giving examples.)

17. Write a computer program that computes binomial coefficients by the iterative method of Pascal's triangle. Use your program to compute $C(30; 5)$ and $C(30; 15)$.

18. If you are familiar with a computer language that permits the use of recursive algorithms, attempt Exercise 17 by means of recursion. Compare the amounts of computer resources used by the two methods.

19. (a) Show the steps used by the algorithm BINOM on an input of 4 and 2.

   (b) Show the steps used by the algorithm BINOM on an input of 5 and 3. (You may shorten the description of each step to a bare minimum.)

(c) Note how, especially in part (b), some instructions of the form EXECUTE BINOM(N,K) are repeated again and again with the same N and K. Suggest an explanation of why this method is less efficient than an iterative method.

SECTION 4   BINOMIAL AND MULTINOMIAL THEOREMS

The Binomial Theorem

Our analysis of subsets and labellings makes it easy to prove a fundamental theorem of elementary algebra, the binomial theorem. It is because of their use in expanding the power of a binomial that the binomial coefficients get their name.

Theorem 4.1. $(x + y)^n = \sum_{i=0}^{n} C(n; i)x^i y^{n-i}$.

Proof. $(x + y)^n$ is a product of n factors, each equal to $x + y$. We multiply them together by repeatedly choosing one summand from each factor, multiplying the choices together, and then adding the results from all possible sequences of choices. For example, by choosing x from all the terms, we get x^n in the product. By choosing x in one term, and y in the rest, we get the product xy^{n-1}; this products occurs n times, once for each possible term from which we select x. Selecting x from k terms and y from the other $n - k$ terms amounts to labelling k of the terms with an x and $n - k$ of the terms with a y. Thus, there are $C(n; k)$ ways in which the product $x^k y^{n-k}$ occurs in the result. Thus the final product is

$$\sum_{k=0}^{n} C(n; k)x^k y^{n-k}. \quad \blacksquare$$

Note that the formula we get for $(x + y)^2$ from Theorem 4.1 is $(x + y)^2 = y^2 + 2xy + x^2$. Ordinarily we would write down $(x + y)^2 = x^2 + 2xy + y^2$. This suggests that an equally valid form of the binomial theorem is

$$(x + y)^n = \sum_{i=0}^{n} C(n; i)x^{n-i} y^i.$$

This may be proved by writing $x + y$ as $y + x$ and using the fact that $C(n; i) = C(n; n - i)$. This formula gives the expansion for $(x + y)^2$ in the order we are used to.

Example 4.1.

$$(2x + 3)^4 = \sum_{k=0}^{4} C(4; k)(2x)^k y^{4-k}$$
$$= C(4; 0)(2x)^0 3^4 + C(4; 1)(2x)^1 3^3$$
$$\quad + C(4; 2)(2x)^2 3^2 + C(4; 3)(2x)^3 3^1 + C(4; 4)(2x)^4$$
$$= 81 + 4 \cdot 2 \cdot 27x + 6 \cdot 4 \cdot 9x^2 + 4 \cdot 8 \cdot 3x^3 + 1 \cdot 16 \cdot x^4$$
$$= 81 + 216x + 216x^2 + 96x^3 + 16x^4. \ \blacksquare$$

Example 4.2.

$$(x - y)^6 = \sum_{k=0}^{6} C(6; k)x^k(-y)^{6-k}$$
$$= \sum_{k=0}^{6} C(6; k)(-1)^{6-k}x^k y^{6-k}$$
$$= y^6 - 6xy^5 + 15x^2y^4 - 20x^3y^3 + 15x^4y^2 - 6x^5y + x^6,$$

using the last row of the Pascal triangle, Table 1.2, rather than computing each number $C(6; k)$ individually. $\blacksquare$

Example 4.3.   Surprising formulas sometimes come from applying the binomial theorem to simple sums.   For example, let's see what happens when we expand $(2 - 1)^{10}$.

$$(2 - 1)^{10} = \sum_{k=0}^{10} C(10, k)2^k(-1)^{10-k}$$
$$1^{10} = \sum_{k=0}^{10} C(10, k)2^k(-1)^k(-1)^k(-1)^{10-k}$$
$$1 = (-1)^{10} \sum_{k=0}^{10} C(10, k)(-2)^k$$
$$1 = (-1)^{10} \left(\binom{10}{0} 2^0 - \binom{10}{1} 2^1 + \binom{10}{2} 2^2 - \cdots + \binom{10}{10} 2^{10} \right)$$
$$1 = (-1)^{10} \left(1 - 2 + \binom{10}{2} 4 - \cdots + 1024 \right),$$

giving us a rather strange looking way to write the number 1. $\blacksquare$

Multinomial Coefficients

We have seen the value of binomial coefficients in expanding a power of binomial. It is natural to ask whether a similar family of coefficients might prove helpful when we wish to expand a power of a "trinomial" $(x + y + z)$ or general "multinomial" $(x_1 + x_2 + \cdots + x_m)$. In the same vein, we might ask whether labellings with more than two labels might also be counted with numbers like binomial coefficients. In fact, these two questions have related answers.

Suppose we have n objects, k_1 labels of type 1, k_2 labels of type 2, . . . , and k_m labels of type m and suppose that $k_1 + k_2 + \cdots + k_m = n$. In how many ways may we assign the labels to the objects? Another way to phrase this question is as follows. We are given a set N with n elements and a second set $M = \{1, 2, \ldots, m\}$. How many functions f are there from N to M that map k_i elements of N to element i of M? We denote the number of labellings by

$$C(n; k_1, k_2, \ldots, k_m) \text{ or } \binom{n}{k_1, k_2, \ldots, k_m}.$$

Theorem 4.2. $\binom{n}{k_1, k_2, \ldots, k_m} = \dfrac{n!}{k_1! k_2! \ldots k_m!}.$

Proof. If f is a function from an n–element set N to $\{1, 2, \ldots m\} = M$, we use the notation $f^{-1}(\{i\})$ to denote the *set* of elements of N to which f assigns i. In terms of a labelling, $f^{-1}(\{i\})$ stands for the set of elements with a label of type i. We may specify a function f by specifying

$$f^{-1}(\{1\}), f^{-1}(\{2\}), \ldots f^{-1}(\{m\}),$$

(that is, by specifying which elements of N are related to i, or labelled with label i for each i). This specification gives a list of m sets, the first of which must have size k_1, the second size k_2, and so on. Of course, the function given by each such list is the type we desire, and so the number of functions in question is the same as the number of possible lists. We have $\binom{n}{k_1}$ choices for $f^{-1}(\{1\})$, and once we have chosen $f^{-1}(\{1\})$, we have $n - k_1$ elements left for the remaining choices. Once we have chosen the first i sets in the list, we have $n - k_1 - k_2 - \cdots - k_i$ elements left for our remaining choices; in particular, the number of choices for $f^{-1}(i + 1)$ is $C(n - k_1 - k_2 - \cdots - k_i; k_{i+1})$. Note that since $k_m = n - k_1 - \cdots - k_{m-1}$, the last binomial coefficient is $C(k_m; k_m)$. Thus, by the multiplication principle, the total number of lists is given by the formula

$$\binom{n}{k_1}\binom{n-k_1}{k_2}\binom{n-k_1-k_2}{k_3}\cdots\binom{n-k_1-k_2\cdots-k_{m-1}}{k_m}$$

$$=\frac{n!}{k_1!(n-k_1)!}\cdot\frac{(n-k_1)!}{k_2!(n-k_1-k_2)}\cdots\frac{(n-k_1-k_2-\cdots-k_{m-1})!}{k_m!\cdot 0!}$$

$$=\frac{n!}{k_1!k_2!\ldots k_m!}\ \blacksquare$$

Example 4.4.   We must paint nine different chairs with green, red and blue paint. We have enough blue for two chairs, enough red for three chairs and enough green for four chairs. In how many different ways may we paint all the chairs?

The number of paint jobs is

$$C(9; 2, 3, 4) = \frac{9!}{2!\cdot 3!\cdot 4!}$$

$$= \frac{9\cdot 8\cdot 7\cdot 6\cdot 5}{2!\cdot 3!} = 9\cdot 4\cdot 7\cdot 5 = 1260.\ \blacksquare$$

The multinomial theorem tells us how to expand a power of a multinomial.

The Multinomial Theorem

Theorem 4.3. $(x_1 + x_2 + \cdots + x_m)^n$ is the sum of all possible terms of the form

$$C(n; k_1, k_2, \ldots k_m)x_1^{k_1}x_2^{k_2}\ldots x_m^{k_m}$$

using nonnegative integers k_i with $k_1 + k_2 + \cdots + k_m = n$. We write this sum in the form

$$\sum_{\substack{(k_1,k_2,\ldots k_m);\\k_1+k_2+\cdots+k_m=n}} C(n; k_1, k_2, \ldots k_m)x_1^{k_1}x_2^{k_2}\ldots x_m^{k_m} = (x_1 + x_2 + \cdots x_m)^n$$

Proof.   The proof is in direct analogy with that of the binomial theorem. ■

We read the sum sign in the theorem above as "the sum over all lists k_1, k_2, . . . , k_m such that $k_1 + k_2 + \cdot \cdot \cdot + k_m = n$."

Example 4.5.   What is the coefficient of x^6yz^2 in $(x + y + z + w)^9$? What is the coefficient of x^6yz^2 in $(x + y + z + w)^{10}$?

The coefficient of x^6yz^2 in $(x + y + z + w)^9$ is, by Theorem 4.3, $C(9; 6, 1, 2, 0)$, and

$$C(9; 6, 1, 2, 0) = \frac{9!}{6! \cdot 1! \cdot 2! \cdot 0!} = \frac{9 \cdot 8 \cdot 7}{2} = 252.$$

The coefficient of x^6yz^2 in $(x + y + z + w)^{10}$ is *zero* because x^6yz^2 *is not* one of the terms shown in Theorem 4.3. ■

Multinomial Coefficients from Binomial Coefficients

It would be helpful to have a method like that of Pascal's triangle for computing multinomial coefficients. However, if we have a Pascal triangle written down, we can use it to compute multinomial coefficients as well by applying the following theorem which was proved as part of the proof of Theorem 4.2.

Theorem 4.4.

$$C(n; k_1, k_2, . . . , k_m) = C(n; k_1) \cdot C(n - k_1, k_2)$$
$$. . . C(n - k_1 - k_2 \cdot \cdot \cdot - k_{m-1}, k_m).$$

In a different notation, we can write

$$C(n; k_1, k_2, . . . , k_m) = \prod_{i=1}^{m} C\left(n - \sum_{j=1}^{i-1} k_j; k_i\right).$$

We read the symbol $\prod_{i=1}^{m}$ as "the product from i equals 1 to m of . . .".

Example 4.6. $\begin{pmatrix} 7 \\ 3 \quad 2 \quad 2 \end{pmatrix} = \binom{7}{3}\binom{4}{2}\binom{2}{2} = 35 \cdot 6 \cdot 1 = 210.$

EXERCISES

1. Write out the expansion of $(x + y)^4$.
2. Write out the expansion of $(x + 2)^5$.
3. Write out the expansion of $(2x - 3)^4$.

4. Using Pascal's triangle, write out the expansion of $(x + 2)^6$.

5. Using Pascal's triangle, write out the expansion of $(2x - y)^6$.

6. What is the coefficient of x^5 in $(6x + \frac{1}{2})^7$?

7. What is the coefficient of $\sqrt{2}$ in $(1 + \sqrt{2})^7$?   (Rewrite $(1 + \sqrt{2})^7$ in the form $a + b\sqrt{2}$ to answer this question.)

8. What is the coefficient of x^4 in $(4x - \frac{1}{2})^9$?

9. The number 2.01 is $2 + 0.01$. To compute $(2.01)^6$ in a straightforward way would require six multiplications. Use the binomial theorem and Pascal's triangle to compute $(2.01)^6$. How close would you have come to the correct answer if you had stopped after you had the sum of only the largest three terms?

10. Repeat the procedure of Exercise 9 with $(1.99)^6$. Why is the sum of the largest three terms (in absolute value) closer to the correct answer than the corresponding sum was to the answer in Exercise 9?

11. (a) Is $(1.001)^{1000}$ greater than 1.1?

   (b) Is $(1.0001)^{10,000}$ greater than 2?

12. Write out the expansion of $(x + y + z)^3$.

13. Write out the expansion of $(1 + \sqrt{2} + \sqrt{3})^3$ and collect like terms.

14. What is the coefficient of x^2y^3 in $(x + 2y + 3)^7$?

15. In how many ways may nine different pieces of candy be given to Sam, Mary and Pat so that Sam gets two pieces, Mary gets three pieces and Pat gets four pieces?

16. In how many ways may nine different pieces of candy be given to three children so that each child gets three pieces?

17. In how many ways may nine different pieces of candy be given to three children so that one child gets two pieces, a second child gets three pieces and the third child gets four pieces?

18. (a) In how many ways may a central processing unit assign nine jobs to three slave computers so that each computer gets three jobs?

   (b) In how many ways may the CPU assign the jobs so that no slave gets just one job or none at all?

19. An *ordered partition* of a set S into k parts is a list $S_1, S_2, \ldots, S_k$ of disjoint subsets of S, such that $S = S_1 \cup S_2 \cup \cdots \cup S_k$. How many ordered partitions of an n-element set into k parts have the property that the ith set S_i has size k_i? (If some of the numbers k_i are 0, the list is sometimes called a "composition" of S, rather than a partition.)

20. What is $\sum\limits_{i=1}^{10} \binom{10}{i} \cdot 3^{10-i}$?

21. Write down the formula for the expansion by the binomial theorem of $(x - 1)^n$. What can you conclude about the alternating sum $\binom{n}{0} - \binom{n}{1} + \binom{n}{2} + \cdots \pm \binom{n}{n}$ if $n > 0$?

22. By taking derivatives, show that

(a) $\binom{n}{1} + 2\binom{n}{2} + 3\binom{n}{3} + \cdots = n2^{n-1}$.

(b) $\binom{n}{1} - 2\binom{n}{2} + 3\binom{n}{3} - 4\binom{n}{4} + \cdots = 0$.

23. Write down the formula you get by computing the coefficient of $x^n y^n$ on both sides of the equation $(x + y)^n (x + y)^n = (x + y)^{2n}$. Apply the binomial theorem individually to both factors on the left side of the equation and apply it to the right side of the equation as well.

SECTION 5   MATHEMATICAL INDUCTION

The Principle of Induction

A number of proofs we have given have been made hard to read because the phrase, "and so on" appears in the middle of the proof. As proofs get more and more complex, this lack of precision becomes more and more confusing. In such situations, a reasonably simple solution is to appeal to a principle called the principle of mathematical induction. There are two closely related versions of the principle.

Version 1. Suppose the set S contains the integer i. Suppose also that for each integer j such that j is in S, the integer $j + 1$ is in S. Then S contains every integer greater than or equal to i.

Version 2. Suppose the set S contains the integer i. Suppose also that for each integer j such that all integers between i and j inclusive are in S, the integer $j + 1$ is in S. Then S contains every integer greater than or equal to i.

Frequently you will find versions of the principle based on the idea that some statement involving an integer i is true for certain values of the integer. These versions may seem more difficult to understand because they involve the idea of a statement about an integer i. They are, however, entirely equivalent to the two versions given here, and the two versions given here are also equivalent to each other. For your own understanding, you should learn to use the principle as stated here. Then you should have no difficulty adapting to some other version. For the remainder of this section, we shall give examples of the use of induction in proving theorems.

Proving Formulas Work

Example 5.1.   Prove that $1 + 3 + 5 + \cdots + 2n - 1 = n^2$ for all $n \geq 1$.

Let S be the set of all integers n such that $1 + 3 + 5 + \cdots + 2n - 1 = n^2$. The integer 1 is in S since $1 = 1^2$. Suppose the integer k is in S. Then $1 + 3 + \cdots + 2k - 1 = k^2$ and

$$1 + 3 + 5 + \cdots + 2k - 1 + 2(k + 1) - 1$$
$$= \underbrace{1 + 3 + 5 + \cdots + 2k - 1}_{k^2 + 2k + 1} + 2k + 1 \qquad = (k + 1)^2.$$

Thus the integer $k + 1$ is in S. Then by the principle of mathematical induction, S contains all positive integers, and so our formula is proved. ■

Informal Induction Proofs

There is no need for proofs by induction to be as formal as they are in this section. The formalism is intended to illustrate exactly what is happening in an inductive proof. We will slowly dispense with the formalism in the text as the reader becomes more familiar with inductive proofs.

Example 5.2.   Do Example 5.1 with no formal reference to a set S.

We wish to show that $1 + 3 + 5 + \cdots + 2n - 1 = n^2$ for all $n > 1$. Clearly this is the case when $n = 1$. Now suppose it is the case when $n = k - 1$, so that

$$1 + 3 + \cdots + 2(k - 1) - 1 = (k - 1)^2.$$

Then by addition,

$$
\begin{aligned}
1 + 3 + \cdots + 2(k - 1) - 1 + 2k - 1 &= (k - 1)^2 + 2k - 1 \\
&= k^2 - 2k + 1 + 2k - 1 \\
&= k^2.
\end{aligned}
$$

Thus by the principle of mathematical induction, the formula holds for all n. (Note, by the way, it doesn't matter whether we derive the case $n = k + 1$ from $n = k$ or the case $n = k$ from $n = k - 1$.) ■

The General Sum Principle

Example 5.3.   The simplest version of the sum principle states that if two sets A and B have no elements in common, then the size of $A \cup B$ is the sum of the sizes of A and B; in symbols $|A \cup B| = |A| + |B|$. Use this principle and the principle of mathematical induction to prove that if $A_1, A_2, \ldots, A_n$ is a list of sets such that for every distinct i and j, $A_i \cap A_j = \emptyset$, that is, A_i and A_j have no elements in common, then

$$\left| \bigcup_{i=1}^{n} A_i \right| = \sum_{i=1}^{n} |A_i|.$$

(The notation $\bigcup_{i=1}^{n} A_i$ stands for the union of all the sets in the list and is read as "the union from i equal 1 to n of A_i.")

Let S be the set of all n such that for mutually disjoint sets $A_1, A_2,$ . . . , A_n,

$$\left| \bigcup_{i=1}^{n} A_i \right| = \sum_{i=1}^{n} |A_i|.$$

Then 2 is in S by the sum principle (and 1 is in S because the statement then reads $|A_1| = |A_1|$).

Now suppose k is in S, and let $A_1, A_2, \ldots A_k, A_{k+1}$ be a list of mutually disjoint sets. Then

$$\bigcup_{i=1}^{k+1} A_i = \left(\bigcup_{i=1}^{k} A_i \right) \cup A_{k+1}.$$

However, since A_{k+1} has no elements in common with any A_i for $i \leq k$, it has no elements in common with $\bigcup_{i=1}^{k} A_i$. Thus by the sum principle for pairs of sets,

$$\left| \bigcup_{i=1}^{k+1} A_i \right| = \left| \bigcup_{i=1}^{k} A_i \right| + |A_{k+1}|.$$

However, k is in S, so that

$$\left| \bigcup_{i=1}^{k} A_i \right| = \sum_{i=1}^{k} |A_i|.$$

Thus

$$\left| \bigcup_{i=1}^{k+1} A_i \right| = \left| \bigcup_{i=1}^{k} A_i \right| + |A_{k+1}| = \sum_{i=1}^{k} |A_i| + |A_{k+1}| = \sum_{i=1}^{k+1} |A_i|.$$

Thus $k + 1$ is in S, so by Version 1 of the induction principle, S contains all positive integers; this proves the formula. ∎

An Application to Computing

Example 5.4.   Show that given a list of 2^n numbers in increasing order, it is possible to determine whether a particular number k is in the list by comparing k to at most $n + 1$ numbers of the list.

Let $S = \{n \mid$ With $n + 1$ comparisons it is possible to tell if a number k is a list of 2^n numbers in increasing order$\}$.

Then 0 is in S, for 2^0 is 1 and in a list of 1 number, only 1 comparison is required to determine if k is in the list. Now suppose we have a list of 2^{m+1} numbers in increasing order and let m be in S. Given the number k, compare it to the number a in position 2^m of the list.

If $k = a$, stop.

If $k < a$, determine whether k is between position 1 and 2^m using at most $m + 1$ comparisons. (This is possible since m is in S.)

If $k > a$, determine whether k is between position $2^m + 1$ and position 2^{m+1} of the list using at most $m + 1$ comparisons. (This is possible since m is in S.)

Thus, we can find whether k is in the list in a total of $1 + (m + 1)$ comparisons, so $m + 1$ is in S. Therefore, by the principle of mathematical induction, all non-negative integers are in S. This proves we can determine whether k is in a list of 2^n numbers using at most $n + 1$ comparisons. ∎

Proving a Recursion Works

Example 5.5. Show that if a function $D(n, k)$ which is defined for pairs n, k with $0 \leq k \leq n$, has the properties that $D(0, n) = D(n, n) = 1$, and $D(n, k) = D(n - 1, k - 1) + D(n - 1, k)$, then $D(n, k) = n!/(n - k)!k!$

Let S be the set of all n such that $D(n, k) = \dfrac{n!}{(n - k)! \cdot k!}$. Since $D(0, 0) = 1$ and $\dfrac{0!}{(0 - 0)!0!} = 1$, then 0 is in S. Suppose $n - 1$ is in S. Then for each k with $0 < k < n$

$$
\begin{aligned}
D(n, k) &= D(n - 1, k - 1) + D(n - 1, k) \\
&= \frac{(n - 1)!}{(k - 1)!(n - k)!} + \frac{(n - 1)!}{k!(n - k - 1)!} \\
&= \frac{k(n - 1)!}{k!(n - k)!} + \frac{(n - 1)!(n - k)}{k!(n - k)!} \\
&= \frac{n(n - 1)!}{k!(n - k)!} = \frac{n!}{k!(n - k)!}.
\end{aligned}
$$

(The reason we wrote the equations for $0 < k < n$ rather than $0 \leq k \leq n$ was the two multiplications of the numerator and denominator of a fraction by k or $n - k$ in the next to last line of the equations.)   Also $D(n, 0) = D(n, n) = 1 = \dfrac{n!}{(n - 0)!0!}$, so for all k with $0 \leq k \leq n$ we have the equation $D(n, k) = \dfrac{n!}{(n - k)!k!}$. Thus n is in S.   By the principle of mathematical induction, S contains all non-negative integers.   Thus $D(n, k) = \dfrac{n!}{k!(n - k)!}$ for all n. ∎

Note that this example shows that the binomial coefficients must be completely determined by the rules we used to generate Pascal's triangle.

A Sample of the Second Form of Mathematical Induction

Example 5.6.   A prime number is a positive integer greater than 1 which has no positive factor other than itself and 1.   Show that every positive number greater than 1 is a product of prime numbers, or is prime.

Let S be the set of positive integers that are prime or products of prime numbers.   Then 2 is in S because 2 is prime.   Suppose now that all numbers between 2 and k are in S.   We shall show that $k + 1$ is in S.   If $k + 1$ has no factors other than 1 or $k + 1$, it is prime and thus is in S.   If $k + 1$ has a positive factor m different from itself and 1, then $k + 1 = mj$ for some positive integer j. Since $m > 1$, j cannot be greater than or equal to $k + 1$.   Thus $j < k + 1$.   Then both m and j are either prime or products of prime numbers.   Thus $mj = k + 1$ is a product of prime numbers, so $k + 1$ is in S.   Thus, by the principle of mathematical induction, S contains all integers greater than 1, and so every integer greater than 1 is a prime number or a product of prime numbers. ∎

This example shows how useful the second form of the principle of mathematical induction can be.   How, though, do we know which form to use?   Generally you do not decide on a proof by induction on the basis of some magical insight that induction is required.   Instead, you may first work out examples for quite a few values of the parameter (the k we've used in our examples) and you see how one example seems to follow naturally from the previous one—or from all the previous ones.   Perhaps instead you see how to divide the problem into several cases, each one with the parameter value being one smaller—or else into several cases each with a smaller parameter value (which is not necessarily just one smaller).   It is on this basis that you decide which form is more useful.

The two forms of the induction principle are equivalent mathematically. In other words, given one version of the principle, we could derive the other as a theorem.   Proving this equivalence is important to the student interested in the foundations of mathematics; however, since the proof is more a matter of logic than combinatorics, we will not include it here.

EXERCISES

1. Use induction to prove that

$$\sum_{i=1}^{n} i = 1 + 2 + \cdots + n = \frac{n(n+1)}{2}.$$

2. Use induction to prove that

$$\sum_{i=1}^{n} i^2 = \frac{n(n+1)(2n+1)}{6}.$$

3. Prove that 6 is a factor of $n^3 + 5n$ for all positive integers n.

4. Use induction to prove that $(1 + x)^n \geq 1 + nx$ if $x \geq 0$ and n is a positive integer.

5. Prove that $n^3 > 2n^2$ if $n \geq 3$.

6. Prove that $2^n > n^2$ for $n > 5$.

7. Prove that for each integer $n \geq 3$, the number of distinct prime factors of n is less than n.

8. Prove by induction that a set with n elements has 2^n subsets. To get started, look at the following questions for several values of n. How many subsets of $\{1, 2, \ldots n\}$ contain n? How many don't contain n?

9. Use induction to reprove Theorem 1.2.

10. Use induction on n to prove there are m^n functions from an n–element set to an m–element set.

11. The simplest version of the multiplication principle states that if n disjoint sets each have m elements, then their union has mn elements. Prove this using induction and the simplest form of the sum principle. (Example 5.3 is a good model because the problem stated here can be regarded as a special case of the problem stated in Example 5.3.)

12. Using induction and the simplest form of the multiplication principle stated in Exercise 11, prove the multiplication principle for lists of length n.

13. Let g be a function from the nonnegative integers to the positive integers with properties that $g(1) = a$ and $g(m + n) = g(m)g(n)$. Prove that $g(0) = 1$. Prove that $g(n) = a^n$ for $n \geq 0$.

14. Show that it is possible to sort a list of 2^n numbers in no special order into a list in increasing order by making no more than $n \cdot 2^n$ comparisons. (Hint: Use a method like that of Example 5.5 and observe that two lists of j numbers each in increasing order may be merged into a single list in increasing order in at most $2j$ steps.)

15. Show that if $m \neq n$, there is no one–to–one correspondence (Section 3) from $\{1, 2, \ldots, n\}$ to $\{1, 2, \ldots, m\}$. This is why we can count the number of elements of a set by placing it in a one–to–one correspondence with a set whose size we already know.

16. Prove that $\sum_{j=0}^{n} \binom{j}{k} = \binom{n+1}{k+1}.$

Suggested Reading

Berge, Claude: *Principles of Combinatorics*, Academic Press 1971 (Chapter 1).

Feller, William: *An Introduction to Probability Theory and Its Applications*, Third Edition, John Wiley 1968 (Chapter 3).

Ryser, Herbert John: *Combinatorial Mathematics*, CARUS MATHEMATICAL MONOGRAPHS *14*, Mathematical Association of America 1963 (Chapter 1).

2 Equivalence Relations, Partitions and Multisets

SECTION 1   EQUIVALENCE RELATIONS

The Idea of Equivalence

Suppose we are writing out seating charts to arrange four people clockwise around a circular table for a game of cards. If we obtain one arrangement from a previous arrangement by shifting each person's place one seat to the right, the relative positions of all four players will remain the same. Thus, the two arrangements are equivalent for the purpose of playing a game of cards. Because of the possibility of equivalence, the number of truly distinct seating patterns is not the same as total number of seating charts. Without a more precise understanding of what we mean by equivalence, it is hard to even say what we mean by "the number of truly distinct seating patterns." It will often happen that two arrangements of objects that are technically different—such as two different lists of places at a table—are so closely related that they are equivalent for our purposes. It is for this reason that we introduce the concept of an equivalence relation.

We gave a precise definition of the word relation in our examination of the definition of a function. Recall that a relation R on a set S is some set of ordered pairs of elements of S. Under one set of conditions, we might see two objects as being equivalent, while under another set of conditions, we might see the two objects as being nonequivalent. However, there are certain properties that the relation expressed by *any* statement of the form "is equivalent to" should satisfy. For example, any object should be equivalent to itself. Also if object A is equivalent to object B, then object B should be equivalent to object A. Further, if object A is equivalent to object B, and object B is equivalent to object C, then objects A and C should be equivalent. We take these properties as the defining properties of the idea of equivalence.

Equivalence Relations

We shall say that a set R of ordered pairs of elements of a set X is an *equivalence relation on X* if

35

(1) (x, x) is in R for all x in X. (*reflexive* law)

(2) If (x, y) is in R, then (y, x) is in R for all x and y in X. (*symmetric* law)

(3) If (x, y) and (y, z) are in R, then (x, z) is in R for all x, y and z in X. (*transitive* law)

On any set X, the relation of "equality" is an equivalence relation. Less obvious examples of equivalence relations arise in many different ways.

Example 1.1.   Let X be the set of college students attending some college. Let R be the relation, "has the same grade point average as." Show that R is an equivalence relation on X.

(1) Any person has the same "GPA" as him or herself.

(2) If person A has the same GPA as person B, then person B has the same GPA as person A.

(3) If person A has the same GPA as B and B has the same GPA as C, then A has the same GPA as C.

Thus R is an equivalence relation.   Because many relations are readily understood, it is often possible to verify the three laws as we have done here without having to talk about ordered pairs. ■

Circular Arrangements

Example 1.2.   Let X be the set of all possible lists of the four names {Bill, Chuck, Maria, Sarah}.   Regard the lists as seating charts for four people around a circular table having four places marked 1, 2, 3, 4.   Define a relation R on X by letting (K, L) be in R if the list K may be obtained from the list L by shifting each person the same number of places to the right or the left.   For example, (using initials) B, C, M, S is related by R to S, B, C, M.   In terms of ordered pairs, we would say that $(\langle B, C, M, S\rangle, \langle S, B, C, M\rangle)$ is in R.   (The angle brackets around the lists have no special meaning, and are chosen just as a convenient form of "punctuation.")   Show that R is an equivalence relation.

(1) For each list L, $(L, L) \in R$ because L may be obtained from L by shifting each person no places (or four places!).

(2) If K may be obtained from L by shifting everyone n places to the right, then L may be obtained from K by moving everyone n places to the left.

(3) We will say a left shift through n places is a right shift through $-n$ places.   A right shift through m places followed by a right shift through n places results in a right shift through $m + n$ places.   (Why is this true even if m or n is negative?)   Thus if $(K, L) \in R$ and $(L, M) \in R$, then $(K, M) \in R$.

Thus R is an equivalence relation. ■

In the example above, suppose we start with the list B, C, M, S and ask which other lists are equivalent to it. By means of right shifts through 1, 2 and 3 places, respectively, we get S, B, C, M; M, S, B, C and C, M, S, B. By means of left shifts through 1, 2 and 3 places, respectively, we get C, M, S, B; M, S, B, C and S, B, C, M. Thus the set of *all* lists equivalent to B, C, M, S is

$$S_1 = \{B, C, M, S;\ \ S, B, C, M;\ \ M, S, B, C;\ \ C, M, S, B\}.$$

We could begin with another list and write the set S_2 of all lists equivalent to it. Continuing, we would get *sets of lists* $S_1, S_2, \ldots, S_k$ such that each list is in one and only one set, with all the lists in any one set being equivalent. Further, if two lists are in different sets, they will not be equivalent. Thus each *set of lists* corresponds to a seating pattern, and different sets of lists correspond to different patterns.

If we ask how many distinct seating patterns we have, then we wish to know how many sets of lists we have; this number is the number k of sets S_1, $S_2, \ldots, S_k$. It is clear that we have four lists in each set, and that each list of initials appears in one and only one set. Thus the union of all of the four–element sets $S_1, S_2, \ldots, S_k$ is the set of *all* lists of the four names, and so has 4! elements. On the other hand, each ordered pair (S_i, L), where L is a list in S_i, corresponds to exactly one list L in the union of the sets S_i. Thus by the multiplication principle

$$4! = k \cdot 4$$

so $k = 3! = 6$. Therefore, there are six distinct seating arrangements.

This kind of analysis leads us to a general principle.

Equivalence Classes

Theorem 1.1.   Let R be an equivalence relation on a set X. Then there is a collection **C** of nonempty subsets of X such that

(1) Each element of X is in some set C_x of **C**.
(2) Any two different sets in **C** are disjoint.
(3) For each set C_y in **C**, all elements of C_y are equivalent relative to R.
(4) If two elements x and y are equivalent relative to R, then they lie in the same set of **C** (in fact $C_x = C_y$).

Proof.   For each x in X, define C_x by

$$C_x = \{y \mid (x, y) \in R\}.$$

Let **C** be the collection of sets C_x. Note that C_x is not empty (it contains the element x). Condition 1 of the Theorem holds because x is in C_x. To check Condition 2, let x and y be in X. If $C_x \cap C_y$ is not empty, then C_x and C_y have an element z in common. Using this common element we conclude that $C_x = C_y$. To see why, note that $(x, z) \in R$ since $z \in C_x$ and $(z, y) \in R$ since $z \in C_y$. Then $(x, y) \in R$ by the transitive law. Thus $x \in C_y$ and $y \in C_x$. To see why $C_x = C_y$, note that if $w \in C_x$, then (w, x) and (x, y) are in R, so that (w, y) is in R. Thus $w \in C_y$. That is, every element of C_x is in C_y. Similarly, every element of C_y is in C_x, so $C_x = C_y$. In other words, for any x and y, either $C_x = C_y$ or C_x and C_y are disjoint.

Condition 3 follows immediately from the definition of C_x and the symmetric and transitive laws. Condition 4 follows from Condition 2, because if x and y are equivalent, then both x and y are in $C_x \cap C_y$ (so then C_x and C_y cannot be different because they are not disjoint). Thus the collection **C** given by

$$\underline{C} = \{C_x \mid x \in X\}$$

is the desired collection. (Note this notation *does not mean* that if C_x and C_y are equal, they appear twice. This notation means that the collection consists of all the distinct sets among the C_x's, each included once.) ∎

The sets C_x in the collection **C** are called *equivalence classes*. A collection **C** of nonempty mutually disjoint sets whose union is X is called a *partition* of X. Thus given an equivalence relation, we have a partition **C** of X into equivalence classes. It is straightforward to show that given a partition **P** of X, there is an equivalence relation whose equivalence classes are just the sets of **P**.

Theorem 1.2.   If **P** is a partition of X, then there is one and only one equivalence relation whose equivalence classes are the classes of **P**.

Proof.   Exercise 7. ∎

Counting Equivalence Classes

In our example of seating arrangements, all the equivalence classes had the same size. This made it easy to compute the number of equivalence classes. As you may see in the exercises, it is possible to construct examples of useful equivalence relations whose equivalence classes *do not have* the same size. On the other hand, many practical problems give rise to equivalence relations whose equivalence classes all have the same size.

Theorem 1.3.   If R is an equivalence relation on a set X with n elements and each equivalence class has m elements, then R has n/m equivalence classes.

Proof.   This is a direct consequence of the multiplication principle. ∎

Example 1.3.   A company has built new corporate headquarters in which five different, but compatible, computers are to be installed for its five divisions. There are five floors linked by a communication cable with a connector on each floor; the cable makes a loop through all five floors and then returns to the first floor.   If communications may go in only one direction along the loop (for example, in the order 1, 2, 3, 4, 5 and back to 1), in how many ways may the computers be arranged along the loop?   (Two ways are different if they differ in communications patterns.)   How many arrangements are there if communications may go in both directions along the loop?

The first question is just like the card player's question.   There are 5! assignments of computers, but shifting each computer a number of spaces to the right or left around the loop gives an equivalent assignment.   In any other change, some machine would send its messages to a different machine than before.   Each assignment belongs to an equivalence class consisting of five different assignments, so there are

$$\frac{5!}{5} = 4! = 24$$

equivalence classes of assignments.   Thus there are 24 different communications patterns possible.

On the other hand, if two–way communications are possible, an arrangement with A on floor 1, B on floor 2, C on floor 3, D on floor 4, and E on floor 5 has the same communications pattern as the reversed arrangement $E\ D\ C\ B\ A$ (because each computer can communicate with exactly the same machines as before).   Thus, taking all shifts and reversals, we discover that the arrangement $ABCDE$ generates the equivalence class

$$\{ABCDE,\ BCDEA,\ CDEAB,\ DEABC,\ EABCD,$$
$$EDCBA,\ AEDCB,\ BAEDC,\ CBAED,\ DCBAE\}.$$

Now if we shift or reverse any arrangement in this set, we get another arrangement in the set.   Thus, the set is indeed the equivalence class generated by $ABCDE$.   We have 10 arrangements per equivalence class and $5!/10 = 120/10 = 12$ different communications patterns possible. ∎

The Inverse Image Relation

There is another reason why equivalence relations will be important to us. Given a function f from a set A to a set B, we can define a relation $R(f)$ by saying $(x, y) \in R(f)$ if $f(x) = f(y)$. That is, x is equivalent to y relative to $R(f)$ if $f(x) = f(y)$. It is straightforward to check that $R(f)$ is indeed an equivalence relation. For reasons we shall see later on, it is natural to call this the "inverse image relation" of f.

Example 1.4. Let f be the function that assigns to each student in a college the postal (zip) code of that person's home town. Describe the inverse image relation $R(f)$ and its equivalence classes.

Two people are related by $R(f)$ if and only if they have the same postal codes. Each equivalence class consists of all students with a given postal code. ■

Example 1.5. Let $f(x, y) = y - x$ be a function defined on the plane with its usual real number coordinate system. The inverse image relation $R(f)$ is an equivalence relation on ordered pairs (x_1, y_1), (x_2, y_2), etc. of real numbers in which (x_1, y_1) is equivalent to (x_2, y_2) if $y_1 - x_1 = y_2 - x_2$. Thus for each real number b, there is an equivalence class of all ordered pairs (x, y) such that $y - x = b$, or $y = x + b$. Thus, each equivalence class is a straight line with slope 1. In particular, when b is zero, we get the straight line through the origin given by $y = x$. ■

EXERCISES

1. Write down the relation determined by "x is related to y if $|x - y| \le 1$," for the set of integers from 1 to 5. Is this relation an equivalence relation?

2. Is the relation "is a brother of" an equivalence relation on the set of all males? On the set of all people? Answer the same questions for "is a brother of or is."

3. Two people are siblings if they are brother or sister. Answer the questions of Exercise 2 for "is a sibling of." Answer the questions for "is a sibling of or is."

4. Two people are cousins if a parent of one is a sibling of a parent of the other. Is the relation, "is a cousin of or is," an equivalence relation?

5. Write down the set of ordered pairs which is the relation "is a factor of" for the set of integers between 1 and 12. (Note, $2 = 2 \cdot 1$ so 2 is a factor of 2.) Is this relation an equivalence relation?

6. For this exercise, write an ordered pair of positive integers as $a|b$. Is the relation given by "$a|b$ is related to $c|d$ if $ad = bc$" an equivalence relation?

7. Let $\mathbf{P} = \{C_1, C_2, \ldots, C_k\}$ be a partition of the set S. Suppose the relation R is defined by $(x,y) \in R$ if x and y are both in the same set of $\mathbf{P}$, i.e. the same C_i. Show that R is an equivalence relation. What are the equivalence classes of R? What part of Theorem 1.2 have you just proved? Finish the proof of Theorem 1.2.

8. In how many ways may k people be arranged around a circular table?

9. In how many ways may five men and five women be arranged around a circular table so that no two neighbors are of the same sex?

In Exercises 10–15 two arrangements of beads are equivalent as necklaces if one may be obtained from the other by picking it up, moving it around and flipping it if necessary.

10. In how many ways may six distinct beads be arranged on a string as a necklace?

11. Suppose we are given two red beads (R) and two black beads (B). Define two lists of two R's and two B's as equivalent if the same necklace is obtained by stringing the red and black beads in the orders given in the two lists. Show that this is an equivalence relation on the set of lists of two R's and two B's. Write down the equivalence classes and note that they have different sizes. How many necklaces are possible?

12. In how many ways may k distinct beads be arranged on a necklace?

13. In how many ways may k distinct red beads and two distinct black beads be arranged on a necklace?

14. We have k distinct red beads and k distinct black beads. In how many different ways may the beads be arranged around the edge of a circular table so that they alternate in color? In how many ways may they be strung on a necklace so that they alternate in color?

15. Answer the two questions of Exercise 14 assuming no bead of one color can be situated between two beads of the opposite color (in place of the assumption that they alternate in color).

16. What is the inverse image relation $R(f)$ associated with $f(x) = x^2$ for all real numbers x and what are its equivalence classes? (Note: You should *not* have to deal with ordered pairs as we did in Example 1.5. Why?)

17. How many equivalence classes will $R(f)$ have for a function f from an n–element set to a k–element set if
      (a) f is one–to–one?
      (b) f is onto?

SECTION 2   DISTRIBUTIONS, MULTISETS AND PARTITIONS

The Idea of a Distribution

We will now begin a general study of ways of distributing objects among recipients. There are a number of important, and perhaps surprising, applications found here. The objects might be jobs to be distributed by a computer to a network of slave computers, or fruit to be distributed into packages in a grocery store, etc. The order in which the recipients receive the objects may or may not be important. The items being distributed, or even the recipients, may or may not be identical.

In fact we already began this study in Chapter 1 when we studied functions. A distribution of k distinct objects among n distinct recipients is nothing but a function from the objects to the recipients which states who is to receive each object. Thus there are n^k such distributions. Since we know the number of such distributions, we now study cases in which the objects, the recipients, or both are not distinct. Our first problem is to compute the number of distributions of

k identical objects, apples, say, or pieces of candy, to n distinct recipients—for example, children. For the sake of intuition, we will often speak of distributing candy to children, though there are quite a number of possible applications to other situations. Feller's book on probability, for example, gives a list of sixteen possible applications.* There are several approaches to the problem; we choose one which can be applied consistently and which makes use of our previous study of equivalence relations.

Distributing Identical Objects to Distinct Recipients

Example 2.1.   In how many ways may k identical candy bars be distributed among n children?

We can number the candy bars (in our imagination) with the numbers 1 through k. Imagine the n children sitting in a row, and imagine the candy is distributed by placing the pieces of candy intended for some child directly in front of this child. Then we have k objects divided into n piles (some of the piles might be empty). To make sure the piles don't get mixed up, we can put $n - 1$ dividers between the piles. (Why do we need $n - 1$?) We regard all the objects as equivalent, and note that the $n - 1$ dividers can be used in any order. Thus we have k pieces of candy $c_1, c_2, \ldots, c_k$ and $n - 1$ dividers $d_1, d_2, \ldots, d_{n-1}$, and determine a division by making a list of these $n + k - 1$ objects. A rearrangement of the pieces of candy will not give a different division, nor will a rearrangement of the dividers. Thus two lists of c_i's and d_j's are equivalent if one may be obtained from the other by rearranging the c_i's among themselves or by rearranging the d_j's among themselves. By the multiplication principle, each list is equivalent to $k! \cdot (n - 1)!$ other lists. Since this equivalence is an equivalence relation on the set X of all $(n + k - 1)!$ lists, Theorem 1.3 tells us there are $\dfrac{(n + k - 1)!}{k!(n - 1)!}$ equivalence classes of lists. ∎

Now suppose we want to know the number of distributions such that each child gets a piece of candy. If the pieces of candy were distinct, we would be asking for the number of functions from the pieces of candy onto the children; in Chapter 1 we remarked that this problem is beyond our current methods. We will find a partial solution to the problem in this chapter and solve it completely at the beginning of Chapter 3. However, with identical pieces of candy, the problem is not hard. We just give each child one piece of candy, leaving us $k - n$ pieces, and apply the method of Example 2.1 to those $k - n$ pieces. (This is a hint for Exercise 3.)

---

* William Feller, *An Introduction to Probability Theory and Its Applications*, Vol. 1, Third Edition, John Wiley and Sons, New York, 1968, p. 10.

Labelling Identical Objects with Distinct Labels

The methods of the above example can be applied in general. Imagine putting a label consisting of a child's name on each piece of candy. The example suggests the following theorem whose proof is just like the method used there.

> **Theorem 2.1.** Suppose we are given n distinct types of labels. The number of ways to assign one of these labels to each of k identical objects is $\binom{n + k - 1}{k}$.

Ordered Compositions

From a distribution of the candy to the children, we can define a function f from the set of children to the non-negative integers by letting $f(x)$ be the number of pieces of candy given to child x. We let the symbols $x_1, x_2, \ldots, x_n$ stand for the names of the children. Then a distribution of k identical pieces of candy to the n children corresponds to a function f such that

$$\sum_{i=1}^{n} f(x_i) = k.$$

Another way to make a general statement of the solution of the distribution problem involves such function and is our next theorem. Again, the method of proof is that of Example 2.1.

> **Theorem 2.2.** The number of functions f from a set $\{x_1, x_2, \ldots, x_n\}$ to the nonnegative integers such that
>
> $$\sum_{i=1}^{n} f(x_i) = k$$
>
> is $C(n + k - 1; k).$

In number theory, the function of given in Theorem 2.2 would be called an *ordered composition* of k with n parts, because it gives a list of n numbers that add up to k.

Multisets

At times a general statement of the solution to one kind of problem can be interpreted to give a solution to another kind of problem. In this case, our study of distributions can be reinterpreted to explain possibly repeated selections of objects out of a set. For example, the set of letters used in the word *roof* is the set {*f*, *o*, *r*} (in alphabetical order). The full selection of letters used, however, is {*f*, *o*, *o*, *r*}. Because this notation suggests a set with multiple occurrences of elements allowed, we call such a selection a multiset. Thus the multiset of letters of the word *feet* is {*e*, *e*, *f*, *t*} while the set of letters used is {*e*, *f*, *t*}. If we throw a pair of dice 15 times, the *multiset* of resulting sums has exactly 15 entries. However, since the sum must be a number between 2 and 12, the *set* of results has no more than 11 entries and could have as few as one entry. Sometimes a multiset is called a "combination with repetitions."

To define a multiset more carefully, we shall define a k–element *multiset chosen from a set S* as an ordered pair $(S, m) = M$ where m is a function from S to the nonnegative integers such that the sum of the values of $m(x)$ for all x in S is k. The number $m(x)$ is called the *multiplicity* of x in M. The idea is that $m(x)$ tells us how often x appears in our multiset. Such a formal definition of a multiset is important primarily because it shows us how to determine precisely when an object is or is not a multiset. We can take advantage of the precise definition to determine exactly how many multisets of size k we can choose from a set S with n elements. It is clear that we have one multiset for each function m. Thus by Theorem 2.2, the number of multisets of size k chosen from an n–element set is $C(n + k - 1, k)$. Multisets of k elements chosen from a set of size n are sometimes called "combinations (with repetitions) of n things taken k at a time."

Example 2.2. How many fruit baskets with 10 pieces of fruit can be completed using any number of oranges, apples, pears and bananas?

We are asking for the number of 10–element multisets chosen from a 4–element set; so our answer is

$$C(10 + 4 - 1, 10) = \binom{13}{10} = 13 \cdot 22 = 286. \ \blacksquare$$

Example 2.3. What is the multiplicity of each letter in the word *excellent*? Since x, c, n and t each occur once, $m(x) = m(c) = m(n) = m(t) = 1$. Since 1 appears twice, $m(1) = 2$ and since e appears three times, $m(e) = 3$. Note m is a multiplicity function for a nine–element multiset chosen from a six–element set. $\blacksquare$

If the function m happens to take on only the values zero and one, then the multiset it describes can in fact be regarded as a subset of S. In fact, in this case m is the function, called the characteristic function, we introduced to represent subsets in our second proof that an n–element set has 2^n subsets.

Distributions with Specified Class Size

Now we shall ask a more precise kind of question about distributions. That is, "in how many ways may we distribute pieces of candy so that child 1 gets j_1 pieces, child 2 gets j_2 pieces, and so on?" If the pieces of candy are all the same, the answer is that there is just one way—pass it out according to the numbers j_i. If the pieces of candy are all different, then a distribution of the type desired is a labelling of the candy so that j_i pieces are labelled as intended for child i. We have seen that there are $C(n; j_1, j_2, \ldots, j_k)$ such labellings.

Essentially the same methods as above may be used to discuss distributions of distinct objects to identical recipients. Since the recipients are identical, we are simply dividing our objects up into non–overlapping sets, one for each of our identical recipients. Thus we are studying partitions of sets.

Example 2.4. A class of 30 college students is to be divided into two discussion sections with 10 students each and two discussion sections with five students each to test the effect of size on the value of discussion sections. In how many ways may the class be divided?

We are asking for the number of partitions of a 30–element set into two classes of size 10 plus two classes of size 5. One way to proceed would be to label the 30 elements with labels 1 through 4. Thus 10 elements get label 1, 10 elements get label 2, 5 elements get label 3 and 5 elements get label 4. For our purposes, two labellings would be equivalent if we got one from the other by interchanging the people with label 1 and label 2, or interchanging those with label 3 and label 4, or both. Thus each labelling is equivalent to three other labellings. We now have an equivalence relation on the set of labellings with four labellings in each equivalence class. All the labellings in an equivalence class are the same for our purposes; the number of divisions of the class is the *number* of classes. The total number of labellings is $C(30; 10, 10, 5, 5)$, and so the number of classes is $C(30; 10, 10, 5, 5)/4$. ■

The partitions of the students are said to have type vector (0000200002), meaning that the partition has two parts of size 5 and two parts of size 10. In general, we say a partition has *type vector* $(j_1, j_2, \ldots, j_n)$ if the partition has j_i classes of size i for $i = 1$ to n. Thus if $j_i = 0$, we have no classes of size i; if $j_i = 1$, we have one class of size i and so on. (As above, we drop trailing zeros, so the length of a type vector is the size of the largest part.)

Theorem 2.3. The number of partitions of a set with n elements into j_1 classes of size 1, j_2 classes of size 2, . . . up to j_n classes of size n is

$$\frac{n!}{\prod_{i=1}^{n} (i!)^{j_i} \cdot j_i!}.$$

Proof. We can form a partition with type vector $(j_1, j_2, \ldots, j_n)$ by listing all the n elements and taking the first j_1 elements to be classes of size 1, then taking the next $2j_2$ elements two at a time to be classes of size 2 listed one after another, then taking the next $3j_3$ elements to be classes of size 3 listed one after another, and so on. We say that two lists are equivalent if they yield the same partition of type $(j_1, j_2, \ldots, j_n)$. Since this is an equivalence relation, the number of partitions of type $(j_1, j_2, \ldots, j_n)$ is the number of equivalence classes. Two lists are equivalent if one is obtained from the other by any rearrangement of the j_1 classes of size 1. There are $j_1!$ ways to do so. Similarly, listing classes of size i in a different order gives an equivalent list. There are $j_i!$ such arrangements. (Note that if j_i is zero, $j_i!$ is one.) Also, a given class of size i may be listed in $i!$ different ways. Since there are j_i classes of size i, there are, by the multiplication principle, $(i!)^{j_i}$ ways to arrange the elements within each class of size i. Since there are an additional $j_i!$ ways to arrange the classes themselves, by the multiplication principle there are $(i!)^{j_i} \cdot j_i!$ ways to arrange the elements of classes of size i among the appropriate positions in the list. By one more application of the multiplication principle, there are

$$(1!)^{j_1} \cdot j_1!(2!)^{j_2} \cdot j_2! \quad \ldots \quad (n!)^{j_n} \cdot j_n! = \prod_{i=1}^{n} (i!)^{j_i} \cdot j_i!$$

ways to arrange all the n elements in a list that represents a given partition of type $(j_1, j_2, \ldots, j_n)$. Thus each equivalence class has this number of lists. The total number of lists is $n!$. Therefore by Theorem 1.3, the number of equivalence classes is

$$\frac{n!}{\displaystyle\prod_{i=1}^{n} (i!)^{j_i} \cdot j_i!} . \quad \blacksquare$$

It is possible to give a proof based on labellings as in the example that precedes the theorem. However, the notation used in this section differs from the notation we used to describe labellings, and a longer proof would have been needed to relate the two notations. In the exercises, we shall reformulate the notation of labellings so that you may relate the two ideas for yourself. Note, however, that $\dfrac{30!}{(5!)^2 \cdot 2! \cdot (10!)^2 \cdot 2!}$ is the same as $C(30; 10, 10, 5, 5)/4$, so our example and our theorem agree.

EXERCISES

1. In how many ways may 10 identical candy bars be distributed to four children?

2. In how many ways may 10 identical candy bars be distributed to four children if each child must get at least one candy bar? (Hint: If you first give each child one piece, how many pieces can you distribute at will?)

3. In how many ways may n identical pieces of candy be distributed among k children if each child is to get a piece?

4. In how many ways may n identical pieces of candy be distributed among k children if each child is to get at least j pieces?

5. Write down the multiplicity function for the letters in the multiset of letters of the phrase *ulterior motive*.

6. Give a sensible definition of a "multisubset" of a "multiset," and explain why you think it is sensible.

7. In how many ways may n identical chemistry books, r identical mathematics books, s identical physics books and t identical astronomy books be arranged on three bookshelves? (Assume there is no limit on the number of books per shelf.)

8. Three dice are thrown. The three tops give a multiset of numbers chosen from $\{1, 2, 3, 4, 5, 6\}$. How many different multisets of three numbers could be facing up?

9. A store sells eight different kinds of candy. In how many ways may you choose a bag of 15 pieces?

10. A fast food restaurant sells four different breakfasts. How many orders for breakfast for a family of six are possible?

11. How many labellings are there for n–element sets with j_1 of the labels used once, j_2 of the labels used twice, . . . , and j_i of the labels used i times?

12. Derive Theorem 2.3 from the result of Exercise 11.

13. In how many ways may 40 students be grouped into four groups of five each for a discussion section led by a graduate student and two groups of 10 each for a section led by a professor?

14. How many partitions of a 15–element set have two classes of size 4, three classes of size 2, and one class of size 1?

15. Two committees of four people each are chosen from a 10-member board of directors. The people on the two committees form an eight–element multiset. How many different multisets of people could be the members of the committees?

16. In how many ways may 100 distinct beads be used to make three necklaces with 20 beads and four necklaces with 10 beads?

17. In how may ways may n distinct books be arranged on k shelves, assuming that each shelf can hold all n books?

18. (De Moivre) Show that the number of ways of obtaining the positive integer m as a sum of a list of n nonnegative integers is $C(m + n - 1, m)$.

19. If m identical dice and n identical coins are thrown, how many results may be distinguished?

20. In game 1 we throw 6 dice and win if we get a 1 at least once. In game 2 we thrown 12 dice and win if we get at least two 1's. What fraction of the outcomes leads to a win in each game?

SECTION 3 PARTITIONS AND STIRLING NUMBERS

Partitions of an n–Element Set into k Classes

We have computed the number of partitions of an n–element set into k_1 classes of size 1, k_2 classes of size 2, and so on. We now take up the computation of the number of partitions of an n–element set into k classes. In terms of distributions, we are asking for the number of distributions of n distinct objects among k identical recipients.

The corresponding distribution problems for identical objects were solved in Exercise 3 and Example 2.1 of the previous section. Although we are not yet in a position to give an exact formula for the number of partitions of an n-element set into k classes, we can say a good deal about the number of such partitions.

We use $S(n, k)$ to stand for the number of partitions of an n–element set into k classes or parts, and we call the numbers $S(n, k)$ *Stirling numbers of the second kind.* (Stirling found two important families of coefficients; now we are interested in the second family **he** discussed.) It is implicit in this definition that $S(n, k) = 0$ unless n and k are integers and $1 \le k \le n$. However, in the case $n = k = 0$, we define $S(0, 0) = 1$, in effect saying that the empty set has a single partition into no parts. These numbers have some interesting analogies with the numbers $C(n; k)$, one of which is Stirling's triangle of the second kind.

Stirling's Triangle of the Second Kind

Theorem 3.1. $S(n, k) = S(n - 1, k - 1) + kS(n - 1, k)$

Proof.   Let $N = \{1, 2, \ldots, n\}$ and $N' = \{1, 2, \ldots, n - 1\}$. Then a partition of N may have one class consisting of n alone; the number of such partitions is the number of partitions of N' with $k - 1$ parts. A partition of N may have n in a class containing some elements of N' as well. Deleting n gives a partition of N' with k parts, and since n may have been in any of the k parts, k different partitions of N yield the same partition of N'. Since n either is or isn't in a class by itself, this proves the theorem. ∎

Example 3.1.   It is customary to write Stirling's triangle in the form of a right triangle, as in the portion on and below the main diagonal in Table 3.1. Note that $S(3, 2) = S(2, 1) + 2S(2, 2) = 1 + 2 \cdot 1 = 3$ says that $S(3, 2)$ is the sum of twice the number above it plus the number immediately to the left of the number directly above it. The remainder of the table is filled in similarly. ∎

There is another recursive relationship that relates the value of $S(n, k)$ to values with smaller n–values. It relates $S(n, k)$ to the numbers $S(m, k - 1)$ for all smaller values of m. Since $S(m, k - 1)$ is zero when m is smaller than $k - 1$, there are cases where this relation can be useful because of the small number of terms in the sum. Note that the relation given in Theorem 3.2 says that the entry in row n and column k of Table 2.1 is a weighted sum of the part of column $k - 1$ that lies above row n.

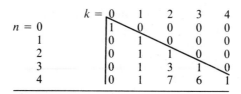

Table 2.1

Theorem 3.2. $\quad S(n, k) = \sum_{m=0}^{n-1} \binom{n-1}{m} S(m, k-1).$

Proof. If P is a partition of $N = \{1, 2, \ldots, n\}$ and if we eliminate from P the class containing n, we get a partition of a *subset* $M \subseteq \{1, 2, \ldots, n-1\}$. The resulting partition has $k-1$ parts. Every partition of every subset M of $\{1, 2, \ldots, n-1\}$ arises exactly once from this kind of construction, so the number of partitions of N into k parts is the number of partitions of subsets of $\{1, 2, \ldots, n-1\}$ into $k-1$ parts. The formula gives the number of all such partitions of a subset. ■

Example 3.2. As you can see by examining column 2 and row 4 of Table 2.1,

$$S(4, 3) = \sum_{m=0}^{3} \binom{3}{m} S(m, 2) = 1 \cdot 0 + 3 \cdot 0 + 3 \cdot 1 + 1 \cdot 3 = 6. \quad ■$$

The Inverse Image Partition of a Function

The numbers $S(n, k)$ are closely related to functions. Suppose we are given a function f from $N = \{1, 2, \ldots, n\}$ onto $K = \{1, 2, \ldots, k\}$. We use the notation

$$f^{-1}(i) = \{x \mid f(x) = i\};$$

that is $f^{-1}(i)$ is the *subset* of N consisting of elements that f maps onto i. $f^{-1}(i)$ is called the inverse image of i. Thus, we have a partition, called the *inverse image partition*,

$$\mathbf{P} = \{f^{-1}(1), f^{-1}(2), \ldots, f^{-1}(k)\}$$

of N into k classes. This is the equivalence class partition of the inverse image relation $R(f)$ associated with f. In addition, we have a function g defined on **P** by $g(f^{-1}(i)) = i$; note that g is a one–to–one function from **P** to K.

Onto Functions and Stirling Numbers

On the other hand, given a partition

$$\mathbf{Q} = \{C_1, C_2, \ldots, C_k\}$$

of N into k classes and a one–to–one function g from **Q** to K, we can define a function f from N onto K by

$$f(x) = i \text{ if } x \text{ is in } C_i.$$

Thus the number $F(N, K)$ of functions from N onto K is, by the multiplication principle, $S(n, k)k!$, the product of the number of partitions and the number of one–to–one functions. This proves

> *Theorem 3.3.* The number of functions from an n–element set onto a k–element set is $S(n, k)k!$.

Stirling Numbers of the First Kind

Stirling did not regard these numbers as representing the number of partitions, but rather as the coefficients of a certain polynomial. To see why, we first examine the Stirling numbers of the first kind. These arise from studying polynomials and at first glance appear to have nothing to do with partitions. Suppose n and k are non-negative integers and let $n \le k$. The number $\dfrac{k!}{(k - n)!}$ of one–to–one functions from an n–element set to a k–element set is denoted by $(k)_n$. As we saw in Section 2 of Chapter 1,

$$(k)_n = k(k - 1) \ldots (k - n + 1)$$

which is a polynomial of degree n in k. Because it is a polynomial, it can be written in the form

$$(k)_n = \sum_{j=0}^{n} s(n, j)k^j.$$

We use this equation to *define* Stirling numbers of the first kind. In other words, the numbers $s(n, j)$ are the coefficients that arise when we expand the product $(k)_n$. The numbers $s(n, j)$ are called *Stirling numbers of the first kind*.

For any real or even complex number x, we can define $(x)_n = x(x - 1)$ ... $(x - n + 1)$. Then $(x)_n$ is a polynomial in x of degree n. We call $(x)_n$ a "factorial power" of x of degree n. Although, technically speaking, we defined $s(n, j)$ in terms of the coefficients of a polynomial with an integral variable, it is not surprising that changing k to x does not change the coefficients $s(n, j)$.

Theorem 3.4.   For any x,

$$(x)_n = \sum_{j=0}^{n} s(n, j)x^j.$$

Proof.   A polynomial in x of degree n is completely determined by its value at any $n + 1$ distinct values of x. However, for each integer k larger than n, the substitution $x = k$ into both sides of the equation yields the equations we used to define the Stirling numbers. Thus the left– and right–hand sides of the equation are equal for infinitely many values of x, and thus for $n + 1$ values of x. Therefore, the left– and right–hand sides are equal as polynomials in x. ∎

Example 3.3.   Write $(x)_3$ as an ordinary polynomial.

$$(x)_3 = x(x - 1)(x - 2) = x(x^2 - 3x + 2)$$
$$= x^3 - 3x^2 + 2x.$$

Note that we can tell from this example that $s(3, 0) = 0$, $s(3, 1) = 2$, $s(3, 2) = -3$, and $s(3, 3) = 1$. ∎

Stirling Numbers of the Second Kind as Polynomial Coefficients

It was Stirling who first studied the relation between factorial powers of n and ordinary powers of n. The numbers $S(n, k)$ (which stand for the number of partitions of an n–element set into k classes) are called Stirling numbers of the second kind because Stirling discovered they also relate the factorial powers of n to ordinary powers of n.

Theorem 3.5. $k^n = \sum_{j=0}^{k} S(n, j)(k)_j.$

Proof.   If $n > 0$, let $F(N, J)$ denote the number of functions from an n–

element set N onto the set J. Then if we add the numbers $F(N, J)$ for *all* subsets J of a set K, we get the total number of functions from N to K. To have a notation consistent with our usual sum notation, we will write $\sum_{J=\emptyset}^{K} F(N, J)$ to stand for the sum of the quantity $F(N, J)$ over all *sets J* that contain (or equal) the empty set and are contained in (or equal to) K. Thus this notation stands for the sum of 2^k numbers, one for each subset of K. In this notation

$$k^n = \sum_{J=\emptyset}^{K} F(N, J)$$

$$= \sum_{j=0}^{k} \binom{k}{j} F(N, J) \qquad \text{(note the change to summing over integers)}$$

$$= \sum_{j=0}^{k} \binom{k}{j} S(n, j)j! \qquad \text{(by Theorem 3.3)}$$

$$= \sum_{j=0}^{k} \frac{k!}{(k-j)!} S(n, j)$$

$$= \sum_{j=0}^{k} S(n, j)(k)_j.$$

Note that our definition that $S(0, 0) = 1$ makes this theorem true for $n = 0$. ∎

Replacing k by x in this theorem (as in Theorem 3.4) presents certain technical difficulties because x would have to be the upper index of the sum. We finesse this difficulty by observing that for an integer k, $(k)_j = 0$ if $j > k$, and $S(n, j) = 0$ if $j > n$, so that

$$\sum_{j=0}^{k} S(n, j)(k)_j = \sum_{j=0}^{n} S(n, j)(k)_j.$$

Then just as we proved Theorem 3.4, we may prove

Theorem 3.6. $\quad x^n = \sum_{j=0}^{n} S(n, j)(x)_j.$

Example 3.4. Write x^3 as a combination of falling factorials.

Applying Theorem 3.6, we get

$$x^3 = \sum_{j=0}^{3} S(3, j)(x)_j$$

$$= S(3, 0)(x)_0 + S(3, 1)(x)_1 + S(3, 2)(x)_2 + S(3, 3)(x)_3$$
$$= 0 + 1(x)_1 + 3(x)_2 + 1(x)_3$$
$$= x + 3(x)(x - 1) + 1(x)(x - 1)(x - 2).$$

To verify the use of the theorem, we multiply out the polynomials and add.

$$x + 3(x^2 - x) + 1 \cdot x(x^2 - 3x + 2) = x + 3x^2 - 3x + x^3 - 3x^2 + 2x = x^3. \ \blacksquare$$

If you have had linear algebra, you may recognize that the Stirling numbers are "change of basis" coefficients. That is, they can be used to switch from one basis for the vector space of polynomials of degree n or less to another basis.

Stirling's Triangle of the First Kind

The numbers $s(n, k)$ also satisfy a recurrence relation which may be used to compute them in a triangular array. Note that $s(n, 0) = 0$ and $s(n, n) = 1$. Because of the theorem below, we can then compute a triangular array containing all the values $s(n, k)$, the Stirling numbers of the first kind.

Theorem 3.7. $s(n, k) = s(n - 1, k - 1) - (n - 1)s(n - 1, k).$

Proof.

$$\sum_{j=0}^{n} s(n, j)k^j = (k)_n = (k)_{n-1}(k - n + 1)$$

$$= \sum_{i=0}^{n-1} s(n - 1, i)k^i(k - n + 1).$$

For a given j other than 0 or n, two terms on the right contain k^j, namely

$$s(n - 1, j - 1)k^j \text{ and } (-n + 1)s(n - 1, j)k^j.$$

Equating coefficients of x^j proves the theorem. $\blacksquare$

It is interesting to note that since some of the values of $s(n, k)$ are negative, we cannot interpret $s(n, k)$ as representing the size of some set of objects as we can interpret $S(n, k)$. (See, however, Exercise 18.)

The Total Number of Partitions of a Set

The total number B_n of partitions of an n-element set is a sum of Stirling numbers of the second kind, namely

$$B_n = \sum_{k=0}^{n} S(n, k).$$

The numbers B_n are called the Bell numbers after E.T. Bell who studied how the value of B_n increases as n increases. Much of the information known about the Bell numbers can be derived from properties of the Stirling numbers.

For example, from the second recursion formula for $S(n, k)$, we get

Theorem 3.8. $B_n = \sum_{m=0}^{n-1} \binom{n-1}{m} B_m.$

Proof.   Since

$$B_n = \sum_{k=0}^{n} S(n, k),$$

$$B_n = \sum_{k=0}^{n} \sum_{m=0}^{n-1} \binom{n-1}{m} S(m, k-1)$$

$$= \sum_{m=0}^{n-1} \sum_{k=0}^{n} \binom{n-1}{m} S(m, k-1)$$

$$= \sum_{m=0}^{n-1} \binom{n-1}{m} \sum_{k=0}^{n} S(m, k-1)$$

$$= \sum_{m=0}^{n-1} \binom{n-1}{m} \sum_{j=0}^{n-1} S(m, j),$$

since $S(m, -1) = 0$. However,

$$\sum_{j=0}^{n-1} S(m, j) = \sum_{j=0}^{m} S(m, j) = B_m$$

because $S(m, j)$ is zero if $j > m$, and we know that $m \leq n - 1$. Substituting B_m for the second sum yields the formula. ∎

Theorem 3.8 may be interpreted as follows: In a partition of $N = \{1, 2, \ldots, n\}$, one class must contain n; the remaining classes must be a partition of some m–element subset of the $(n - 1)$–element set $N' = \{1, 2, \ldots, n - 1\}$.

EXERCISES

1. Write out rows 0 through 8 of Stirling's triangle for Stirling numbers of the second kind.

2. Write down the values of the Bell numbers B_n for n between 1 and 8. Observe the ratio of each value to the preceding one. Can you make any intuitive statement about how fast B_n grows as a function of n?

3. Write out rows 0 through 8 of the triangular array of Stirling numbers of the first kind.

4. Express $(x)_4$ as an ordinary polynomial.

5. Express x^4 in terms of falling factorial polynomials.

6. Express the polynomial $3x^4 + 2x^2 + 1$ in terms of factorial polynomials $(x)_4$, $(x)_3$, etc.

7. Write down the partitions of $\{1, 2, 3, 4\}$ into two parts. Did you get $S(4, 2)$ partitions?

8. What are the possible type vectors of partitions of $\{1, 2, 3, 4, 5, 6\}$ into three parts? How many partitions do you have for each type vector? Use this information to compute $S(6, 3)$.

9. A computer has eight jobs to divide among five different slave computers. Assuming each slave gets at least one job, in how many ways may this be done? Answer the same questions assuming the five slave computers are indistinguishable to the master computer.

10. In how many ways may nine distinct pieces of candy be placed in three identical paper bags so that there is some candy in each bag? What if we don't require that each bag must have some candy?

11. How many functions are there from an eight–element set onto a four–element set?

12. Express $\binom{n}{5}$ as a polynomial in n.

13. The rising factorial polynomial $(n)^k$ is defined as $(n)^k = n(n + 1) \ldots (n + k - 1)$, and is thus a polynomial of degree k in n, so $(n)^k = \sum_{j=0}^{k} t(k, j)n^j$. Find a recurrence relation that is analogous to that for $s(n, k)$ and use it to write out the first five rows of a triangular array of t–values.

14. Show that $S(n, n - 1)$ and $|s(n, n - 1)|$ are both $\binom{n}{2}$.

15. Show that $S(n, 2) = 2^{n-1} - 1$.

*16. The delta function $\delta(m, n)$ is 0 if $m \neq n$ and is 1 if $m = n$. Show that if r is the smaller of m and n,

$$\sum_{k=0}^{r} s(n, k)S(k, m) = \delta(m, n)$$

and

$$\sum_{k=0}^{r} S(n, k)s(k, m) = \delta(m, n)$$

17. Use the interpretation of Theorem 3.8 given after the proof of the theorem to find another (shorter) proof of the theorem analogous to the proof of Theorem 3.1.

*18. Recall that a one–to–one function from the set $N = \{1, 2, \ldots, n\}$ onto N is called a permutation of N; it is also called a permutation on n letters. If $f: N \rightarrow N$ is a permutation, then the sequence $a, f(a), f(f(a)), f(f(f(a))), \ldots$ must repeat because it has at most n different values. Suppose we write the nonrepeating beginning of the sequence as $(a, f(a), f^2(a), f^3(a), \ldots, f^{i-1}(a))$ where $f^i(a)$ is the first repeated element.

(a) Show that $f^i(a) = a$.

(b) We call the i–tuple in parentheses a *cycle* of the permutation f. Two cycles are equal if they correspond to the same function defined on the same subset of N. Show that each element of N is in one and only one cycle of f.

(c) Let $c(n, k)$ be the number of permutations of n letters with exactly k cycles. Show that $c(n, k) = c(n - 1, k - 1) + (n - 1)c(n - 1, k)$.

(d) How does $c(n, k)$ relate to the absolute values of the Stirling numbers of the first kind?

SECTION 4   PARTITIONS OF INTEGERS

Distributing Identical Objects to Identical Recipients

We have studied distributions of both identical and distinct objects to distinct recipients and partitions of sets of distinct objects, that is, distributions of distinct objects to identical recipients. We have not yet discussed the problem of distributing identical objects to identical recipients. For example, if we have 10 identical pieces of candy and three identical bags, in how many ways may we place some candy in each of the three bags and use up all the candy? We are really asking in how many ways may we specify three numbers that add to 10. The order in which we specify the numbers does not matter because the bags are identical, so our specification is not a list. We may repeat a number in our specification, so our specification is not a set of numbers, but a multiset. We formalize the idea in the definition of a partition of an integer.

A *partition* of a positive integer n is a multiset of positive numbers that add up to n. (A partition of an integer is thus a different concept from a partition of a set.)

Type Vector of a Partition and Decreasing Lists

There are two different ways to describe such a multiset of integers which prove useful in working with partitions. First we can specify the multiset by specifying the multiplicity of each element in it. A list of non-negative numbers $(m_1, m_2, \ldots, m_n)$ such that

$$\sum_{i=1}^{n} im_i = \mathrm{n}$$

tells us the multiplicity of each integer i in the multiset. (Note that n is *not* the number of elements in the multiset; it is the sum of the elements of the multiset.) Such a list could also be a type vector of a partition of an n–element set, specifying the number of classes of size i for each i. We call such a list a *type representation* of a partition. In effect, when we study partitions of integers, we are studying type vectors of partitions of sets.

Our second description of partitions uses the idea of a "decreasing list" of integers. A decreasing list is a list $(i_1, i_2, \ldots, i_k)$ with $i_1 \geq i_2 \geq \cdots \geq i_k$. A decreasing list whose entries add up to n describes a unique partition of n because a multiset of integers may be listed in decreasing order in exactly one way. A *decreasing list representation* of a partition of n is a decreasing list whose entries add up to n.

Example 4.1. The multiset $\{1, 1, 3, 3, 4\}$ is a partition of 12. Its type vector is $(2, 0, 2, 1, 0, 0, 0, 0, 0, 0, 0, 0)$. It is customary to delete trailing zeros and write the type vector as $(2, 0, 2, 1)$ or $(2, 0, 2, 1, 0, \ldots)$. Its decreasing list representation is 4, 3, 3, 1, 1. ∎

The Number of Partitions of n into k Parts

By the number of *parts* of a partition, we mean the size of the multiset, i.e. either the sum of the multiplicities or the length of the decreasing list in our two representations. We use $P(n, k)$ to denote the number of partitions of the integer n into k parts. The theorem below allows us to compute $P(n, k)$ recursively in a fashion somewhat similar to our recursive computations of $C(n; k)$ and $S(n, k)$.

Theorem 4.1. $\displaystyle\sum_{i=1}^{k} P(n, i) = P(n + k, k).$

Proof.   The left–hand side of the equation is the number of partitions of n into k or fewer parts. Suppose we are given a partition P of n into i parts with type vector $(k_1, k_2, \ldots, k_n)$. From this partition, we can construct a partition of $n + k$ with type vector $(k - i, k_1, k_2, \ldots, k_n, 0, \ldots, 0)$ obtained by adding 1 to each of the i nonzero parts of P and then adding $k - i$ parts of size 1. Since

$$k - i + \sum_{j=1}^{n} k_j(i + 1) = k - i + n + i = n + k$$

and

$$k - i + \sum_{j=1}^{n} k_j = k - i + i = k,$$

the partition constructed is a partition of $n + k$ into k parts. Also, given a partition of $n + k$ into k parts, we can construct a partition of n into k or fewer parts by deleting all parts of size 1 and subtracting 1 from the remaining parts. Since these two constructions reverse each other, the number of partitions of $n + k$ into k parts is the number of partitions of n into k or fewer parts. ∎

By letting $n + k$ be m (and thus letting $n = m - k$), we obtain a corollary which simplifies the recursive computation.

Corollary 4.2. $P(m, k) = \sum_{i=1}^{k} P(m - k, i).$

Note that $P(n, 1) = P(n, n) = 1$. Thus if we write a table of values of $P(n, k)$ with the rows corresponding to various values of n, then we have 1's as the first and last nonzero entries in each row. Further, to get the k-th entry in row m, we go up k rows and add up the first k entries of that row. Since the number of partitions of n into k parts is 0 if $k > n$, we have zeros in the upper right–hand portion of our table. We don't have a zero row or column because we have only defined partitions of positive integers. Figure 2.1 shows these properties.

	$k = 1$	$k = 2$	$k = 3$	$k = 4$	$k = 5$	$k = 6$	$k = 7$
$n = 1$	1	0	0	0	0	0	0
$n = 2$	1	1	0	0	0	0	0
$n = 3$	1	1	1	0	0	0	0
$n = 4$	1	2	1	1	0	0	0
$n = 5$	1	2	2	1	1	0	0
$n = 6$	1	3	3	2	1	1	0
$n = 7$	1	3	4	3	2	1	1

Figure 2.1

Ferrer's Diagrams

The idea of a partition of n as a division of a multiset of n identical objects into k parts suggests a convenient geometric visualization. We start with n identical squares, write our partition in decreasing list form as $(n_1, n_2, \ldots, n_k)$ and then place n_1 squares in a row, each touching the previous one. Next we make a row of n_2 squares with the first square of row 2 below and touching the first square of row 1. In general, row i has n_i squares and begins below the first square of row $i - 1$. The geometric figure that results is called a Ferrer's diagram of the partition. The decreasing list $(5, 3, 3, 1)$ represents the partition $5 + 3 + 3 + 1$ of 12; its Ferrer's diagram is in Figure 2.2.

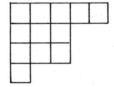

Figure 2.2

Conjugate Partitions

By flipping a Ferrer's diagram over the 45° line, i.e. over its "main diagonal," we get a new diagram whose rows are the columns of the old one. The partition to which this new diagram corresponds to is the *conjugate* of the old one. We use $(5, 3, 3, 1)^*$ to denote the decreasing list of the partition conjugate to $(5, 3, 3, 1)$ shown in Figure 2.3. By studying Figure 2.3, we see that a natural formal definition of a conjugate would read "the decreasing list *conjugate* to $(n_1, n_2, \ldots, n_k)$ is the list $(r_1, r_2, \ldots, r_s)$ in which r_i is the number n_j's larger than or equal to i." (In geometric terms, the number of squares in row i of P^* is equal to the number of rows of P with i or more squares.) The notion of a conjugate gives us additional information about the number of partitions of an integer into k parts.

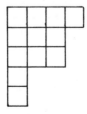

Figure 2.3

Theorem 4.2.   The number of partitions of n into k parts is equal to the number of partitions of n into parts the largest of which is k.

Proof.   A partition has k parts if and only if its conjugate has its largest part equal to k.   Since each partition is the conjugate of some partition, the theorem follows. ■

The Total Number of Partitions of n

There are no known formulas that compute $P(n, k)$ directly from n and k as there are for $C(n; k)$ and $S(n, k)$.   As a result, the numbers $P(n, k)$ have been the subject of intensive study.   Of equal interest is the total number $P(n)$ of partitions of n. Explicit formulas for $P(n)$ are also not known, but there are techniques we will study later that provide information about $P(n)$.   There are elementary results connecting the total number of partitions with the number of partitions whose smallest part is some integer.   For example,

Theorem 4.3.   The number of partitions of n is equal to the number of partitions of $n + 1$ whose smallest part is 1.

Proof.   Suppose we are given a partition of n of type $(k_1, k_2, \ldots, k_n)$. The partition of type $(k_1 + 1, k_2, \ldots, k_n)$ is a partition of $n + 1$ whose smallest part has size 1.   Conversely, if $(h_1, h_2, \ldots, h_{n+1})$ is a partition of $n + 1$ with $h_1 > 0$, then $h_{n+1} = 0$ and $(h_1 - 1, h_2, \ldots, h_n)$ is the type representation of a partition of n. ■

We will find that the numbers $P(n, k)$ and $P(n)$ can be studied fruitfully by means of the technique of generating functions introduced in the next chapter. We have now carried our study of distributions of objects among recipients as far as is practical without the use of these algebraic methods.

EXERCISES

1. In how many ways may seven identical pieces of candy be placed in three identical bags so that no bag is empty?   What if empty bags are allowed?

2. Complete the table of values of $P(n, m)$ through $m = 10$ and $n = 10$.

3. Draw the Ferrer's diagrams for partitions with the following type vector representations (in which ellipses denote zeros).   What integer is being partitioned?

     (a) $(2, 1, 3, 0, 1, 0, 0, 2, 0, \ldots)$
     (b) $(0, 0, 2, 2, 2, 0, \ldots)$
     (c) $(2, 0, 1, 0, 2, 0, \ldots)$

4. Draw the Ferrer's diagrams of partitions with the following decreasing list representations.   What integer is being partitioned?

     (a) $(6, 5, 5, 3)$
     (b) $(5, 4, 3, 2, 1)$
     (c) $(8, 6, 6, 4, 2, 2)$

5. Find the conjugates of the partitions in Exercise 3 and draw their Ferrer's diagrams.

6. Find the conjugates of the partitions in Exercise 4 and draw their Ferrer's diagrams.

7. Show that the number of partitions of n into at most k parts is equal to the number of partitions of n into parts of size at most k.

8. Find $P(20, 10)$. (Hint: Exercise 2 simplifies this problem.)

9. Find $P(15, 8)$.

10. Show that the number of partitions of n into 2 parts is $\frac{n}{2}$ if n is even and $\frac{n-1}{2}$ if n is odd.

11. In how many ways may 15 identical apples be placed in five bags so that each bag has at least two apples?

12. Show that $P(n) - P(n - 1)$ is the number of partitions of n into parts greater than 1.

13. Using the result of Exercise 12, show that

$$P(n + 2) + P(n) \geq 2P(n + 1).$$

14. Show that $P(n, k) \geq \frac{1}{k!} \binom{n-1}{k-1}$.

15. Prove that the number of partitions of n into unequal odd parts is equal to the number of partitions of n conjugate to themselves. (Such partitions are called self-conjugate partitions.) (Hint: Draw the Ferrer's diagram for the following pairs of partitions of 9: (9), (5, 1, 1, 1, 1), (5, 3, 1), (3, 3, 3). For each pair, show that the Ferrer's diagram of the first may be obtained by lining up all squares from the first row *and* column of the second, then all *remaining* squares from the second row and column of the second, then all remaining squares of the third row and column, etc.)

∗ 16. Show that the number of partitions of n into unequal parts is the number of partitions of n into odd parts.

17. By subtracting $p(n - 1, k - 1)$ from $p(n, k)$ and applying Corollary 4.2, derive a recursion formula for $p(n, k)$ with two terms, one involving $p(n - 1, k - 1)$ and the other involving $p(n - k, k)$.

SUGGESTED READING

Berge, Claude: *Principles of Combinatorics*, Academic Press 1971, Chapters 1 and 2.

Knuth, Donald E.: *The Art of Computer Programming, Fundamental Algorithms*, Vol. 1 (Second Edition), Addison Wesley 1973, especially Section 1.2.6.

Lovász, László: *Combinatorial Problems and Exercises*, North-Holland 1979, Chapter 1.

Niven, Ivan and Zuckerman, Herbert S.: *An Introduction to the Theory of Numbers*, Wiley 1960, Chapter 10.

Riordan, John: *An Introduction to Combinatorial Analysis*, Wiley 1958, Chapters 1, 5, and 6.

3 Algebraic Counting Techniques

SECTION 1   THE PRINCIPLE OF INCLUSION AND EXCLUSION

The Size of a Union of Three Overlapping Sets

The sum principle tells us that the size of the union of a family of disjoint sets is the sum of the sizes of the sets. To determine the size of a union of overlapping sets, we must clearly use information about how they overlap—but how? Figure 3.1 shows the Venn diagram of two overlapping sets.

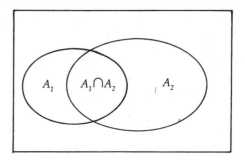

Figure 3.1

Note that if we add $|A_1|$ and $|A_2|$, we include the size of $A_1 \cap A_2$ twice in this sum, so

$$|A_1| + |A_2| = |A_1 \cup A_2| + |A_1 \cap A_2|.$$

Thus we get the formula

$$|A_1 \cup A_2| = |A_1| + |A_2| - |A_1 \cap A_2|.$$

Example 1.1.   Find a formula similar to the one above for the size of a union of three sets.   Figure 3.2 shows a picture of three overlapping sets.

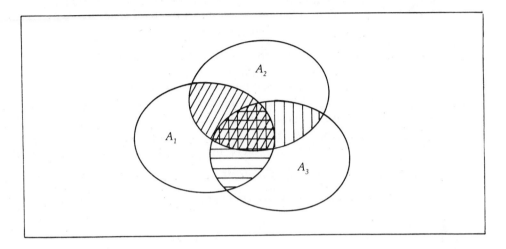

Figure 3.2

In the sum $|A_1| + |A_2| + |A_3|$, an element of $A_1 \cap A_2$ is included in at least two terms; an element of $A_1 \cap A_2 \cap A_3$ is included in all three terms.   In the sum

$$|A_1| + |A_2| + |A_3| - |A_1 \cap A_2| - |A_1 \cap A_3| - |A_2 \cap A_3| \qquad (1.1)$$

an element in A_1, but not A_2 or A_3, is included in only one term.   An element in A_1 and A_2, but not in A_3, is included in two positive terms and one negative term, so it is counted exactly once in this difference.   An element in A_1, A_2, and A_3 is included in three positive terms and three negative terms, so it is not counted at all in the difference (1.1).   Therefore, the expression in (1.1) counts once each element of $A_1 \cup A_2 \cup A_3$ except for the elements in $A_1 \cap A_2 \cap A_3$.   Thus,

$$|A_1 \cup A_2 \cup A_3| = |A_1| + |A_2| + |A_3| - |A_1 \cap A_2| - |A_1 \cap A_3| \qquad (1.2)$$
$$- |A_2 \cap A_3| + |A_1 \cap A_2 \cap A_3|. \quad \blacksquare$$

A formula such as (1.2) is called an inclusion–exclusion formula and the argument that led to it is called an inclusion–exclusion argument.   We shall formalize the principle of inclusion and exclusion as a general counting principle, but first let us do another example.

The Number of Onto Functions

Recall that a function $f : M \to K$ is *onto* if each element of K occurs as the image $f(x)$ of some element x of M. We have not yet been able to compute the number of functions from an m–element set M onto a k–element set K. When K has only a few elements, the computation can be made directly.

  Example 1.2. Find a formula for the number of functions from an m–element set onto a k–element set.

  If, for example, $K = \{a\}$, then there is one function from M to K and it is onto. If $K = \{a_1, a_2\}$, there are 2^m functions from M to K. Of these, one skips a_1 because it maps to a_2, and one skips a_2 because it maps to a_1. Thus there are $2^m - 2$ functions from M onto K.

  Now if $K = \{a_1, a_2, a_3\}$, there are 3^m functions from M to K. Of these, 2^m functions skip a_1 because they map into $\{a_2, a_3\}$. In fact, for each of the $\binom{3}{1}$ sets $\{a_i\}$, there are 2^m functions that skip the element of $\{a_i\}$. However the difference

$$3^m - \binom{3}{1} 2^m, \tag{1.3}$$

is smaller than the number of onto functions because some functions that skip $\{a_1\}$ skip $\{a_2\}$ as well, and thus are subtracted twice. In fact, each of the $\binom{3}{2}$ sets $\{a_i, a_j\}$ is skipped by exactly one function. Thus we have subtracted $\binom{3}{2} \cdot 1$ too many functions in the difference (1.3). The number of functions from M onto K is then

$$3^m - \binom{3}{1} 2^m + \binom{3}{2} \cdot 1. \tag{1.4}$$

The same kind of argument with $K = \{a_1, a_2, a_3, a_4\}$ gives the following. There are 4^n functions from N to K. These functions could skip 1, 2, or 3 (but not all 4) elements of K. The number skipping each of the $\binom{4}{1}$ sets $\{a_i\}$ is 3^n. In the difference $4^n - \binom{4}{1} 3^n$, we subtract the functions that skip each of the $\binom{4}{2}$ two-element sets $\{a_i, a_j\}$ at least twice. Let us examine the quantity

$$4^n - \binom{4}{1} 3^n + \binom{4}{2} 2^n. \tag{1.5}$$

The functions that skip exactly one element are excluded from the quantity exactly once. The functions that skip exactly two elements are excluded from the quantity twice, but are included back in the $\binom{4}{2} \cdot 2^n$ term exactly once, and so are excluded exactly once by the quantity in (1.5). However, the functions that skip the three–element set $\{a_i, a_j, a_k\}$ are excluded three times in the negative term and are included three times (once for each two–element subset of $\{i, j, k\}$) in the term $\binom{4}{2} 2^m$. Thus they have yet to be excluded from the sum. Only functions that skip three elements have yet to be taken care of; there are $\binom{4}{3} \cdot 1$ such functions still to be excluded. Thus we have

$$4^m - \binom{4}{1} 3^m + \binom{4}{2} 2^m - \binom{4}{3} 1^m \qquad (1.6)$$

functions from an m–element set onto a four–element set. Formulas (1.4) and (1.6) suggest immediately that the number of functions from an m–element set onto a k–element set is

$$k^m - \binom{k}{1} (k-1)^m + \cdots \pm \binom{k}{k-1} 1^m = \sum_{i=0}^{k-1} (-1)^i \binom{k}{i} (k-i)^m$$
$$= \sum_{i=0}^{k} (-1)^i \binom{k}{i} (k-i)^m.$$

(The change from $k-1$ to k in the last equality is simply because $(k-k)^m = 0^m = 0$, corresponding to the fact that no functions skip all of K.) We shall prove the formula works later. ■

Counting Arrangements with or without Certain Properties

Rather than trying to prove each formula we get by an inclusion–exclusion argument, let us look for a general description of the kind of problem we are solving and state a general principle that can be applied to such problems. Each problem is a problem of counting arrangements. In Example 1.1, our arrangements were just the elements of the sets A_i. In Example 1.2, our arrangements were the functions from M to K. Our arrangements had certain properties that were interesting to us. In Example 1.1, we had a property P_i for each set A_i. That is, P_i was the property

"The element x is in set A_i".

In Example 1.2, we had a property P_j for each element j of K. P_j was the property

"The function f skips the element j of K".

In Example 1.1, we wanted to know how many arrangements have at least one of the properties. (Note that the union of A_1, A_2, and A_3 is the set of elements in at least one of the sets.) In Example 1.2, we wanted to know how many arrangements had exactly none of the properties.

What kind of data were we given in the two examples? In Example 1.1, the sizes of A_i, $A_i \cap A_j$, and $A_1 \cap A_2 \cap A_3$ were known. The size of A_i is the number of elements with property P_i (and maybe some more properties as well), the size of $A_i \cap A_j$ is the number of elements with properties P_i and P_j, etc. In Example 1.2, we computed the number of functions skipping i; this is the number of functions with at least property P_i. We also computed the number of functions skipping i and j; this is the number of functions with at least properties P_i and P_j.

The Basic Counting Functions $N_\geq$ and $N_=$

Thus the ingredients we should expect to deal with in our principle will be a set of arrangements, a set of properties the arrangements might or might not have, and for each set S of properties the number $N_\geq(S)$ of arrangements having the properties in S (and perhaps other properties as well). (The notation $N_\geq$ was suggested by Bender and Goldman to remind us that the numbers we are dealing with are numbers of arrangements with a certain set of properties and *perhaps other* properties as well.) In Example 1.1, $N_\geq(\{P_i\}) = |A_i|$, $N_\geq(\{P_i, P_j\}) = |A_i \cap A_j|$, etc. In Example 1.2, with $K = \{a_1, a_2, a_3, a_4\}$,

$$N_\geq(\emptyset) = 4^m, \ N_\geq(\{P_i\}) = 3^m,$$
$$N_\geq(\{P_i, P_j\}) = 2^m, \ N_\geq(\{P_i, P_j, P_k\}) = 1^m,$$
$$N_\geq(\{P_1, P_2, P_3, P_4\}) = 0.$$

The principle of inclusion tells us how to find the number of arrangements that have exactly a certain set of properties. For example, in Example 1.2, we want to know how many functions have exactly the empty set of skips. We use $N_=(S)$ to stand for the number of arrangements with exactly the properties in S. From our examples, we expect the principle to tell us that to find $N_=(S)$, we start with $N_\geq(S)$, subtract the number of arrangements with at least one more property besides the properties in S, add the number with at least two more properties, and so on. That is exactly what the summation notation in Theorem 1.1 tells us to do. The symbol J is a dummy variable that takes on as its values all possible sets which are subsets of P and contain S as a subset. Except for the fact that J, S and P are sets, we use and read the summation like ordinary summation notation.

The Principle of Inclusion and Exclusion

Theorem 1.1. (Principle of Inclusion and Exclusion). Suppose A is a set of arrangements and suppose **P** is a set of properties the arrangements may have. For each $S \subseteq \mathbf{P}$, suppose $N_{\geq}(S)$ is the number of arrangements with at least the properties in S and suppose $N_{=}(S)$ is the number of arrangements with the properties in S and no others. Then for each $S \subseteq \mathbf{P}$,

$$N_{=}(S) = \sum_{J=S}^{P} (-1)^{|J|-|S|} N_{\geq}(J).$$

Proof. Suppose an arrangement has only the properties in S. Then it appears in only one term of the sum. If it has one more property P_j in addition to the properties in S, it is counted once in $N_{\geq}(S)$ and once in $N_{\geq}(S \cup \{P_j\})$ with opposite sign, and so is not counted. If an arrangement has i properties not in S, it appears $\binom{i}{j}$ times in terms of the form

$$N_{\geq}(S \cup \{P_{i_1}, P_{i_2}, \ldots, P_{i_j}\}),$$

and the sign alternates according to whether j is even or odd. Thus, the total number of times this arrangement is counted is

$$1 - \binom{i}{1} + \binom{i}{2} - \binom{i}{3} + \cdots + (-1)^i \binom{i}{i} = (-1 + 1)^i$$

by the binomial theorem. Therefore, an arrangement with more properties than those in S is counted 0 times. Thus, the sum counts exactly the desired arrangements. ∎

As an example of how to use Theorem 1.1, we prove the formula from Example 1.2.

Onto Functions and Stirling Numbers

Theorem 1.2. The number of functions from an m-element set onto a k-element set is equal to

$$\sum_{i=0}^{k} (-1)^i \binom{k}{i} (k-i)^m = k^n - \binom{k}{1} (k-1)^m + \cdots + (-1)^{k-1} \binom{k}{k-1} 1^m.$$

Proof. As described above, property P_j is "the function f skips the element j of K". $N_\geq(P_{j_1}, P_{j_2}, \ldots, P_{j_i})$ is the number of functions skipping at least the i elements in $\{j_1, j_2, \ldots, j_i\}$. This is the number of functions mapping into a $(k - i)$–element set, which is the number $(k - i)^m$. As noted above, we want to compute $N_=(\emptyset)$. By Theorem 1.1,

$$N_=(\emptyset) = \sum_{I=\emptyset}^{K} (-1)^{|I|}(k - i)^m.$$

Using the fact that there are $\binom{k}{i}$ subsets I of K of size i, we get

$$N_=(\emptyset) = \sum_{i=0}^{k} (-1)^i \binom{k}{i} (k - i)^m. \qquad \blacksquare$$

Corollary 1.3. The number of partitions of an m–element set into k classes is given by

$$S(m, k) = \frac{1}{k!} \sum_{i=0}^{k} (-1)^i \binom{k}{i} (k - i)^m.$$

Proof. See Theorem 3.3 of Chapter 2. $\blacksquare$

Examples of Using the Principle of Inclusion and Exclusion

Example 1.3. A used car dealer has 18 cars on the lot. Nine of them have an automatic transmission, 12 have power steering, and eight have power brakes. Seven have both automatic transmission and power steering, four have automatic transmission and power brakes, and five have power steering and power brakes. Three cars have power steering and brakes and automatic transmission. How many cars have automatic transmission only? How many cars are "stripped?"

To illustrate the principle of inclusion and exclusion, we let our set of arrangements be the cars and our set of properties be $\{AT, PS, PB\}$. For any set X of properties, let $N_\geq(X)$ be the number of vehicles with the properties in the set X, and perhaps some other properties as well. Let $N_=(X)$ be the number of vehicles with exactly the properties in X. The problem gives us the values of $N_\geq$ and asks for $N_=(\{AT\})$ and $N_=(\emptyset)$. By the principle of inclusion and exclusion

$$N_=(\{AT\}) = N_\geq(\{AT\}) - N_\geq(\{AT, PS\}) - N_\geq(\{AT, PB\}) + N_\geq(\{AT, PS, PB\})$$
$$= 9 - 7 - 4 + 3 = 1$$
$$N_=(\emptyset) = N_\geq(\emptyset) - N_\geq(\{AT\}) - N_\geq(\{PS\}) - N_\geq(\{PB\}) + N_\geq(\{AT, PS\})$$
$$+ N_\geq(\{AT, PB\}) + N_\geq(\{PS, PB\}) - N_\geq(\{AT, PB, PS\})$$
$$= 18 - 9 - 12 - 8 + 7 + 4 + 5 - 3 = 2.$$

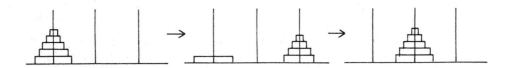

Figure 3.3

This computation is especially easy to visualize by examining the Venn diagram in Figure 3.3. When we subtract the number of cars with each property from the total number of cars, we subtract the areas labelled as *AT, PB*; *PB, PS*; and *AT, PS* out twice. Further, we subtract the area labelled as *AT, PB, PS* three times. When we correct for the oversubtraction by adding in the numbers corresponding to the areas *AT, PB*; *PB, PS*; and *AT, PS*, we unfortunately add in the area labelled *AT, PB, PS* just as often as we subtracted it before. By subtracting out the number corresponding to this last area one more time, we find the number of items corresponding to the area inside the rectangle, but none of the circles. ■

Example 1.4. A group of couples sits around a circular table for a group discussion of marital problems. In how many ways may the group be seated so that no husband and wife sit together?

In this case, our arrangements are seating arrangements and our properties are "husband and wife *i* sit together." Note that there is no requirement that the seating arrangements alternate sex. We assume the couples are numbered as couple 1 through couple *n*. We let $N = \{1, 2, \ldots, n\}$. Note that this N is a set and not the $N_=$ or $N_\geq$ of Theorem 1.1. We let $N_\geq(I)$ be the number of arrangements in which the couples numbered by the integers in I sit together as couples, and let $N_=(I)$ be the number of arrangements in which couples numbered by integers in I—*and* no other couples—sit together. We wish to know $N_=(\emptyset)$. To compute $N_\geq(I)$, we note that by seating the i couples first and allowing the other $2n - 2i$ people to sit down at the remaining places, we are in effect arranging $2n - i$ units (think of $2n - 2i$ skinny people and i fat people each of whom require two chairs) around a circular table. We know that there are $(2n - i - 1)!$ such arrangements. Now each couple can occupy its two chairs in two ways. Therefore, after the $2n - i$ units are assigned to places, there are 2^i ways for the units to be placed in their assigned places. Thus, there are $2^i(2n - i - 1)!$ ways to seat the people so that couples labelled by elements of the i-element set I sit together. Thus $N_\geq(I) = 2^i(2n - i - 1)!$, and

$$N_=(\emptyset) = \sum_{I=\emptyset}^{N} (-1)^{|I|} N_\geq(I) = \sum_{i=0}^{N} (-1)^i \binom{n}{i} 2^i(2n - i - 1)! \quad \blacksquare$$

Example 1.5. A bookcase has five shelves each with 10 books on it. Each shelf contains books on one of five different subjects. In how many ways may the books be removed for dusting and returned to the shelves so that each subject still has a shelf of its own, even though no shelf has a book previously on it?

The arrangements of interest here are arrangements of books on shelves. If we take as property j, "shelf j gets the same subject it had last time," then for a given set of I of i shelves, there are 10! ways to return the books of a given subject to each shelf in I, giving $(10!)^i$ ways to fill these shelves. Next there are $(5 - i)$ other shelves and there are $(5 - i)!$ ways to assign subjects to shelves so that the shelves in I—and perhaps some others—get their original subjects back. Then there are $(10!)^{5-i}$ ways to assign the books to these shelves. Thus we have $(10!)^5(5 - i)!$ arrangements having at least the properties in I, so $N_\geq(I) = (10!)^5(5 - i)!$. Since we want $N_=(\emptyset)$, we apply the principle of inclusion and exclusion to get (with $K = \{1, 2, 3, 4, 5\}$)

$$N_=(\emptyset) = \sum_{I=\emptyset}^{K} (-1)^i(10!)(5 - i)!$$

$$= \sum_{i=0}^{5} (-1)^i(10!)^5 \binom{5}{i} (5 - i)!$$

$$= 5!(10!)^5 \sum_{i=0}^{5} \frac{(-1)^i}{i!}.$$

Had we k shelves rather than 5, each occurrence of 5 would be replaced by k. $\blacksquare$

Derangements

A classic problem in the application of the principle of inclusion and exclusion is the "problem of derangements." A derangement of the set $N = \{1, 2, \ldots , n\}$ is a one–to–one function $f : N \rightarrow N$ such that $f(i) \neq i$ for any i. In more colorful language, the derangement problem is called the hat check problem (see Exercise 14). To solve the derangement problem, we use as our objects the one–to–one functions from N to N and as our property P_i the property that $f(i) = i$. Then the number of derangements is $N_=(\emptyset)$.

* Level Sums and Inclusion-Exclusion Counting

In Example 1.4, we might instead be asking ourselves, "What is the probability that no husband and wife sit together?" Since there are $(2n - 1)!$ possible seating arrangements, we divide $N_=(\emptyset)$ by $(2n - 1)!$ to find this probability. In Example 1.5, we might have wanted the probability that "no shelf has a book previously on it." From this point of view, it is equally interesting to ask, "What is the probability that exactly i couples sit together?" or "What is the probability that exactly i shelves have the books which were previously on them?" To compute the probability that i couples sit together, we need the number of arrangements in which exactly i husbands and wives are side by side: note that this is *not* $N_=(I)$ for some set I, but rather the sum over *all sets* I of size i of $N_=(I)$. Similarly for the bookshelves, we would want the sum over all sets I of size i of $N_=(I)$. This kind of sum makes sense for any inclusion–exclusion problem, and represents the number of arrangements that have exactly i properties. In Examples 1.4 or 1.5, we could have computed this sum by multiplying one value of $N_=(I)$ by an appropriate binomial coefficient (which one?), since all values $N_=(I)$ for all sets of size i are equal. On the other hand, in a problem like Example 1.3, $N_=(I)$ depends on more than just the size of I. It turns out that this sum of $N_=$ values is related to the sum of the $N_{\geq}$ values in a way that reduces the sum over subsets in the inclusion–exclusion theorem to a sum over integers.

In order to see what this relationship is in any inclusion–exclusion situation, we introduce two new functions related to the two functions $N_=$ and $N_{\geq}$. These two new functions are called *level sums* of the old functions. They are defined by

$$N_{=}^{\pm}(i) = \sum_{I:|I|=i} N_=(I)$$

and

$$N_{\geq}^{\pm}(i) = \sum_{I:|I|=i} N_{\geq}(I).$$

In other words, $N_=^+(i)$ is the sum of all values of $N_=(I)$ for sets I of size i and $N_\geq^+(i)$ is the sum of all values of $N_\geq(I)$ for sets I of size i. Then by substitution of the formula of Theorem 1.1 for $N_=(I)$,

$$N_=^+(i) = \sum_{I:|I|=i} N_=(I) = \sum_{I:|I|=i} \sum_{S=I}^{p} (-1)^{|S|-|I|} N_\geq(S)$$

$$= \sum_{S:|S|\geq i} \binom{|S|}{i} (-1)^{|S|-i} N_\geq(S)$$

$$= \sum_{j=i}^{p} \sum_{S:|S|=j} \binom{j}{i} (-1)^{j-i} N_\geq(S)$$

$$= \sum_{j=i}^{p} \binom{j}{i} (-1)^{j-i} \sum_{S:|S|=j} N_\geq(S)$$

$$= \sum_{j=i}^{p} (-1)^{j-i} \binom{j}{i} N_\geq^+(j).$$

This proves

Theorem 1.4. If $N_=$ and $N_\geq$ are related as in Theorem 1.1 and $N_=^\pm$ and $N_\geq^+$ are the "level sums" given by

$$N_=^\pm(i) = \sum_{I:|I|=i} N_=(I)$$

$$N_\geq^\pm(i) = \sum_{I:|I|=i} N_\geq(I),$$

then

$$N_=^\pm(i) = \sum_{j=i}^{p} (-1)^{j-i} \binom{j}{i} N_\geq^+(j).$$

Proof. Given above. ■

Examples of Level Sum Inclusion and Exclusion

Example 1.6. In Example 1.1, how many cars have exactly one feature? We will need to know $N_\geq^+(1)$, $N_\geq^+(2)$ and $N_\geq^+(3)$. By inspection and addition, $N_\geq^+(1) = 9 + 12 + 8 = 29$, $N_\geq^+(2) = 7 + 4 + 5 = 16$ and $N_\geq^+(3) = 3$. Then

$$N_=^+(1) = \sum_{j=1}^{3} (-1)^{j-1} \binom{j}{1} N_{\geq}^+(j)$$

$$= (-1)^0 \binom{1}{1} \cdot 29 + (-1)^1 \binom{2}{1} \cdot 16 + (-1)^2 \binom{3}{1} \cdot 3$$

$$= 29 - 32 + 9 = 6. \quad \blacksquare$$

Example 1.7.   Use Theorem 1.4 to do Example 1.5.

For any integer i, $N_{\geq}^+(i)$ is the sum over all sets I of size i of $N_{\geq}(I)$.   Thus $N_{\geq}^+(i) = \binom{5}{i} (10)!^5 (5-i)! = \frac{5!}{i!} (10)!^5$.   Therefore

$$N_=^+(0) = \sum_{i=0}^{5} (-1)^i \frac{5!}{i!} (10!)^5 = 5!(10!)^5 \sum_{i=0}^{5} \frac{(-1)^i}{i!}. \quad \blacksquare$$

EXERCISES

1. How many functions are there from a five–element set onto a three–element set?

2. What is the Stirling number $S(5, 3)$?

3. In how many ways may 12 distinct pieces of candy be passed out to six children so that each child gets a piece?

4. In how many ways may 12 distinct pieces of candy be placed into six identical bags with no bag left empty?

5. In Example 1.1, how many cars have only power brakes?   How many cars with power brakes do not have automatic transmission?

6. In an experiment on the effects of fertilizer on 27 plots of a new breed of lawn grass, eight plots are given nitrogen, phosphorous, and potash fertilizers, 12 plots are given at least nitrogen and phosphorous, 12 plots are given phosphorous and potash, and 12 plots are given nitrogen and potash.   Eighteen plots receive nitrogen, 18 receive phosphorous and 18 receive potash.   How many plots are left unfertilized?

7. In Exercise 6, how many plots are given exactly one of the three nutrients?

8. In a college class of 28 people, nine are women from the East, five are Easterners over 20 and seven are women over 20.   Sixteen people in the class are Easterners, 16 are women and 12 are over 20.   There are two Eastern women over 20 in the class.   How many men of age 20 or under are in the class?   How many of these men are from the East?   How many of these men are not Easterners?   (This problem is probably best done by Venn diagrams like Figures 3.1 and 3.2 because the questions are not all direct applications of formulas.)

9. The number of members of $A_1 \cup A_2 \cup A_3$ that lie in *none* of the sets A_1, A_2 or A_3 is zero.   Use the principle of inclusion and exclusion to show that

$$|A_1 \cup A_2 \cup A_3| - |A_1| - |A_2| - |A_3| + |A_1 \cap A_2|$$
$$+ \cdots - |A_1 \cap A_2 \cap A_3| = 0.$$

How does this relate to Example 1.1?

10. Show that

$$
\begin{aligned}
|A_1 \cup A_2 \cup \cdots \cup A_n| = {} & |A_1| + |A_2| + \cdots + |A_n| \\
& - |A_1 \cap A_2| - \cdots - |A_i \cap A_j| - \cdots \\
& + |A_1 \cap A_2 \cap A_3| + \cdots + |A_i \cap A_j \cap A_k| + \cdots
\end{aligned}
$$

$$
\cdot
$$
$$
\cdot
$$
$$
\cdot
$$
$$
\cdot
$$

$$
- (-1)^n |A_1 \cap A_2 \cap \cdots \cap A_n|.
$$

(Hint: Exercise 9 suggests the simplest method for doing this problem.)

11. In Example 1.4, in how many seating arrangements are exactly i couples seated side by side?

12. You are to make a necklace with n different pairs of beads. The beads in a pair have the same shape, but different colors. In how many ways can you make the necklace so that no identically shaped beads are side by side?

* 13. In Exercise 12, what is the number of necklaces with exactly one pair of identically shaped beads side by side?

14. A hat check person discovers that n people's hats have been mixed up and returns these hats to the owners at random. In how many ways may the hats be returned so that all of the owners get someone else's hat? What proportion of the total number of distributions do these represent? If you have had calculus, what limiting value does this proportion have? (Hint: the limiting value involves the base e of the natural logarithms and is best found by using what you know about power series.)

15. In how many ways may the hats of Exercise 14 be passed out so that exactly i people receive their own hats?

16. In how many ways may eight pieces of identical candy be passed out to three children so that each child gets at least one piece but no child gets more than 4 pieces?

17. In how many ways may k identical pieces of candy be passed out to n children so that no child gets m or more pieces?

18. In how many ways may n distinct books be arranged on k shelves so that no shelf gets m or more books?

19. In a nursery school, m children take off their wet socks and set them on the radiator to dry. By the end of the day, after the socks have become hopelessly mixed up, each child takes two socks. In how many ways may the children prepare to go home in such a way that each child has at least one sock belonging to someone else? In how many ways may they go home with each child wearing two socks belonging to someone else?

* 20. In Example 1.4, if we require that the couples be seated in the order man, woman, man, woman, . . . , i.e. alternating sex, then we have a classic problem in combinatorics called the "menage problem." The main difference in the two problems lies in the computation of $N_{\geq}(I)$. Of course if I is empty, $N_{\geq}(I) = N_{\geq}(\emptyset)$ is just $(2n - 1)!$. If I has some couples in it, pick the first couple (say, in alphabetical order) and seat them. There are $2n$ ways to choose places for them and two ways to seat them in these two

places. Since the seating pattern must alternate in sex, this determines for the entire table which places are for men and which are for women. Once you choose two side–by–side places for another couple, you can seat them in these places in just one way. Now to fill in the rest of the table, make the same kind of argument we made before using $i - 1$ "fat people" and $2n - 2i$ "skinny people." Observe that each arrangement is equivalent to $2n$ other arrangements this time. From here on, the argument is parallel.

  * 21. Do Exercises 12 and 13 for the case where the identically shaped beads have the same color. (There is a troublesome symmetry problem you must get around here.)

SECTION 2   THE CONCEPT OF A GENERATING FUNCTION

Polya's Change–making Example

Suppose we have a pile of nickels, dimes and quarters and we wish to make change for some amount of money, for example, a dollar.* In how many ways may we do this? Let us use the symbolic notation $QQQQ$ to stand for four quarters, the notation $QQQDDN$ to stand for three quarters, two dimes and a nickel. To shorten the notation, we may write Q^4 or $Q^3 D^2 N$; this abbreviation is particularly useful when we want to represent twenty nickels; we write N^{20} rather than a string of twenty N's. We can visualize the process by constructing three symbolic strings (using a plus sign as if it were the word "or")

$$N^0 + N^1 + N^2 + N^3 + \cdots + N^i + \cdots$$
$$D^0 + D^1 + D^2 + D^3 + \cdots + D^j + \cdots$$
$$Q^0 + Q^1 + Q^2 + Q^3 + \cdots + Q^k + \cdots$$

to stand for the fact that we can take 0, 1, 2, 3, and so on nickels; 0, 1, 2, 3, and so on dimes; and 0, 1, 2, 3, and so on quarters.

  The selection of i nickels, j dimes and k quarters yields the expression $N^i D^j Q^k$. This expression can be derived by multiplying N^i from the first symbolic series, D^j from the second and Q^k from the third. Selecting one term from each of the three series and multiplying the resulting three terms together gives us a typical monomial in the product of the three series. Thus, we can visualize every possible way of selecting nickels, dimes and quarters as terms in the product of the three series. Of course, not all terms $N^i D^j Q^k$ in the product correspond to change for a dollar; we get a dollar if and only if $5i + 10j + 25k = 100$. Suppose we use n^5 in place of N, d^{10} in place of D and q^{25} in place of Q. Then we may rewrite our symbolic product as

---

* This example is based on George Polya's highly readable and illuminating paper, "Picture Writing." (See suggested reading).

$$(N^0 + N^1 + N^2 + \cdots)(D^0 + D^1 + D^2 + \cdots)(Q^0 + Q^1 + Q^2 + \cdots)$$
$$= \sum_{i,j,k} N^i D^j Q^k = \sum_{i,j,k} n^{5i} d^{10j} q^{25k},$$

and then pick out terms in the symbolic product whose exponents add up to 100 to represent the different ways of making change for a dollar. Now we really want to know how *many* terms there are whose exponents add up to 100. If we replace each of the n, d and q by x, then each time we have an $n^{5i} d^{10j} q^{25k}$ with $5i + 10j + 25k = 100$, we will get $x^{5i} x^{10j} x^{25k} = x^{5i+10j+25k} = x^{100}$ after the substitution. Further, if we have a term that leads to

$$N^p D^r Q^s = n^{5p} d^{10r} q^{25s} = x^{5p} x^{10r} x^{25s} = x^{100},$$

then $5p + 10r + 25s = 100$, so the term corresponds to change for a dollar. Thus the number of ways to make change for a dollar is the number of times we get x^{100} after the substitutions; therefore it is the coefficient of x^{100} in the series we get after multiplying out our symbolic series and substituting x^5 for N, x^{10} for D, and x^{25} for Q. But we use the same rules of arithmetic to expand products involving three symbols, N, D, and Q as we use for products involving one symbol; so we may also substitute the x before multiplying out the series. In other words, the number of ways to make change for a dollar is the coefficient of x^{100} in the product of the power series

$$(x^0 + x^5 + x^{10} + \cdots)(x^0 + x^{10} + x^{20} + \cdots)(x^0 + x^{25} + x^{50} + \cdots)$$
$$= \sum_{i,j,k} x^{5i} x^{10j} x^{25k} = \sum_{i,j,k} x^{5i+10j+25k}. \quad (2.1)$$

Power Series

The reader acquainted with the notion of power series will notice that the symbolic multiplications we perform here are ordinary multiplication of power series. A *power series* is a formal sum of the form

$$\sum_{i=0}^{\infty} a_i x^i = a_0 x^0 + a_1 x^1 + a_2 x^2 + \cdots$$
$$= a_0 + a_1 x + a_2 x^2 + \cdots$$

Power series are multiplied in the same way as polynomials. For example, the power series

$$1 + y + y^2 + y^3 + \cdots + y^i + \cdots$$

times the polynomial $1 - y$ gives the product

$$1 + y + y^2 + y^3 + \cdots + y^i + \cdots$$
$$\underline{1 - y}$$
$$-y - y^2 - y^3 - y^4 - \cdots - \qquad y^{i+1} - \cdots$$
$$\underline{1 + y + y^2 + y^3 + y^4 + \cdots + y^i + y^{i+1} + \cdots}$$
$$1$$

Thus, $(1 - y)(1 + y + y^2 + \cdots) = 1$, or after division by $1 - y$,

$$\sum_{i=0}^{\infty} y^i = 1 + y + y^2 + y^3 + \cdots = \frac{1}{1 - y}.$$

This is the formula for the sum of a geometric series often presented in algebra courses.

By making the substitutions x^5, x^{10} and x^{25} for y, we see that

$$1 + x^5 + x^{10} + \cdots = \sum_{i=0}^{\infty} x^{5i} = (1 - x^5)^{-1}$$

$$1 + x^{10} + x^{20} + \cdots = \sum_{j=0}^{\infty} x^{10j} = (1 - x^{10})^{-1}$$

and

$$1 + x^{25} + x^{50} + \cdots = \sum_{k=0}^{\infty} x^{25k} = (1 - x^{25})^{-1}.$$

Substituting these into the expression 2.1 we see that the number of ways to make change for a dollar is the coefficient c_{100} of x^{100} in

$$\sum_{i=0}^{\infty} c_i x^i = (1 - x^5)^{-1} \cdot (1 - x^{10})^{-1} \cdot (1 - x^{25})^{-1} = \frac{1}{(1 - x^5)(1 - x^{10})(1 - x^{25})}.$$

Tables of Values from Systems of Linear Recurrences

Although we have translated our problem into a problem that looks quite different, we have still not solved it. How, after all, are we to go about finding the coefficient c_{100}? It turns out that we can work out a table of values of c_i from what we already know. Let us try to develop an approach to finding the values of c_i by starting with a simpler problem. If we had only a pile of nickels and dimes, then the number of ways to make change for a dollar would be the coefficient b_{100} in the product of two series

$$(1 - x^5)^{-1}(1 - x^{10})^{-1} = \sum_{i=0}^{\infty} b_i x^i, \tag{2.2}$$

and if we had only nickels, the number of ways to make change for a dollar is the coefficient a_{100} in

$$(1 - x^5)^{-1} = \sum_{i=0}^{\infty} a_i x^i.$$

We already know that $a_i = 1$ if $i = 0, 5, 10, \ldots$ and $a_i = 0$ if i is not a multiple of 5. If we multiply Equation 2.2 for $\Sigma b_i x^i$ by $1 - x^{10}$, we get

$$(1 - x^5)^{-1} = (1 - x^{10}) \sum_{i=0}^{\infty} b_i x_i$$

or

$$\sum_{i=0}^{\infty} a_i x^i = \sum_{i=0}^{\infty} b_i(x^i - x^{i+10})$$

$$= \sum_{i=0}^{\infty} b_i x^i - \sum_{i=0}^{\infty} b_i x^{i+10}$$

$$= \sum_{i=0}^{\infty} b_i x^i - \sum_{i=0}^{\infty} b_{i-10} x^i \tag{2.3}$$

if we agree to let $b_{-10} = b_{-9} = \cdots b_{-1} = 0$. Now we already know that b_0 must be 1, because $a_0 b_0$ is the coefficient of x^0 in the product $(1 - x^5)^{-1}(1 - x^{10})^{-1}$ and 1 is the coefficient of x^0 in the product $(1 + x^5 + x^{10} + \cdots)(1 + x^{10} + x^{20} + \cdots)$. By combining the two summations on the last line (2.3) of our equations we see that $a_i = b_i - b_{i-10}$ for $i \geq 0$. Thus

$$b_i = a_i + b_{i-10}.$$

This equation is a linear recurrence (or difference equation) in two variables.

How does the equation help us? First, b_{i-10} is 0 if i is less than 10, so $b_i = a_i$ if $i < 10$. To get b_{10}, note that

$$b_{10} = a_{10} + b_0 = 1 + 1 = 2.$$

To get b_{11}, note that

$$b_{11} = a_{11} + b_1 = 0 + 0 = 0.$$

to get b_{20}, note that

$$b_{20} = a_{20} + b_{10} = 1 + 2 = 3.$$

Now whenever i is not a multiple of 5, a_i is 0. Thus

$$b_i = a_i + b_{i-10} = b_{i-10}$$

and by induction, it is clear that $b_i = 0$. In other words, we need only compute b_i by our technique when i is a multiple of 5. This is how row 2 was created in Table 3.1.

What does this suggest for computing the numbers c_i? Recall that

$$(1 - x^5)^{-1}(1 - x^{10})^{-1}(1 - x^{25})^{-1} = \sum_{i=0}^{\infty} c_i x^i,$$

so that, by substitution from (2.4),

$$\left(\sum_{i=0}^{\infty} b_i x^i\right)(1 - x^{25})^{-1} = \sum_{i=0}^{\infty} c_i x^i.$$

By multiplying both sides by $1 - x^{25}$, we get

$$\sum_{i=0}^{\infty} b_i x^i = (1 - x^{25})\sum_{i=0}^{\infty} c_i x^i.$$

If we let $c_{-25} = c_{-24} = \cdots = c_{-1} = 0$, then we get

$$\sum_{i=0}^{\infty} b_i x^i = \sum_{i=0}^{\infty} (c_i - c_{i-25})x^i,$$

so that $c_i - c_{i-25} = b_i$ or $c_i = b_i + c_{i-25}$. Thus, $c_0 = b_0 + 0 = 1$ and $c_{25} = b_{25} + c_0 = 3 + 1 = 4$. Of course, $c_i = 0$ if i is not a multiple of 5 (why?) and we can compute line 3 of Table 3.1 from line 2 by applying the formula $c_i = b_i + c_{i-25}$.

$i =$	0	5	10	15	20	25	30	35	40	45	50	55	60	65	70	75	80	85	90	95	100
$a_i =$	1	1	1	1	1	1	1	1	1	1	1	1	1	1	1	1	1	1	1	1	1
$b_i =$	1	1	2	2	3	3	4	4	5	5	6	6	7	7	8	8	9	9	10	10	11
$c_i =$	1	1	2	2	3	4	5	6	7	8	10	11	13	14	16	18	20	22	24	26	29

Table 3.1

From the table, we see that there are 29 ways to make change for a dollar using nickels, dimes and quarters.

Now suppose we had asked for the number of ways to make change using nickels, dimes, quarters and half-dollars. The same considerations would lead us to a product of four geometric series rather than three. If this product were rewritten as $\sum_{i=0}^{\infty} d_i x^i$, we would get the d_i's from the c_j's in the way we get the c_i's from the b_j's. This would add a new row to Table 3.1. If we wanted to include pennies as well, then the series $\Sigma a_i x^i$ would be the product of a geometric series for pennies and a geometric series for nickels. We would have to compute the a_i's much as we computed the b_i's and then we would end up entirely rebuilding Table 3.1 in order to find the number of ways to make change for a dollar.

What is a Generating Function?

When we have a sequence c_n of numbers, the power series $\sum_{i=0}^{\infty} c_i x^i$ is called the *generating series* or *generating function* for the sequence c_i. In algebra and calculus we learn how a power series in x can often be used to represent a function of x for appropriate x values. This is why generating series have come to be called generating functions. Generating functions are useful for several reasons. First, they provide a compact notation for a sequence; as an example, we saw that the sequence $100001000010000 \ldots$ could be represented by $(1 - x^5)^{-1}$. Second, manipulations of generating functions can lead to additional information about the sequences of interest; for example, we were able to construct our "change table" using the generating function for the c_i's. Also we shall see that various calculus theorems about power series can be of value. For example, if we know that the generating function for a sequence of numbers is an infinitely differentiable function f, we can apply Taylor's theorem (from calculus) to compute the numbers c_i. Finally, arithmetic operations on series correspond to combinatorial constructions that are often useful. It was this principle that lay behind our analysis of change–making.

Typically, we have a sequence a_i such that a_i is the number of objects in a certain set that have a "value" of some sort equal to i. For example, our objects might be piles (multisets) of nickels. Then the value of interest is the monetary value of the pile. If we assume we have an empty pile, a pile with 1 nickel, a pile with 2, and so on, then a_i is 1 if i is a multiple of 5 and 0 otherwise. Multiplying generating functions corresponds to adding values in the following sense.

The Multiplication Principle for Generating Functions

Theorem 2.1. Let v and w be numerical functions defined on sets S and T. Let a_i be the number of objects x of S with $v(x) = i$ and b_i be the number of objects y in T with $w(y) = i$. Then

$$\left(\sum_{i=0}^{\infty} a_i x^i \right)\left(\sum_{i=0}^{\infty} b_i x^i \right)$$

is the generating function for the sequence c_j, where c_j is the number of ordered pairs $(x, y) \in S \times T$ with $v(x) + w(y) = j$.

Proof. By the usual rules for multiplying power series,

$$c_j = \sum_{i=0}^{j} a_i b_{j-i}.$$

By the multiplication principle, $a_i b_{j-i}$ is the number of ordered pairs (x, y) with $v(x) = i$ and $w(y) = j - i$. Since the ordered pairs (x, y) with $v(x) = i$ and $w(y) = j$ are disjoint from those with $v(x) = i'$ and $v(y) = j'$, the sum principle says that adding $a_i b_{j-i}$ over all conceivable values of i will give us the total number of ordered pairs with total value j. However, the only values of i that could yield $v(x) + w(y) = j$ are those for which $v(x)$ is $0, 1, 2, \ldots, j$. The equation for c_j, and thus the assertion of the theorem, follows. ∎

Example 2.1. In how many ways can we select a multiset of eight candy bars which are either pure milk–chocolate or milk–chocolate almond bars?

We can choose 0, 1, 2, . . . , etc. pure chocolate bars. Thus, if we let $a_i = 1$ for each i, $\sum_{i=0}^{\infty} x^i$ is the generating function for the a's; a_i then represents the number of ways to choose i pure chocolate candy bars. The generating function $\sum_{i=0}^{\infty} x^i$ is also the generating function for the number of ways to select i almond candy bars. Thus, by Theorem 2.1

$$\sum_{i=0}^{\infty} x^i \cdot \sum_{i=0}^{\infty} x^i = \sum_{j=0}^{\infty} c_j x^j$$

is the generating function for the number of ways to select a total of j candy bars, and the coefficient of x^8 is $\sum_{i=0}^{8} 1 \cdot 1 = 9$.

Thus, there are nine ways to choose our eight–element multiset of two kinds of candy bars, and $9 = \binom{8 + 2 - 1}{8} = \binom{9}{8}$, as we learned in Chapter 2. ∎

The Generating Function for Multisets

Theorem 2.2.   The generating function for the number $C(n + k - 1, k)$ of k–element multisets of an n–element set is $(1 - x)^{-n}$.

Proof.   A k–element multiset of an n–element set is described by a multiplicity function which is essentially an n–tuple of nonnegative integers that add up to k. Thus the generating function for the number of n–tuples of nonnegative integers whose values add up to k is the function $\sum_{k=0}^{\infty} C(n + k - 1, k)x^k$. Using mathematical induction, we may convert Theorem 2.1 to a theorem about n–tuples rather than pairs. Then $C(n + k - 1, k)$ is the coefficient of x^k in the product of the n equal generating functions $1 + x + x^2 + \cdots = (1 - x)^{-1}$ for the number of ways to select a single nonnegative integer whose value is j. This product of n equal series is $(1 - x)^{-n}$. ∎

Corollary 2.3. $(1 - x)^{-n} = \sum_{k=0}^{\infty} C(n + k - 1, k)x^k$ for each nonnegative integer n.

Note the similarity between Corollary 2.3 and the binomial theorem which states that $(x + 1)^n = \sum_{k=0}^{n} C(n, k)x^k$.

Generating Functions for Integer Partitions

Example 2.2.   As defined in Chapter 2, a *partition* of an integer n is a multiset of integers that add up to n. What is the generating function for partitions of integers into parts of size 1, 2, 3, 4 and 5?

The number of ways of choosing a multiset of 1's that add up to n is 1, so the generating function for partitions of n with all parts of size 1 is

$$1 + x + x^2 + x^3 + \cdot \cdot \cdot = \frac{1}{1 - x}.$$

The number of ways of choosing a multiset of 2's that add up to n is 1 if n is even and 0 if n is odd. Thus the generating function for partitions of n with all parts of size 2 is

$$1 + x^2 + x^4 + x^6 + \cdot \cdot \ = \frac{1}{1 - x^2}.$$

Similarly, the generating functions for the number of partitions of n with all parts of size 3, 4 and 5 are, respectively,

$$\frac{1}{1 - x^3}, \frac{1}{1 - x^4}, \text{ and } \frac{1}{1 - x^5}.$$

Thus by Theorem 2.1, the generating function for the number of partitions of n into parts of size 1, 2, 3, 4 and 5 is

$$\frac{1}{1 - x} \cdot \frac{1}{1 - x^2} \cdot \frac{1}{1 - x^3} \cdot \frac{1}{1 - x^4} \frac{1}{1 - x^5}. \blacksquare$$

If we apply Theorem 2.1 in the same way to partitions into parts of size 1, 2, ... and so on up to n, we obtain

Theorem 2.4.   The coefficient of x^n in the product

$$\frac{1}{1-x} \cdot \frac{1}{1-x^2} \cdot \frac{1}{1-x^3} \cdots \frac{1}{1-x^m} = \prod_{i=1}^{m} \frac{1}{1-x^i}$$

is the number of partitions of the integer n so long as $m \geq n$.

Proof.   As in our example above, the generating function for partitions of integers into parts of size 1, 2, . . . , m is the product shown. However, if $m \geq n$, then all parts of every partition of n are of size less than or equal to m. ∎

Now recall that $\dfrac{1}{1-x^i} = 1 + x^i + x^{2i} \ldots$, so that even the symbolic product

$$\prod_{i=1}^{\infty} \frac{1}{1-x^i} = (1 + x + x^2 + \cdots)(1 + x^2 + x^4 + \cdots) \cdots$$

$$(1 + x^i + x^{2i} + \cdots) \cdots$$

of infinitely many power series may be interpreted as a power series. This power series is the sum of the products $t_1 t_2 \ldots t_i \ldots$, with one term from each series, in which only finitely many terms t_i are different from 1. In this product, the only terms that can be multiplied together to yield x^m will be terms of the form x^j with $j \leq m$. Thus the coefficient of x^m in this infinite product is the same as the coefficient of x^m in the product of the first m series. This gives us the generating function for all partitions of integers.

Theorem 2.5.   The generating function in which x^n is the coefficient of the number of partitions of the integer n is

$$\prod_{i=1}^{\infty} \frac{1}{1-x^i}.$$

EXERCISES

1. What is the generating function in which the coefficient of x^n is the number of ways of choosing an even number of nickels, an even number of dimes and an even number of quarters with total value n cents?

2. What is the generating function for the number of ways to make n cents using at least one nickel, at least one dime and at least one quarter?

3. Find the coefficient of x^7 in $(1 + x + x^2 + \cdot \cdot \cdot)^9$. (Try Corollary 2.3.)

4. Find the coefficient of x^k in $(1 + x + x^2 + \cdot \cdot \cdot)^9$.

5. Find the coefficient of x^7 in $(1 + x + x^2 + \cdot \cdot \cdot)^n$.

6. Find the coefficient of x^7 in $(x + x^2 + \cdot \cdot \cdot)^n$.

7. Find the coefficient of x^{10} in $(1 + x^2 + x^4 + x^6 + \cdot \cdot \cdot)^6$.

8. Find the coefficient of x^{10} in $\dfrac{1}{1 - x} \cdot \dfrac{1}{1 - x^2}$.

9. Extend Example 2.1 by allowing bittersweet chocolate candy bars as well. In how many ways may we select eight candy bars? (Hint: The sum in the example was really the sum of $1^i 1^j$ over all i and j such that $i + j = 8$. The condition $i + j = 8$ converted it into a single sum; here you can convert to a double sum. Using Corollary 2.3 might be easier!)

10. Extend the change–making example with which we began the section by adding fifty–cent pieces. In how many ways may we make change for a dollar in this case?

11. Extend the change–making example (with which we began the section) by allowing pennies. In how many ways may we make change for a dollar in this case?

12. Extend the change–making example (with which we began the section) by allowing both pennies and fifty–cent pieces. In how many ways may we make change for a dollar in this case?

13. What is the generating function for the number of partitions of an integer such that each part is used an even number of times?

14. What is the generating function for the number of partitions of an integer into parts all of which are even numbers?

15. What is the generating function for the number of ways to pass out apples (from a potentially infinite supply) to five children? What if we require that each child gets at least one apple? In both cases, how many distributions use 15 apples?

16. What is the generating function for the number of ways to pass out an even number of oranges to k children so that each child gets something? How many distributions use $2n$ oranges?

17. What is the generating function for the number of multisets that may be formed using the letter a an even number of times and the letter b any number of times?

18. What is the generating function for the number of multisets that may be chosen from an m–element set A and a (disjoint) k–element set B so that elements chosen from B are used an even number of times?

19. Show by means of generating functions that the number of multisets of k objects chosen from a set of n such that each object appears at least once is $C(k - 1, k - n)$.

20. If $f(x)$ is the generating function for the sequence a_i, and f is infinitely differentiable in a neighborhood of $x = 0$, what does Taylor's theorem tell us about the value of the a_i's in terms of f?

21. What is the generating function for the number of ways of making an n–ounce candy assortment (with n an integer) using three different kinds of candy weighing one–half ounce, four different kinds of candy weighing one ounce, two different kinds weighing two ounces, and one kind of candy weighing four ounces? What about an n–pound candy assortment? Why is there a difference between the processes of finding the generating function for an integral number of ounces and the generating function for an integral number of pounds?

22. Using calculus, take r derivatives of $\dfrac{1}{1-x}$ to get the generating function for the falling factorial numbers $(n)_r$.

SECTION 3 POLYNOMIALS AND OTHER SPECIAL GENERATING FUNCTIONS

Why Polynomials Make Good Generating Functions

To avoid introducing too many new ideas in the last section, we only gave examples of generating functions that were "truly infinite" series. In practice, rather than having an infinite supply of coins as in our change–making example, we have only a certain number of nickels, a certain number of dimes and a certain number of quarters. When we are passing out candy bars, there is some number of candy bars of each type available. When we are considering partitions of integers, it might be desirable to constrain the total number of times a given integer could be used. In all these cases, the generating function that occurs turns out to be simply a polynomial—a particularly elementary kind of power series.

Example 3.1. If in Example 2.1 there are five almond bars and six pure chocolate bars, then in how many ways can we choose a total of eight candy bars?

In this case we can choose 0, 1, 2, 3, 4 or 5 pure chocolate bars. Thus the generating function for pure chocolate bars is $1 + x + x^2 + x^3 + x^4 + x^5$ and the generating function for almond bars is $1 + x + x^2 + x^3 + x^4 + x^5 + x^6$. Finally, the generating function for ordered pairs of selections of candy bars is

$$(1 + x + x^2 + x^3 + x^4 + x^5)(1 + x + x^2 + x^3 + x^4 + x^5 + x^6)$$
$$= 1 + 2x + 3x^2 + 4x^3 + 5x^4 + 6x^5 + 6x^6 + 5x^7 + 4x^8 + 3x^9 + 2x^{10} + x^{11}.$$

Thus, there are four ways to select eight candy bars from among the two types. ∎

Extending the Definition of Binomial Coefficients

In the same way, if we had n different candy bars and wanted to select k of them, choosing zero or one of each kind, our generating function for the number of choices would be a product of n factors:

$$(1 + x)(1 + x) \ \ldots \ (1 + x) = (1 + x)^n.$$

However, by making such a selection, we select k different elements from our n–element set. Since, by the binomial theorem, $(1 + x)^n = \displaystyle\sum_{k=0}^{n} C(n, k)x^k$ we have in essence rederived the fact that there are $C(n; k)$ ways to select k elements from our n-element set.   By extending the definition of $C(n; k)$ to negative values of n, we can relate this generating function to the generating functions of Section 2.

One way to write $\binom{n}{k}$ is $\dfrac{(n)_k}{k!}$, where $(n)_k = n(n - 1) \ldots (n - k + 1)$.

It is interesting to note that when n is a negative number, $n = -m$, if we *define*

$$\binom{-m}{k} = \frac{(-m)_k}{k!} = \frac{(-m)(-m - 1) \ldots (-m - k + 1)}{k!},$$

then

$$\binom{-m}{k} = (-1)^k \frac{(m + k - 1)_k}{k!}$$
$$= (-1)^k C(m + k - 1; k).$$

Thus applying Corollary 2.3 to the power series

$$(1 + x)^{-m} = (1 - (-x))^{-m},$$

we get

$$(1 + x)^{-m} = \sum_{k=0}^{\infty} C(m + k - 1; k)(-1)^k x^k$$
$$= \sum_{k=0}^{\infty} \binom{-m}{k} x^k.$$

The Extended Binomial Theorem

In other words, with the natural definition of $\binom{n}{k}$ for negative values of n, we now have for all integers n

$$(1 + x)^n = \sum_{k=0}^{\infty} \binom{n}{k} x^k,$$

and for any positive or negative integer n, $(1 + x)^n$ is the generating function for the binomial coefficients $\binom{n}{k}$.

Generating Functions can Sometimes Replace Inclusion–Exclusion

As in Exercise 17 of Section 1 of this chapter, it is possible to apply the principle of inclusion and exclusion to determine the number of ways to hand out k identical pieces of candy to n children so that no child gets m or more pieces. (We use the properties, "child i gets more than $m - 1$ pieces"; to compute the number of distributions with exactly j of the properties, we compute $\binom{n - jm + k - 1}{n - 1}$.)

Generating functions make such a problem easier.

Example 3.2. In how many ways may we distribute 10 pieces of candy to three children so that no child gets more than four pieces?

Each child could get 0, 1, 2, 3 or 4 pieces, so the generating function for the number of ways to give n pieces of candy to child i is $1 + x + x^2 + x^3 + x^4$. (Since all the pieces of candy are identical, there is one way to pass out each possible number of pieces.) Thus the generating function for the number of ways to pass out n pieces of candy to three children is $(1 + x + x^2 + x^3 + x^4)^3$. The number of ways to pass out 10 pieces is the coefficient of x^{10} in this power of a multinomial. This coefficient may be computed using the multinomial theorem. It may also be computed by using some clever algebra. One of the standard factorizations we learn in algebra is

$$(1 - x)(1 + x + x^2 + x^3 + x^4) = 1 - x^5,$$

so

$$(1 + x + x^2 + x^3 + x^4) = \frac{1 - x^5}{1 - x}.$$

Thus

$$(1 + x + x^2 + x^3 + x^4 + x^5)^3 = \frac{(1 - x^5)^3}{(1 - x)^3}$$

$$= (1 - x^5)^3(1 - x)^{-3}$$

$$= (1 - 3x^5 + 3x^{10} - x^{15}) \sum_{i=0}^{\infty} \binom{3 + i - 1}{i} x^i$$

$$= (1 - 3x^5 + 3x^{10} - x^{15}) \sum_{i=0}^{\infty} \binom{2 + i}{i} x^i.$$

In this product, the term involving x^{10} is

$$\binom{2 + 10}{10} x^{10} - 3x^5 \binom{2 + 5}{5} x^5 + 3x^{10} \binom{2 + 0}{0} x^0$$

$$= \left\{ \binom{12}{10} - 3\binom{7}{5} + 3 \right\} x^{10}$$

$$= (66 - 63 + 3)x^{10} = 6x^{10}.$$

Thus there are six ways to pass out the candy. ■

The method used in the example above could be generalized to deal with an arbitrary number n of children—in which case we would have $(1 - x^5)^n(1 - x)^{-n}$; or to deal with an arbitrary upper bound m on the number of pieces of candy—5 would be replaced by $m + 1$. The example already essentially shows us how to deal with an arbitrary number of pieces of candy—this number just tells us what power of x to examine. Note how the example mirrors the use of inclusion–exclusion: the coefficients of $1 - 3x^5 + 3x^{10} - x^{15}$ are binomial coefficients and they alternate in sign.

* The Relationship Between Generating Functions and Inclusion–Exclusion on Level Sums

These are exactly the sort of coefficients that we found in the relationship between $N_=^+$ and $N_\geq^+$ in Theorem 1.4. In fact, Theorem 1.4 can be used to give a relationship between the generating functions for the numerical functions N^+ and $N_\geq^+$.

In particular, assume as in Section 1 that $N_=$ and $N_\geq$ count arrangements that have exactly the properties and at least the properties, respectively, in one of the subsets of a set P of properties. We define

$$N_=^+(i) = \sum_{T: |T|=i} N_=(T)$$

and

$$N_\geq^+(i) = \sum_{T: |T|=i} N_\geq(T).$$

According to Theorem 1.4,

$$N_=^+(i) = \sum_{j=i}^{p} (-1)^{j-i} \binom{j}{i} N_\geq^+(j).$$

Note that $N_{\geq}^{\pm}(j) = 0$ if $j > p$ and $\binom{j}{i} = 0$ if $j < i$. Thus we can rewrite the sum as

$$N_{=}^{\pm}(i) = \sum_{j=0}^{\infty} (-1)^{j-i} \binom{j}{i} N_{\geq}^{\pm}(j).$$

This form is well suited for relating the power series for $N_{=}^{\pm}$ and $N_{\geq}^{\pm}$. In particular,

$$\sum_{i=0}^{\infty} N_{=}^{\pm}(i)x^i = \sum_{i=0}^{\infty} \left(\sum_{j=0}^{\infty} (-1)^{j-i} \binom{j}{i} N_{\geq}^{\pm}(j) \right) x^i$$

$$= \sum_{j=0}^{\infty} \left(\sum_{i=0}^{\infty} (-1)^{j-i} \binom{j}{i} x^i \right) N_{\geq}^{\pm}(j)$$

$$= \sum_{j=0}^{\infty} N_{\geq}^{\pm}(j) \sum_{i=0}^{\infty} (-1)^{j-i} \binom{j}{i} x^i$$

$$= \sum_{j=0}^{\infty} N_{\geq}^{\pm}(j)(x - 1)^j.$$

In summary,

Theorem 3.1. Suppose $N_{=}$ and $N_{\geq}$ are defined on subsets of a set P and $N_{\geq}(T) = \sum_{S=T}^{P} N_{=}(S)$. Then the generating function for the associated numerical function $N_{=}^{\pm}$ may be obtained from the generating function for $N_{\geq}^{\pm}$ by replacing x^j by $(x - 1)^j$.

Proof. Given above. ∎

Example 3.3. Rework Example 3.2 using Theorem 3.1. The property of interest here is "child i gets at least five pieces of candy." Thus, for a given set I of children (properties), we give each child in I five pieces of candy (in one way) and then pass out the remaining pieces in

$$\binom{10 - 5i + 3 - 1}{3 - 1}$$

ways. Thus, $N_{\geq}(I)$ depends only on the size of I, so

$$N_{\geq}^{\pm}(i) = \binom{3}{i} N_{\geq}(I) = \binom{3}{i}\binom{12 - 5i}{2}.$$

Because we don't want any child to get more than four pieces, we must find the number of arrangements in which no (i.e. zero) children get more than four pieces of candy. However, this number of arrangements is $N^+(0)$, which is the coefficient of x^0 in:

$$\sum_{i=0}^{\infty} N^+(i)x^i = \sum_{j=0}^{\infty} \binom{3}{j}\binom{12 - 5j}{2}(x - 1)^j.$$

However, in a power series, the coefficient of x^0 is $f(0)$. Thus,

$$N_{\pm}^{\pm}(0) = \sum_{j=0}^{\infty} \binom{3}{j}\binom{12 - 5i}{2}(-1)^j = \sum_{j=0}^{3} \binom{3}{j}\binom{12 - 5i}{2}(-1)^j$$

$$= \binom{3}{0}\binom{12}{2} - \binom{3}{1}\binom{7}{2} + \binom{3}{2}\binom{2}{2} - 0$$

$$= 6. \quad \blacksquare$$

Note that we have the same alternating series as before, although it arose from a much different sequence of operations with power series.

Other Kinds of Generating Functions

The binomial theorem is not the only way in which polynomial generating functions arise. Recall that the Stirling numbers relate falling factorial polynomials to ordinary polynomials. In particular, since

$$(x)_n = x(x - 1) \ldots (x - n + 1) = \sum_{k=0}^{n} s(n, k)x^k,$$

the falling factorial polynomial of degree n is the generating function for the Stirling numbers of the first kind.

It is natural in this situation to ask if the similar polynomial equation involving the Stirling numbers of the second kind gives us a generating function for them. This polynomial equation is

$$x^n = \sum_{k=0}^{n} S(n, k)(x)_k.$$

Note that we no longer have powers of x in the sum, and so this is not a generating function in the usual sense. For this reason, we define generating functions relative to a set of "indicator functions." For our purposes, a set $\mathbf{I}$ of polynomials will be called a set of *indicator functions* if for each nonnegative integer n, there is exactly one polynomial $p_n(x)$ of degree n in $\mathbf{I}$. A straightforward proof by induction shows that any polynomial can then be represented in one and only one way as a sum of numerical multiples of indicator polynomials in $\mathbf{I}$. (In the lanuage of linear algebra, $\mathbf{I}$ is a basis for the vector space of polynomials.) At times even more general indicator functions may be described, but our description includes all the usual types of such polynomials.

We define the *generating function for the sequence a_n relative to $\mathbf{I}$* to be the series

$$\sum_{n=0}^{\infty} a_n p_n(x).$$

Thus x^n is the generating function for the Stirling numbers of the second kind relative to the indicator family of falling factorial polynomials.

One other indicator family is suggested by an analysis of the binomial theorem. Recall that $(n)_j = \dfrac{n!}{(n-j)!}$ is the number of one-to-one functions from a j-element set to an n-element set. The binomial theorem can be rewritten to give $(n)_j$ explicitly:

$$(x+1)^n = \sum_{j=0}^{n} \binom{n}{j} x^j = \sum_{j=0}^{n} \frac{(n)_j}{j!} x^j = \sum_{j=0}^{n} (n)_j \frac{x^j}{j!}.$$

Thus, the generating function for the number of one-to-one functions from a j-element set into an n-element set *relative to* the indicator functions $\dfrac{x^j}{j!}$ is $(x+1)^n$. Note that $(n)_j$ is 0 if $j > n$, so

$$\sum_{j=0}^{n} \binom{n}{j} x^j = \sum_{j=0}^{\infty} (n)_j \frac{x^j}{j!}.$$

The importance of one-to-one functions in our work so far suggests that generating functions relative to the indicator family $\dfrac{x^j}{j!}$ might arise in other situations as well, and in fact they do. Because the exponential function e^x is the generating function for the all ones $(1, 1, 1, 1 \ldots)$ sequence relative to $\dfrac{x^j}{j!}$ it is customary to call

$$\sum_{j=0}^{\infty} a_j \frac{x^j}{j!}$$

the *exponential generating function* for the sequence a_i. This is the subject of Section 5.

EXERCISES

1. Two dice (a red one and a white one) are thrown. Write a generating function for the number of outcomes in which the two faces showing on top add up to i. Using Theorem 2.1, explain why this generating function can be factored and what its factors mean.

2. A penny, nickel, dime and quarter are tossed. Write the generating function for the number of ways that i heads can occur. Explain why this polynomial can be factored and explain what the factors mean.

3. Repeat Exercise 1 with three dice.

4. What is the generating function for the sum of the faces if n distinct dice are thrown?

5. Extend Example 3.1 by allowing four bittersweet candy bars. In how many ways may we select eight candy bars in this case?

6. What is the generating function for the number of partitions of the integer n into distinct parts? (No two parts may be equal.)

7. What is the generating function for the number of partitions of the integer n into distinct even parts?

8. What is the generating function for the number of partitions of the integer n into parts, at most i of which are of size i?

9. What is the generating function for the number of ways of distributing identical pieces of candy to Joe and Mary so that Joe gets a nonzero even number of pieces and Mary gets an odd number of pieces? What if no more than 11 pieces are to be handed out?

10. How many six–element multisets may be formed by using up to 3 a's, up to 5 s's, up to 2 e's and up to 3 c's? What if each letter must be used once?

11. In how many ways may you make change for a dollar using no more than 10 nickels, no more than five dimes and no more than four quarters?

12. In how many ways can you pass out 10 identical pieces of candy to three children so that each child gets between two and four pieces? Between three and five pieces?

∗13. Redo the "hat check" problem (Exercise 14, Section 1) using Theorem 3.1.

14. Redo Exercise 17, Section 1, using generating functions.

15. Expand $(1 + x)^n(1 + x^{-1})^n$ and use the resulting expression to prove that

$$\binom{2n}{n} = \sum_{k=0}^{n} \binom{n}{k}^2.$$

16. Write the generating function for the number of ways to choose a snack of n items chosen to include 0 or one each of s distinct candy bars, 0 or two each of r different kinds of fruit and one or two soft drinks chosen from among t varieties. (Hint: First do the problem without the soft drinks.)

17. From their generating function, what can you conclude about the sum of the Stirling numbers of the first kind?

18. Using calculus and generating functions, determine the value of the sum

$$\binom{n}{1} + 2\binom{n}{2} + 3\binom{n}{3} + \cdots + n\binom{n}{n}.$$

∗ 19. Using generating functions, show that

$$\sum_{i=0}^{k} \binom{m}{i}\binom{n}{k-i} = \binom{m+n}{k}.$$

∗ 20. Fractional powers also satisfy the binomial theorem. In particular, for any real number r we define

$$\binom{r}{k} = \frac{(r)_k}{k!} = \frac{r(r-1)\cdots(r-k+1)}{k!}$$

so

$$(1+x)^r = \sum_{K=0}^{\infty} \binom{r}{k} x^k$$

makes sense.

(a) Write explicit expressions for $\left(\dfrac{-\frac{1}{2}}{5}\right)$ and $\left(\dfrac{-\frac{1}{2}}{5}\right)$.

(b) Show that fractional powers of the binomial $(1 + x)$ obey the binomial theorem.

(c) Show that $(1 - 4x)^{\frac{-1}{2}}$ is the generating function for $\binom{2n}{n}$.

SECTION 4 RECURRENCE RELATIONS AND GENERATING FUNCTIONS

The Idea of a Recurrence Relation

One of the reasons why generating functions are an important tool is that they allow us to manipulate and sometimes explicitly find sequences of numbers that satisfy rules like

$$a_n = 2a_{n-1},$$

or

$$a_n = 2a_{n-2} + a_{n-1},$$

or

$$a_n = a_{n-1} + 2.$$

Such rules are called *linear recurrence relations* or *linear difference equations*. We have already used such recurrences to compute tables of values. Equations like those above are rather analogous to differential equations, and the use of generating functions to find numbers a_i satisfying these equations is similar to the use of power series in solving differential equations. The equation $a_n = 2a_{n-1}$ arises in the study of subsets of a set—if a_n is the number of subsets of $\{1, 2, \ldots, n\}$, then a_n is the number of subsets not containing n plus the number of subsets containing n, and both of these numbers are a_{n-1}. Of course, we have already shown that since $a_0 = 1$ it follows that $a_n = 2^n$. Let us also derive this result using generating functions, showing each detail explicitly.

How Generating Functions are Relevant

Example 4.1. Find the generating function $\sum_{i=0}^{\infty} a_i x^i$ in which a_i is the number of subsets of an i–element set.

Since we know that $a_{i+1} = 2a_i$, we know that by substitution we may write

$$\sum_{i=0}^{\infty} a_{i+1} x^i = \sum_{i=0}^{\infty} 2a_i x^i \qquad (4.1)$$

Now multiply both sides by x to get

$$x \sum_{i=0}^{\infty} a_{i+1} x^i = x \sum_{i=0}^{\infty} 2a_i x^i \qquad (4.2)$$

and

$$\sum_{i=0}^{\infty} a_{i+1} x^{i+1} = 2x \sum_{i=0}^{\infty} a_i x^i. \tag{4.3}$$

(The reason for the algebraic manipulation above consisting of multiplying by x and applying the distributive law was to get the $a_{i+1} x^{i+1}$ term into the left hand side of Equation 4.3. If we let $j = i + 1$ then we have $a_j x^j$ on the left hand side and $a_i x^i$ on the right hand side. This lets us "solve" the equation for the power series $\sum_{i=0}^{\infty} a_i x^i$; we now proceed to do so.)

$$\sum_{i=0}^{\infty} a_i x^i - a_0 = 2x \sum_{i=0}^{\infty} a_i x^i,$$

or

$$\sum_{i=0}^{\infty} a_i x^i - 2x \sum_{i=0}^{\infty} a_i x^i = a_0,$$

so

$$(1 - 2x) \sum_{i=0}^{\infty} a_i x^i = a_0,$$

or

$$\sum_{i=0}^{\infty} a_i x^i = \frac{a_0}{1 - 2x} = a_0(1 + 2x + (2x)^2 + (2x)^3 + \cdots)$$

$$= a_0 \sum_{i=0}^{\infty} 2^i x^i.$$

Thus $a_i = a_0 2^i$, and since $a_0 = 1$, $a_i = 2^i$. ∎

Second-order Linear Recurrence Relations

A classic example of how recurrence relations arise is Fibonacci's problem; we present three variations on Fibonacci's problem, culminating in Fibonacci's original example. The problems all deal with an imaginary population of rabbits.

Similar problems arise in the study of leaf and tip growth in plants, in the analysis of algorithms and in geometry.

Example 4.2.   In variation 1, we have a population of rabbits who reproduce in pairs. Each pair of rabbits born in a particular month produces a pair of offspring in each of the following two months and dies by the end of that second month. Thus if a_n denotes the number of rabbits present at the end of n months, the equation

$$a_{n+2} = 2a_{n+1} - a_n$$

states that each pair of rabbits has reproduced and that the two–month–old rabbits perished.

Substituting this equation into the generating function for a_n yields

$$\sum_{n=0}^{\infty} a_{n+2}x^n = 2 \sum_{n=0}^{\infty} a_{n+1}x^n - \sum_{n=0}^{\infty} a_n x^n$$

Multiplying by x^2 gives

$$\sum_{n=0}^{\infty} a_{n+2}x^{n+2} = 2x \sum_{n=0}^{\infty} a_{n+1}x^{n+1} - x^2 \sum_{n=0}^{\infty} a_n x^n$$

$$\sum_{n=2}^{\infty} a_n x^n = 2x \sum_{n=1}^{\infty} a_n x^n - x^2 \sum_{n=0}^{\infty} a_n x^n$$

$$\sum_{n=0}^{\infty} a_n x^n - a_1 x - a_0 = 2x \left(\sum_{n=0}^{\infty} a_n x^n - a_0 \right) - x^2 \sum_{n=0}^{\infty} a_n x^n.$$

Rearranging terms and placing all occurrences of the generating function on the left gives

$$\sum_{n=0}^{\infty} a_n x^n (1 - 2x + x^2) = (a_1 - 2a_0)x + a_0$$

or

$$\sum_{n=0}^{\infty} a_n x^n = \frac{a_0 + (a_1 - 2a_0)x}{(1 - x)^2}.$$

However by the extended binomial theorem (Theorem 2.2),

$$(1 - x)^{-2} = \sum_{k=0}^{\infty} C(2 + k - 1; k)x^k$$

$$= \sum_{k=0}^{\infty} C(k + 1; k)x^k = \sum_{k=0}^{\infty} (k + 1)x^k,$$

so by substitution,

$$\sum_{n=0}^{\infty} a_n x^n = (a_0 + (a_1 - 2a_0)x) \sum_{k=0}^{\infty} (k + 1)x^k$$

$$= \sum_{k=0}^{\infty} a_0(k + 1)x^k + (a_1 - 2a_0)(k + 1)x^{k+1}$$

$$= a_0 + \sum_{k=1}^{\infty} \{(k + 1)a_0 + k(a_1 - 2a_0)\}x^k$$

$$= a_0 + \sum_{k=1}^{\infty} \{k(a_1 - a_0) + a_0\}x^k.$$

Thus $a_n = n(a_1 - a_0) + a_0$. Note that if we started out so that $a_1 = a_0$, then the end–of–the–month population would always be a_0; if a_1 were greater than a_0 the population would grow linearly, and if a_1 were less than a_0, our population would die out (the assumptions we use in describing the recurrence relation would be invalid, though, once a_i became 0). The values of a_0 and a_1 represent the initial conditions of the experiment. If in month 0 we have a pair of rabbits that have had babies once, then a_1 would be a_0. However, if in month 0 we have rabbits born during month 0, then $a_1 = 2a_0$. (Note: By month 0, we mean the month at the end of which we begin recording the population.) This problem is one in which a solution could have been found without generating functions; however, the methods used are quite general. ■

In outline, to solve a linear recurrence relation of degree k—that is a linear recurrence of the form

$$a_{n+k} = \sum_{i=0}^{k-1} b_i a_{n+i},$$

we substitute the relation for a_n in the generating function. We then re–express each infinite series in terms of $\sum_{n=0}^{\infty} a_n x^n$ and solve for this power series. The result always turns out to be a quotient of two polynomials and depends on the numbers $a_0, a_1, \ldots, a_{k-1}$. If we know these numbers in advance, they may be used instead of the symbols a_i; this simplifies the arithmetic. As we shall see, a quotient of two polynomials can be re–expressed as a product of a polynomial and a power series, in particular, as a product of a polynomial and some number of geometric series. Let us illustrate with a second rabbit problem.

Example 4.3. Our rabbits now have a long life span. In fact, they all live longer than the length of the experiment, and so are assumed to have infinite life spans. A pair of rabbits requires a maturation period of one month before they can produce offspring; each pair of mature rabbits present at the end of one month produces two new pairs of rabbits by the end of the next month. The following recurrence relation states that the rabbit population at the end of one month consists of all rabbits present at the end of the previous month plus new offspring produced by rabbits that are mature.

$$a_{n+2} = a_{n+1} + 2a_n.$$

Substituting this into the power series gives us

$$\sum_{n=0}^{\infty} a_{n+2}x^n = \sum_{n=0}^{\infty} a_{n+1}x^n + 2\sum_{n=0}^{\infty} a_n x^n$$

which, after appropriate manipulation, yields

$$\left(\sum_{n=0}^{\infty} a_n x^n\right)(1 - x - 2x^2) = a_1 x + a_0 - a_0 x,$$

so that

$$\sum_{n=0}^{\infty} a_n x^n = \frac{(a_1 - a_0)x + a_0}{1 - x - 2x^2} = \frac{a_0 + (a_1 - a_0)x}{(1 - 2x)(1 + x)} \tag{4.1}$$

$$= (a_0 + (a_1 - a_0)x)\left(\sum_{i=0}^{\infty} (2x)^i\right)\sum_{j=0}^{\infty} (-x)^j$$

$$= (a_0 + (a_1 - a_0)x)\left(\sum_{i=0}^{\infty} 2^i x^i\right)\left(\sum_{j=0}^{\infty} (-1)^j x^j\right)$$

$$= (a_0 + (a_1 - a_0)x)\sum_{k=0}^{\infty}\left(\sum_{i=0}^{k} 2^i(-1)^{k-i}\right)x^k.$$

Now it is possible to show that the value of the inner sum is $\dfrac{2^{k+1}}{3} + \dfrac{(-1)^k}{3}$;

however, this kind of computation is not always so easy. Instead we use the method of partial fractions. This method, which lets us replace a product by a sum, is sometimes used in calculus for integration of quotients of polynomials. The basic idea is that we can find numbers r and s with

$$\frac{1}{(ax + b)(cx + d)} = \frac{r}{ax + b} + \frac{s}{cx + d}$$

if $(ax + b)$ and $(cx + d)$ aren't multiples of each other.

Also, we can find numbers r, s and t with

$$\frac{1}{(ax + b)^2(cx + d)} = \frac{r}{(ax + b)^2} + \frac{s}{ax + b} + \frac{t}{cx + d}.$$

The idea can be extended to larger numbers of terms or higher powers. In our example we write the equation

$$\frac{1}{(1 - 2x)(1 + x)} = \frac{r}{(1 - 2x)} + \frac{s}{(1 + x)}$$

which gives us

$$\frac{1}{(1 - 2x)(1 + x)} = \frac{r + rx + s - 2sx}{(1 - 2x)(1 + x)}$$

so that $r + s = 1$ and $rx - 2sx = 0$. Dividing the second equation by x (or replacing x by the possible value 1) yields $r - 2s = 0$, so $3s = 1$ and $3r = 2$. Thus

$$\frac{1}{(1 - 2x)(1 + x)} = \frac{2}{3} \cdot \frac{1}{1 - 2x} + \frac{1}{3} \cdot \frac{1}{1 + x}$$

$$= \frac{2}{3} \sum_{i=0}^{\infty} (2x)^i + \frac{1}{3} \sum_{i=0}^{\infty} (-x)^i$$

$$= \frac{2}{3} \sum_{i=0}^{\infty} 2^i x^i + \frac{1}{3} \sum_{i=0}^{\infty} (-1)^i x^i.$$

Collecting terms and multiplying by $a_0 + (a_1 - a_0)x$ yields from Equation (4.1)

$$\sum_{i=0}^{\infty} a_i x^i = \frac{a_0 + (a_1 - a_0)x}{(1 - 2x)(1 + x)}$$

$$= \frac{a_0 + (a_1 - a_0)x}{3} \sum_{i=0}^{\infty} (2^{i+1} + (-1)^i)x^i,$$

so that for $i > 0$

$$a_i = \frac{a_1 - a_0}{3} (2^i + (-1)^{i-1}) + \frac{a_0}{3} (2^{i+1} + (-1)^i)$$

$$= \frac{a_1 + a_0}{3} \cdot 2^i + \frac{2a_0 - a_1}{3} \cdot (-1)^i.$$

Thus since a_0 and a_1 are nonnegative, if either is nonzero the population will grow essentially exponentially. If we start with mature rabbits at time 0, then $a_1 = 3a_0$; if we start with immature rabbits at time 0, then $a_1 = a_0$. ∎

The Original Fibonacci Problem

Fibonacci's original problem is no more difficult theoretically; however, a slightly unexpected square root pops up.

Example 4.4. Now, as in example 4.3, our rabbits can reproduce after one month maturation; however, each pair of mature rabbits present at the end of one month produces exactly one pair of baby rabbits during (and before the end of) the next month. Again the rabbits live forever. (This also describes a simple model of plant growth and branching; the rabbits correspond to branch tips.) Then if a_n is the number of rabbits at the end of month n, our recurrence relation will be:

$$a_{n+2} = a_{n+1} + a_n.$$

Substituting into the generating function as above we get

$$\sum_{i=0}^{\infty} a_i x^i = \frac{a_0 + (a_1 - a_0)x}{1 - x - x^2}.$$

From the quadratic formula (or completing the square), we can see that

$$1 - x - x^2 = -(x^2 + x - 1) = -\left(x - \frac{-1 + \sqrt{5}}{2}\right)\left(x - \frac{-1 - \sqrt{5}}{2}\right)$$

For the sake of brevity, let us use G (for "golden ratio") to stand for $\dfrac{1 + \sqrt{5}}{2}$ and H to stand for $\dfrac{1 - \sqrt{5}}{2}$. Then $G + H = 1$, $GH = -1$, and $G - H = \sqrt{5}$. Since

$$\frac{1}{1 - x - x^2} = -\frac{1}{(x + G)(x + H)} = \frac{r}{x + G} + \frac{s}{x + H},$$

then $rx + sx = 0$ and $rH + sG = -1$, so $r = -s$ and thus

$$r = \frac{1}{G - H} = \frac{1}{\sqrt{5}} \text{ and } s = \frac{1}{H - G} = -\frac{1}{\sqrt{5}}.$$

This gives us

$$\frac{a_0 + (a_1 - a_0)x}{1 - x - x^2} = \frac{a_0 + (a_1 - a_0)x}{G - H}\left(\frac{1}{x + G} - \frac{1}{x + H}\right)$$

$$= \frac{a_0 + (a_1 - a_0)x}{G - H}\left(\frac{1}{G} \cdot \frac{1}{1 + x/G} - \frac{1}{H} \cdot \frac{1}{1 + x/H}\right)$$

$$= \frac{a_0 + (a_1 - a_0)x}{G - H}\left(\frac{1}{G}\sum_{i=0}^{\infty}\left(\frac{-x}{G}\right)^i - \frac{1}{H}\sum_{i=0}^{\infty}\left(\frac{-x}{H}\right)^i\right)$$

$$= \frac{a_0 + (a_1 - a_0)x}{G - H}\sum_{i=0}^{\infty}\left(\frac{1}{G^{i+1}} - \frac{1}{H^{i+1}}\right)(-1)^i x^i$$

$$= \frac{a_0 + (a_1 - a_0)x}{G - H}\sum_{i=0}^{\infty}(-1)^i\frac{H^{i+1} - G^{i+1}}{H^{i+1}G^{i+1}}x^i$$

$$= \frac{a_0 + (a_1 - a_0)x}{G - H}\sum_{i=0}^{\infty}(G^{i+1} - H^{i+1})x^i.$$

Substituting for G and H, we get

$$\sum_{n=0}^{\infty}a_n x^n = \frac{a_0 + (a_1 - a_0)x}{\sqrt{5}}\sum_{i=0}^{\infty}\left\{\left(\frac{1 + \sqrt{5}}{2}\right)^n - \left(\frac{1 - \sqrt{5}}{2}\right)^n\right\}x^i.$$

By applying the binomial theorem to $\left(\dfrac{1 + \sqrt{5}}{2}\right)^n$ and $\left(\dfrac{1 - \sqrt{5}}{2}\right)^n$, we may show that their difference is an integral multiple of $\sqrt{5}$. Thus so long as a_0 and a_1 are integers, a_n will be an integer as well. Note that for large values of n, the G^n term will be far larger in absolute value than the H^n term; thus the "Fibonacci numbers" a_n grow at essentially the same rate as G^n.

As before, we could give a general formula for a_n. In practice, we would substitute our initial conditions (a_0 and a_1) first. It is traditional to start with one pair of baby rabbits; thus $a_0 = a_1 = 1$ (pair). Then

$$a_n = \frac{G^n - H^n}{\sqrt{5}} = \frac{1}{\sqrt{5}} \left(\frac{1 + \sqrt{5}}{2}\right)^n - \frac{1}{\sqrt{5}} \left(\frac{1 - \sqrt{5}}{2}\right)^n. \quad \blacksquare$$

General Techniques

The techniques we introduced work equally well when the polynomial to be factored has complex factors. By now it should be clear why this method works for all linear recurrence relations. In particular, the methods used may be applied to an arbitrary recurrence relation of degree 2 of the form

$$a_{n+2} = ba_{n+1} + ca_n.$$

In this case, we will have to factor the polynomial

$$1 - bx - cx^2.$$

Before we factor this polynomial, we get a generating function for a_n of the form

$$\sum_{n=0}^{\infty} a_n x^n = \frac{a_0 + (a_1 - ba_0)x}{1 - bx - cx^2}.$$

We factor the polynomial, expand it in partial fractions and get a sum of two geometric series in which the coefficients are essentially powers of the roots of the polynomial. Finally we can write a_n as a (linear) combination of the n-th powers of the roots of the quadratic polynomial; the only undetermined quantities are a_0 and a_1. Of course, the method extends to linear difference equations of higher degree; more surprisingly, it also gives generating functions for solutions of *non-homogeneous* linear recurrence relations of the form

$$c_0 a_n + c_1 a_{n-1} + \cdots + c_k a_{n-k} = f(n).$$

These generating functions will typically include a product of the generating function $\Sigma f(n)x^n$ and a solution to the corresponding homogeneous problem. In all cases, if we are not searching for a general solution, we may substitute the values of a_0 and a_1 for our problem at the beginning of our manipulations. This often simplifies the arithmetic.   For example, Example 4.4 would have been far less formidable–looking had we begun with $a_0 = a_1 = 1$.

EXERCISES

1. What is $\sum_{i=0}^{\infty} (\tfrac{1}{3})^i x^i$, $\sum_{i=0}^{\infty} (-\tfrac{1}{3})^i x^i$?   What recurrence relations do these generate the solutions to?

2. Use generating functions to solve the recurrence relation $a_n = 3a_{n-1}$.

3. Redo Example 4.2 assuming *in advance* that $a_0 = 1$ and $a_1 = 2$.

4. Redo Example 4.3 assuming *in advance* that $a_1 = a_0 = 1$.

5. Redo Example 4.4 assuming *in advance* that $a_1 = a_0 = 1$.

6. Use generating functions to solve the recurrence relation $a_{n+2} = 4a_{n+1} - 4a_n$. First consider $a_0 = a_1 = 1$; next consider $a_0 = 1$, $a_1 = 0$ ànd $a_0 = 0$, $a_1 = 1$.

7. Use generating functions to solve the recurrence relation $a_{n+3} = 3a_{n+2} - 3a_{n+1} + a_n$ in the case with $a_0 = a_1 = a_2 = 1$.

8. Solve the following variant of Fibonacci's problem.   Each mature pair of rabbits present at the end of one month produces three more pairs of rabbits during the month following; further, the baby rabbits take a month to mature.   Assuming no rabbits die, how many rabbits are present after n months assuming we start with 10 baby rabbits?

9. Solve the following variant of Fibonacci's problem.   A pair of rabbits produces one pair of baby rabbits between their first and second months, but then dies in the next month after reproducing one more pair.   (This gives a linear recurrence of degree 3.)   How many rabbits are present at the end of n months?

10. The linear recurrence $a_{n+1} = a_n + 1$ has the (almost obvious) solution $a_n = a_0 + n$.   Using generating functions, show how this arises.

11. Find the generating function for the solution of the recurrence relation $a_{n+1} = ra_n + b$.

12. The "tower of Hanoi" is a puzzle consisting of three vertical posts (mounted on a board) and some number n of rings of different diameters.   In standard form, the rings are all stacked on one post in decreasing order of size from bottom to top (Figure 3.4).   A solution to the puzzle consists of first choosing a second post on which the rings are to be stacked; then moving rings from post to post in such a way that a larger ring is never placed on a smaller ring with the goal of setting all the rings on the second post.   If a_n is the minimum number of moves to solve a puzzle with n rings, then $a_{n+1} = 2a_n + 1$.   (First solve the problem of getting all but the biggest ring onto the third post, next move the biggest ring to the chosen post, and then solve the problem of moving all but the biggest ring to the chosen post.)   Find the number of moves needed with n rings.   In particular, what if $n = 5$?

13. Write a recursive program that uses the idea underlying the recurrence relation of Exercise 12 to give a computer solution of the tower of Hanoi problem for any number of rings.

14. Solve the recurrence relation $a_{k+1} = a_k + 2^k$ with $a_0 = 2$.

∗15. Solve the recurrence relation $a_{n+2} = 4a_{n+1} - 4a_n + 2^n$ with $a_0 = 2$, $a_1 = 4$.

∗16. Find the generating function for the solution to $a_{n+2} = 3a_{n+1} + 2^n$ for $a_0 = 1$.

∗17. Find the generating function for the solution to $a_{n+2} = a_{n+1} + 2a_n + 2^n$ for $a_0 = 2$, $a_1 = 4$.

∗18. A merge sort of a list of numbers can be described as follows.   Split the list in half, apply merge sort to each half and then merge the two sorted lists in increasing order.

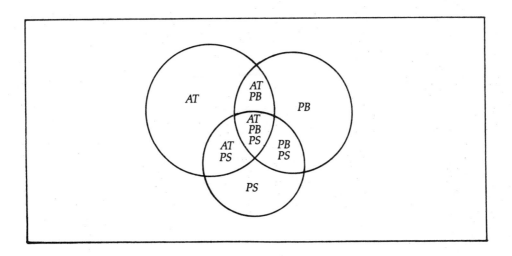

Fig. 3.4

If the list has only one element, do nothing to that one element list. Let a_n be the number of comparisons made by a merge sort on an n–element list. For $n = 1, 2, 4$, figure out by experiment how many comparisons you use (note that you must make some comparisons as you merge).. Assuming n is a power of 2, write a recurrence relation for the numbers a_n. Since this recurrence involves $a_{n/2}$, it is not linear, and the merging keeps it from being homogeneous. There is a solution to this recurrence involving $n\log_2(n)$. Try to find it. This kind of recurrence arises in "divide and conquer" algorithms used to solve computing problems.

* 19. Find a recurrence relation for the number of ways to divide a regular n–gon into triangles by means of nonintersecting diagonals. Solve this recurrence relation.

* 20. Let F_n be the sequence of solutions to the Fibonacci recurrence with $a_0 = a_1 = 1$. Find all n with $F_n = n$.

21. Show that $\sum_{k=0}^{n} \binom{n-k}{k}$ is a solution to the Fibonacci recurrence relation.

* 22. Show that if $F_1, F_2, \ldots, F_n$ is a sequence of solutions to the Fibonacci recurrence, then for each m

$$\sum_{k=0}^{n} \binom{n}{k} F_{m+k}$$

is also a Fibonacci number.

*   **SECTION 5   EXPONENTIAL GENERATING FUNCTIONS**

What Happens When We Multiply Exponential Generating Functions?

Many of our important results on generating functions were based on multiplying generating functions. We began our study of generating functions with Pólya's change–making example.   From this example, we were led to understand what happens when we multiply generating functions.   First, if we have two generating functions $\sum_{i=0}^{\infty} a_i x^i$ and $\sum_{i=0}^{\infty} b_i x^i$, we interpret a_i as being the number of arrangements $\underline{s}$ in some set S whose "value" $v_1(\underline{s})$ is the integer i and let b_j be the number of arrangements $\underline{t}$ in some set T whose "value" $v_2(\underline{t})$ is the integer j.   Then the coefficient c_k of x^k in the product

$$\sum_{i=0}^{\infty} c_i x^i = \sum_{i=0}^{\infty} a_i x^i \cdot \sum_{j=0}^{\infty} b_j x^j$$

is the number of ordered pairs $(\underline{s}, \underline{t})$ with $\underline{s}$ in S and $\underline{t}$ in T and $v_1(\underline{s}) + v_2(\underline{t}) = k$.

This result does not apply to the a_i, b_i and c_i in exponential generating functions, that is, generating functions of the form $\sum_{i=0}^{\infty} a_i \dfrac{x^i}{i!}$ defined at the end of Section 3.   However, there is a similar interpretation of the coefficients c_k of $\dfrac{x^k}{k!}$ in the product

$$\sum_{i=0}^{\infty} c_i \frac{x^i}{i!} = \sum_{i=0}^{\infty} a_i \frac{x^i}{i!} \cdot \sum_{j=0}^{\infty} b_j \frac{x^j}{j!}$$

of exponential generating functions.

To lead into this interpretation, we begin with another example.   Suppose we have two bookshelves, shelf S and shelf T.   We will arrange books on these shelves according to some prescribed rules such as, perhaps, no shelf is empty or the books are in alphabetical order by title or, perhaps, arranged totally at random.

A given arrangement $\underline{s}$ of books on shelf S uses a certain set $v_S(\underline{s}) = B_S$ of books, while a given arrangement $\underline{t}$ of books on shelf T uses a certain set $v_T(\underline{t}) = B_T$ of books.   Of course B_S and B_T have no books in common, so $B_S \cap B_T$ is empty and $B_S \cup B_T$ is the entire set B of books used on both shelves.   We will assume that the number of prescribed arrangements of a set I of books on shelf S depends only on the size i of the set I, and thus is some number a_i.   We assume the same thing about the number of arrangements b_j of a set J of j books on shelf T.   (Note, we aren't assuming that $a_i = b_i$; for example, for some reason we might be allowed to place the books on shelf S at random, but on shelf T only in the single arrangement of alphabetical order by title.)

In this context, is there a natural interpretation of the coefficients of c_k of the product

$$\sum_{k=0}^{\infty} c_k \frac{x^k}{k!} = \sum_{i=0}^{\infty} a_i \frac{x^i}{i!} \cdot \sum_{j=0}^{\infty} b_j \frac{x^j}{j!} \, ?$$

First, notice that $\dfrac{c_k}{k!}$ will be the *sum* of all products of the form $\dfrac{a_i b_j}{i! j!}$ in which $i + j = k$. Thus $j = i - k$ and c_k is the sum over all values of i of the product

$$\frac{k!}{i!(k - i)!}\, a_i b_{k-i} = C(k; i, k - i) a_i b_{k-i}.$$

If we have a specific set K of k books, then we may interpret $C(k; i, k - i)$ as the number of ways to choose i books for shelf S and $k - i$ books for shelf T.   Then we may interpret $a_i b_{k-i}$ as the number of ways of arranging this ordered pairs of sets of books on the two shelves.   To compute c_k, we must add

$$C(k; i, i - k) a_i b_{k-i}$$

over all values of i and j, that is, over all values of i and $i - k$.   Thus

$$c_k = \sum_{i=0}^{k} C(k; i, i - k) a_i b_{k-i}$$

which is the total number of ways of dividing the set K of books into an ordered pair of sets, one for shelf S and one for shelf T, and then arranging these books. This is just the number of permissible arrangements of books on the two shelves.
To summarize, suppose we are given the exponential generating function $\sum_{i=0}^{\infty} a_i \dfrac{x^i}{i!}$ for the number of permissible arrangements of a set I of size i on shelf S and the exponential generating function $\sum_{j=0}^{\infty} b_j \dfrac{x^j}{j!}$ for the number of permissible arrangements of a set J of size j on shelf T.   The coefficients c_k of $\dfrac{x^k}{k!}$ in the product of the two generating functions is the number of permissible arrangements of a set K of size k on the two shelves S and T.
We now give an example of what the actual generating functions might be under the condition that both shelf S and shelf T must receive at least one book. We will then try to formulate these ideas concisely enough to state a theorem which summarizes them.

Example 5.1.   What is the exponential generating function for the number of ways to arrange k books on two shelves so that each shelf receives at least one book?

There are no ways to arrange no books on shelf S and $i!$ ways to arrange i books. Thus, the exponential generating function $\sum_{i=0}^{\infty} a_i x^i$ is $\sum_{i=1}^{\infty} i! \frac{x^i}{i!}$ and the expo-

nential generating function $\sum_{j=0}^{\infty} b_j x^j$ is $\sum_{j=1}^{\infty} j! \frac{x^j}{j!}$. However

$$\sum_{i=1}^{\infty} i! \frac{x^i}{i!} = \sum_{i=1}^{\infty} x^i = \frac{x}{1 - x}.$$

Thus, the exponential generating function for the number of ways to arrange the books on the two shelves is

$$\frac{x^2}{(1 - x)^2} = x^2 \sum_{i=0}^{\infty} C(2 + i - 1, i)x^i = \sum_{i=0}^{\infty} \binom{i + 1}{i} x^{i+2}$$

$$= \sum_{k=2}^{\infty} \binom{k - 1}{k - 2} x^k = \sum_{k=2}^{\infty} (k - 1)x^k$$

$$= \sum_{k=2}^{\infty} k!(k - 1) \frac{x^k}{k!}.$$

Thus there are $k!(k - 1)$ ways to arrange k books on two shelves so that each shelf receives at least one book. ■

From our example, let us try to find a general principle that explains what is happening. For ordinary generating functions, we had *numerically* valued functions v_1 and v_2 defined on sets S_1 and S_2 as our "value functions" and we interpreted the coefficient of x^n in the product of the two relevant power series as the number of ordered pairs (a, b) in $S_1 \times S_2$ of value $v_1(a) + v_2(b) = n$. In the case of exponential generating functions, we assign to each element a of S_1 a *subset* $V_1(a)$ of some relevant set (such as the set of books above), and to each element b of S_2 a *subset* $V_2(b)$. Further we assume the number of elements a with $V_1(a) = I$ is equal to the number of elements a with $V_1(a) = J$ if I and J are of the same size. We make the same assumption about V_2. We let a_i be the number of elements of S_1 whose value is a fixed set I of size i; we let b_j be the number of elements of S_2 whose value is a fixed set J of size j. Then if

$$\sum_{n=0}^{\infty} c_n \frac{x^n}{n!} = \sum_{i=0}^{\infty} a_i \frac{x^i}{i!} \sum_{j=0}^{\infty} b_j \frac{x^j}{j!},$$

we may interpret c_n as the number of ordered pairs whose values are disjoint sets with union a specific set N of size n.

The coefficient of $\dfrac{x^n}{n!}$ in a product of k generating functions may similarly be interpreted as the number of k–element lists whose k values are disjoint sets with union a specific set N of size n.

The Exponential Generating Function for Onto Functions

Example 5.2. Compute the exponential generating function for the number of functions from an n–element set *onto* the set $\{1, 2\}$.

Let S_1 and S_2 both be the set of all nonempty subsets of the integers. Let the "value" of an element of S_1 or S_2 be that set itself. Then the number of ordered pairs of elements whose values are disjoint sets with union N is the number of *ordered* partitions of N into two parts, or equivalently the number of functions from N onto the set $\{1, 2\}$. In this case, $a_i = b_i = 1$ if $i > 0$, $a_0 = b_0 = 0$. Thus

$$\sum_{i=0}^{\infty} a_i \frac{x^i}{i!} = \sum_{i=0}^{\infty} b_i \frac{x^i}{i!} = e^x - 1,$$

so that

$$\epsilon c_n x^n = (e^x - 1)^2 = e^{2x} - 2e^x + 1$$
$$= \sum_{i=0}^{\infty} 2^i \cdot \frac{x^i}{i!} - \sum_{i=0}^{\infty} 2 \frac{x^i}{i!} + 1$$
$$= \sum_{i=1}^{\infty} (2^i - 2) \frac{x^i}{i!}.$$

Thus the number of functions from an n–element set N onto the set $\{1, 2\}$ is $2^i - 2$, as we would expect from an elementary analysis of the problem. However, a more impressive fact is that with the same kind of computation, we can see that the exponential generating function for the number of functions that map onto a k–element set is $(e^x - 1)^k$. ∎

We have not yet proved the result just used. A precise statement of this result and its proof follows. Though not difficult, the proof is an intricately condensed version of the computations we made in the bookshelf example. Thus, there is no harm in passing over the proof.

The General Multiplication Principle for Exponential Generating Functions

Theorem 5.2. Let $S_1, S_2, \ldots, S_k$ be sets and let V_i be a function from S_i to the subsets of some set T. Suppose further that for all j–element sets J, the number of elements a of S_i such that $V_i(a) = J$ is the same. Let $f_i(x)$ be the exponential generating function in which the coefficient of $\dfrac{x^j}{j!}$ is the number of elements a of S_i with $V_i(a) = J$ for one particular j–element set J. Then the coefficient of $\dfrac{x^n}{n!}$ in the product of the k generating functions f_1, f_2, . . . , f_k is the number of lists $(a_1, a_2, \ . \ . \ . \ , a_k)$ of elements $a_i \in S_i$ such that the sets $V(a_1), V(a_2), \ . \ . \ . \ , V(a_k)$ are disjoint and their union is a particular n–element subset N of the set T.

Proof.   Suppose $f_i(x) = \displaystyle\sum_{j=0}^{\infty} b_{ij} \dfrac{x^j}{j!}$. Then the coefficient of $\dfrac{x^n}{n!}$ in their product is the sum of the products

$$b_{1j_1} b_{2j_2} \ . \ . \ . \ b_{kj_k} \cdot \frac{n!}{j_1! j_2! \ . \ . \ . \ j_k!}$$

where the sum runs over all choices of $j_1, j_2, \ . \ . \ . \ , j_k$ that add up to n. However $\dfrac{n!}{j_1! j_2! \ . \ . \ . \ j_k!}$ is $C(n; j_1, j_2, \ . \ . \ . \ , j_k)$, the number of labellings of n objects (chosen from T) using label 1 j_1 times, label 2 j_2 times and so on. Thus the coefficient of $\dfrac{x^n}{n!}$ is the sum of $b_{1j_1} b_{2j_2} \ . \ . \ . \ b_{kj_k}$ over all such labellings. Given such a labelling, there are b_{1j_1} elements a of S_1 whose value $V_1(a)$ is the set of objects labelled with label 1, and in general, there are b_{ij} elements a of S_i whose value $V_i(a)$ is the set of objects labelled with label i. Thus by the multiplication principle, there are $b_{1j_1} b_{2j_2} \ . \ . \ . \ b_{kj_k}$ lists of elements a_i chosen from the sets S_i whose values $V_i(a_i)$ are disjoint such that the set of n objects being labelled is their union. Thus the sum for the coefficient of $\dfrac{x^n}{n!}$ counts each such list exactly once. ∎

Informal Version of the Multiplication Principle

We can restate the multiplication theorem for exponential generating functions in a more intuitive form. Note that each arrangement of books on a shelf uses a certain set of books. Similarly if we study arrangements of letters in words, letters will use a certain set of positions in the word; if we study functions from a set N into a set K, each function uses a certain subset of K as an image. Without attempting a formal definition of the word "uses" we can state

Corollary 5.2.   Let A and B be multisets of objects. Suppose $\sum\limits_{i=0}^{\infty} a_i \dfrac{x^i}{i!}$ is

the exponential generating function for the number of objects in A using any

given set I of size i, and let $\sum\limits_{j=0}^{\infty} b_j \dfrac{x^j}{j!}$ be the exponential generating func-

tion for the number of objects in B using any given set J of size j. If

$$\sum_{i=0}^{\infty} c_k \frac{x^k}{k!} = \sum_{i=0}^{\infty} a_i \frac{x^i}{i!} \sum_{j=0}^{\infty} b_j \frac{x^j}{j!}$$

then c_k is the number of pairs of objects in A $\times$ B using disjoint sets whose
union is a set K of size k.

Putting Lists Together and Preserving Order

As our example of the bookshelves suggests, exponential generating functions frequently are useful in situations where the arrangements we are counting are ordered in some way. In lists of objects, the order in which the objects appear is important, whereas in sets or multisets the order is unimportant. Thus results we
developed for sets or multisets using ordinary generating functions may have analogs for lists that we could develop using exponential generating functions. In
particular, if we multiply the generating functions for multisets chosen from a set
A (with certain restrictions) and multisets chosen from a set B (with perhaps other
restrictions) and if A and B are disjoint, then we get the generating functions for
multisets chosen from $A \cup B$ (subject to all the restrictions). There are many
ways to put two lists, one chosen from A and one chosen from B, together to form a
new list. We will say that we *interleave* or *shuffle* together the lists $a_1, a_2, \ldots,$
a_i and $b_1, b_2, \ldots, b_j$ when we form the lists $c_1, c_2, \ldots, c_{i+j}$ in which each c_k
is from a different position in the list of a's or list of b's and the order of the a's
and b's in the list of c's is the same as in the original lists.

> ***Theorem 5.3.***   Suppose **A** is a set of lists with a_n lists of length n chosen
> from a set S and **B** is a set of lists with b_m lists of length m chosen from a set
> T with $S \cap T = \emptyset$. Then the exponential generating function for the
> number of lists of length k chosen from $S \cup T$ that may be formed by
> shuffling together lists from **A** and **B** is
>
> $$\sum_{i=0}^{\infty} a_i \frac{x^i}{i!} \cdot \sum_{j=0}^{\infty} b_j \frac{x^j}{j!}.$$

Proof. We may think of a list of length i as occupying a certain set of i positions chosen from $\{1, 2, \ldots, k\}$—for example, the first i positions or the last i positions. Each way of shuffling a list of length i from **A** together with a list of length $k - i$ from **B** uses a certain i–element subset I of $\{1, 2, \ldots, k\} = K$ for the lists from **A** and the complementary set $K - I$ for the list from **B**. There are a_i lists that use positions in I and b_{k-i} lists that use positions in $K - I$. The total number c_k of ways to shuffle together lists from **A** and **B** to get a list of length k is the number of pairs of lists, one from **A** and one from **B**, that together use positions in the set $K = \{1, 2, \ldots, k\}$. By Corollary 5.2, the exponential generating function for c_k is the product of the exponential generating function for a_i and the exponential generating function for b_j. ■

Exponential Generating Functions for Words

Example 5.3. How many nonsense words of length n may be made using the letters a, b, c and d?

For each letter, the number of ways to make a list of j copies of that letter is just one, so that generating function for the number of lists of each individual letter is

$$\sum_{i=0}^{n} 1 \frac{x^i}{i!} = e^x.$$

Multiplying the four generating functions together gives

$$\sum_{i=0}^{\infty} a_i x^i = e^{4x} = \sum_{i=0}^{\infty} 4^i \frac{x^i}{i!}$$

for the generating function for words using the four letters, so there are 4^n words of length n. ■

Solving Recurrence Relations with Exponential Generating Functions

Exponential generating functions will help in solving certain kinds of recurrence relations that are not linear, that is recurrence relations of the form

$$a_n = ba_{n-1} + ca_{n-2},$$

in which b and c are functions of n. The classical example is the problem of derangements (the hat check problem). Let D_n be the number of lists of $\{1, 2, \ldots, n\}$ in which integer i is not in position i for any i. Then some integer k between 1 and $n - 1$ is in position n. (Thus there are $n - 1$ choices for k.) Now we examine two cases. In case one, n is in position k; in this case we have D_{n-2} arrangements of the remaining integers. In case two, n is not in position k. Thus, among positions 1 up to $n - 1$, we see a list of the numbers $1, 2, \ldots, k-1, k+1, \ldots, n$ such that i is not in position i if $i < n$ and n is not in position k. In this case we have D_{n-1} such arrangements. Since in both cases we had $n - 1$ choices for k, we have

$$D_n = (n - 1)D_{n-1} + (n - 1)D_{n-2}.$$

(From this it is possible to derive $D_n = nD_{n-1} + (-1)^n$.)
   This may be rewritten as

$$D_{n+2} = (n + 1)D_{n+1} + (n + 1)D_n.$$

To remove the $(n + 1)$ from the recursion, we multiply by $\dfrac{x^{n+1}}{(n + 1)!}$ and sum over all $n \geq 0$. (Note $D_2 = 1$, $D_1 = 0$, so if we set $D_0 = 1$, we may start the sum at $n = 0$ and satisfy the recurrence relation.) We get

$$\sum_{n=0}^{\infty} D_{n+2} \frac{x^{n+1}}{(n + 1)!} = \sum_{n=0}^{\infty} D_{n+1} \frac{x^{n+1}}{n!} + \sum_{n=0}^{\infty} D_n \frac{x^{n+1}}{n!}.$$

Setting $D(x) = \sum_{n=0}^{\infty} D_n \dfrac{x^n}{n!}$ and using primes to denote derivatives, we get

$$\left(\sum_{n=0}^{\infty} D_{n+2} \frac{x^{n+2}}{(n + 2)!} \right)' = x \left(\sum_{n=0}^{\infty} D_{n+1} \frac{x^{n+1}}{(n + 1)!} \right)' + xD(x)$$

$$(D(x) - D_1 x - D_0)' = x(D(x) - D_0)' + xD(x)$$

so that

$$D'(x) - D_1 = xD'(x) + xD(x)$$
$$D'(x)(1 - x) = xD(x) \qquad\qquad \text{because } D_1 = 0$$
$$\frac{D'(x)}{D(x)} = \frac{x}{1 - x} = \frac{1}{1 - x} - 1$$

$$\ln D(x) = -\ln(1 - x) - x + c \qquad \text{by integration}$$

$$D(x) = \frac{1}{1 - x} \, e^{-x} e^c \qquad \text{by exponentiation.}$$

Now $D(0) = 1$, so $c = 0$, giving

$$D(x) = \frac{e^{-x}}{1 - x} = e^{-x}(1 + x + x^2 + \ldots)$$

$$= \sum_{i=0}^{\infty} (-1)^i \frac{x^i}{i!} \cdot \sum_{j=0}^{\infty} x^j$$

$$= \sum_{i=0}^{\infty} \left(\sum_{j=0}^{i} \frac{(-1)^j}{j!} \right) x^i.$$

Thus since $D(x)$ is the exponential generating function for D_n, we get

$$D_n = n! \sum_{j=0}^{n} \frac{(-1)^j}{j!}.$$

EXERCISES

1. Show that the ordinary generating function for the number of ways to place a set of n books on two shelves is $\sum_{n=0}^{\infty} (n + 1)! x^n$. What if each shelf must receive at least one book?

2. Show that for any real number a other than 0, $\lim_{n \to \infty} n! a^n$ does not exist because $n! a^n$ tends to infinity. What does this tell you about the convergence (or lack of convergence) of the ordinary generating function in Exercise 1?

3. For what values of x does the exponential generating function for the arrangements in Exercise 1 converge?

4. What is the exponential generating function for the number of ways to place n books on one shelf so that the shelf does not receive more than ten books?

5. What is the exponential generating function for the number of ways to arrange n books on two shelves so that neither shelf gets more than 10 books?

6. Do Exercise 5 assuming each shelf must receive at least one book.

7. Do the bookshelf example (Example 5.1) at the beginning of this section for k shelves.

8. Do Exercise 5 for k shelves.

9. Do Exercise 5 for k shelves, each of which must receive a book.

10. What is the exponential generating function for the number of functions from a set N onto a set K such that each element of K is the image of at most two elements of N?

11. What does Taylor's theorem tell us about the number a_k in the exponential generating function

$$f(x) = \sum_{k=0}^{\infty} a_k \frac{x^k}{k!}?$$

12. How many "nonsense" words of six letters may be made using up to three a's, five s's, two e's and three c's if each letter must be used at least once? What is the generating function for the number of i–letter "words" that may be made from this selection of letters?

13. What is the exponential generating function for Stirling numbers $S(n, k)$ of the second kind with fixed k? (Hint: How do they relate to onto functions?)

14. What is the exponential generating function for the number of ways to pass out n distinct pieces of candy to three children so that each child gets a piece? In how many ways may 10 pieces be passed out so that each child gets a piece?

15. Redo Exercise 14 assuming each child gets at least two pieces.

16. Redo Exercise 14 assuming no child gets more than four pieces.

17. A function from N to K is "doubly onto" if each element of K is the image of at least two elements of N. What is the exponential generating function for the number of doubly onto functions onto a k–element set?

18. For what distribution problem is a power of the hyperbolic cosine the relevant exponential generating function?

19. What is the exponential generating function for words from the 26–letter alphabet in which the vowels (a, e, i, o or u) must be used an even number of times?

20. Why is $D_n - nD_{n-1} = -\{D_{n-1} - (n - 1)D_{n-2}\}$? Conclude from this that

$$D_n = nD_{n-1} + (- 1)^n.$$

21. Use the recurrence relation $D_n = nD_{n-1} + (- 1)^n$ to derive the exponential generating function $D(x)$ for D_n without using derivatives.

* 22. Let R_n be the number of lists of $\{1, 2, \ldots, n\}$ such that $i + 1$ is not immediately to the right of i. Show that R_n satisfies a recurrence relation much like that of D_n, find the exponential generating function for R_n and find a formula for R_n.

Suggested Reading

Hall, Marshall, Jr.: *Combinatorial Mathematics*, Blaisdell 1967, Chapters 2, 3 and 4.

Knuth, Donald E.: *The Art of Computer Programming*, Vol. 1 (2nd Ed.), *Fundamental Algorithms*, Addison Wesley 1973, Chapters 1 and 2.

Liu, C.L.: *Introduction to Combinatorial Mathematics*, McGraw-Hill 1968, Chapters 2, 3 and 4.

Lovász, László: *Combinatorial Problems and Exercises*, North-Holland 1979, Chapters 1 and 2.

Niven, Ivan and Zuckerman, Herbert: *An Introduction to the Theory of Numbers,* Wiley 1960, Chapter 10.

Pólya, George: "Picture Writing", *American Mathematical Monthly* December, 1956, p. 689.

Riordan, John: *An Introduction to Combinatorial Analysis,* Wiley 1958, Chapters 2, 3, 7 and 8.

Stanley, Richard P.: *Generating Functions* in *Studies in Combinatorics, MAA Studies in Mathematics,* Vol. 17, Gian-Carlo Rota, ed., Mathematical Association of America 1978.

4 *Graph Theory*

SECTION 1 EULERIAN WALKS AND THE IDEA OF GRAPHS

The Concept of a Graph

In our work so far, we have concentrated on formulas that are useful in computing the number of arrangements of a certain kind. In this chapter, we begin another kind of study, the study of the *properties* of certain kinds of arrangements. In the past, a number of different kinds of arrangements have been called graphs. There is no general agreement among mathematicians on exactly which "graphical arrangements" should be called graphs; however, the terminology introduced here seems to be gaining the greatest acceptance.

To begin with a concrete example, suppose a company has a distributed computer network with machines in seven cities. Not all of these computers are able to communicate with one another; Table 4.1 gives, for each of the seven cities, the locations of the computers which can communicate with the computer in a given city.

Computer Location	Potential Communication Links				
Boston	Albany	Atlanta	Cleveland	Denver	
Albany	Boston	Cleveland	Atlanta		
Atlanta	Boston	Albany	Cleveland	Dallas	
Cleveland	Atlanta	Albany	Boston	Denver	Dallas
Dallas	Atlanta	Denver	Sacramento	Cleveland	
Denver	Boston	Dallas	Cleveland	Sacramento	
Sacramento	Denver	Dallas			

Table 4.1

Each day when the system is started up, part of the start-up procedure involves testing the communications links. The testing consists of sending mes-

117

sages back and forth and comparing transmitted and received messages. If a message could start in one city and be transmitted through the network in such a way as to use each link exactly once and end up in the starting city this would be the least costly way to check all the links. Is there such a routing? In Figure 4.1 we show a simplified geographic picture of the network, with the circles representing cities and the lines representing communications links. The "map" in Figure 4.1 could just as well stand for possible airline routes between cities—in this context it is natural to ask if an airplane could shuttle from city to city, passing through each connection on the route map exactly once. If the figure were a map of a subdivision and the circles were intersections, we might wish to ask whether a mail delivery truck could enter the subdivision at one point, deliver mail along each street and then leave after traversing each street exactly once.

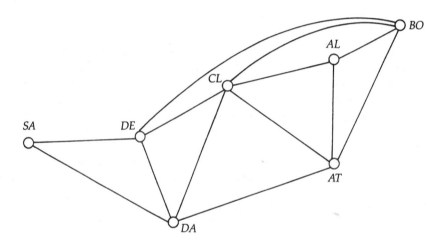

Figure 4.1

The picture in Figure 4.1 is generally called a *graph*. Informally, a graph is a collection of points, called vertices, together with lines between the points, called edges. A more precise definition is that a *graph* consists of a set V, called a *vertex set,* and a set E of two−element subsets of V, called an *edge set.* The elements of V are called *vertices* and the elements of E are called *edges.* We use the shorthand notation $G = (V, E)$ to say "G is a graph with vertex set V and edge set E."

Example 1.1. Write down the vertex set and edge set of the graph shown in Figure 1.1.

By inspection we write

$$V = \{BO, AL, AT, CL, DA, DE, SA\}$$

and

$$E = \{\{BO, AT\}, \{BO, AL\}, \{BO, CL\}, \{BO, DE\}, \{AL, AT\}, \{AL, CL\}, \{AT, CL\},$$
$$\{AT, DA\}, \{CL, DE\}, \{CL, DA\}, \{DE, DA\}, \{DE, SA\}, \{DA, SA\}\}. \quad \blacksquare$$

Multigraphs and the Königsberg Bridge Problem

One of the questions that gave rise to modern graph theory was a question that Euler (pronounced "oiler") attributed to the citizens of the old town of Königsberg in Prussia.   Königsberg consisted of an island where two rivers came together, along with some land along each riverbank; it also had quite a few bridges.   A schematic map of Königsberg is shown in Figure 4.2.

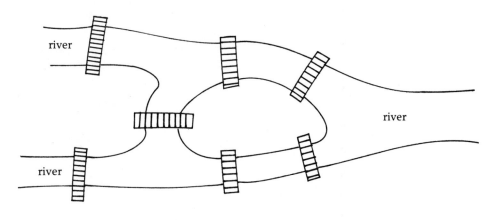

Figure 4.2

The solid bars represent bridges crossing rivers; it is possible to walk between any two bridges touching a given land mass.   Euler attributed to the townspeople the question, "Is it possible to to take a walk through town, starting and ending at the same place, and cross each bridge exactly once?"   Euler recognized that the shape of the land masses made no difference to the problem and that the more abstract diagram in Figure 4.3, in which circles represent land masses and lines represent bridges, contained the essence of the problem.

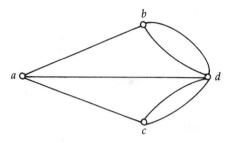

Figure 4.3

The picture represents a type of arrangement we shall call a multigraph. A *multigraph* consists of a set V, called a vertex set, and a multiset of two–element multisets, (chosen from V) called edges. We use the notation $G = (V, E)$ and refer to the "multigraph G." Another multigraph is shown in Figure 4.4. The graph shown in Figure 4.1 is also a multigraph, i.e., graphs are just special kinds of multigraphs.

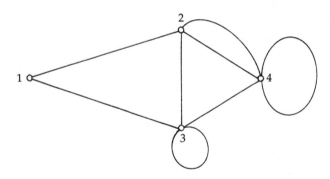

Figure 4.4

The edge multiset of the multigraph in Figure 4.3 is $\{\{a, b\}, \{b, d\}, \{b, d\},$ $\{c, d\}, \{c, d\}, \{a, c\}, \{a, d\}\}$. The edge multiset for Figure 4.4. is $\{\{1, 2\}, \{2, 4\},$ $\{2, 4\}, \{4, 4\}, \{3, 4\}, \{3, 3\}, \{2, 3\}, \{1, 3\}\}$. Although our definition of a multigraph is expressed in terms of sets, we may visualize multigraphs in much the same way we visualize graphs. We start by drawing points on a piece of paper, one for each vertex, connecting two vertices with a line whenever these two vertices are the vertices of an edge, and drawing a line from a vertex to itself whenever the multiset containing that vertex twice is an edge. An edge connecting a vertex to itself is called a *loop*.   Notice that a graph has no loops because a loop is not representable as a two–element set.

Walks, Paths and Connectivity

The questions we have asked in our examples can be asked for any multigraph. Such questions arise so often that special terminology has been developed to help make these ideas easy to discuss. A *walk* in a multigraph is an alternating sequence of vertices and edges

$$v_1 e_1 v_2 e_2 \; . \; . \; . \; e_{n-1} v_n$$

such that e_i contains v_i and v_{i+1}.

A *path* is a walk in which no vertex appears twice. (Unfortunately what we call a walk is often called a path; then what we call a path is called a simple path.) If the first and last vertex of a walk are equal, the walk is called a *closed* walk; a closed walk in which only the first and last vertex are equal is a *cycle*. (Unless the edge $\{x, y\}$ has multiplicity more than 1, however, we do not call the trivial closed walk $x\{x, y\}y\{x, y\}x$ a cycle.) (See Figure 4.5 for examples.) A walk that includes each edge of the multigraph $G = (V, E)$ (exactly as many times as it is in E) is called an *Eulerian* walk.

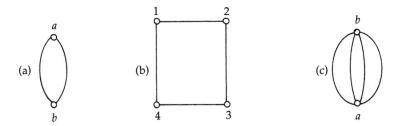

Figure 4.5 The multigraphs (a) and (b) are cycles; in (c) the four–edge closed walk from a to b and back is not a cycle. However, the four-edge closed walk in (c) is an Eulerian walk.

In our two examples, we asked whether a multigraph had a closed Eulerian walk. It was Euler who noticed that the question of the existence of such a walk could be answered by determining how many edges touch each vertex. The *degree* of a vertex v, denoted by $d(v)$, is the number of two–element edges that touch v plus twice the number of loops that touch v. Thus the degree of v is the total number of times v appears in edges of the multigraph. Intuitively speaking, it is the number of lines "sticking out of" v.

Note that if a multigraph has an Eulerian walk, then (with the possible exception of the first and last vertex in the walk) each vertex appears an even number of times in edges—once in each edge preceding it in the walk and once more in each edge following it in the walk. Thus each vertex (with the possible exception of the first and last in the walk), has even degree. Further, if the first and last vertex are the same, it has even degree, and if the first and last vertex are different, then they have odd degree.

Note also that if a multigraph has an Eulerian walk, then it is *connected* in the sense that given any two vertices, there is a walk from one to the other. We say that the vertex u *is connected to* the vertex v if there is a walk from u to v. It is simple to verify that the relation "is connected to" is an equivalence relation on the vertices of a multigraph. Thus the vertices are divided into equivalence classes (called *connectivity classes* or *connected components*), so that vertices in the same class are connected while vertices in different classes are not. Every edge is between two vertices in some connectivity class. Figure 4.6 shows two graphs, one connected and one not connected.

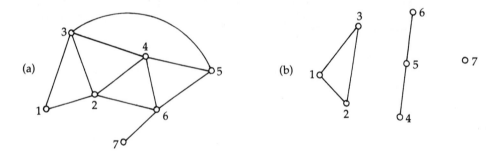

Figure 4.6 Graph (a) is connected, but graph (b) has three connected components.

Eulerian Graphs

The two ideas of "degree" and "connectivity" can be used to describe the *Eulerian multigraphs*, those multigraphs with an Eulerian walk.

> *Theorem 1.1.* A multigraph is Eulerian if and only if it is connected and either all or all but two vertices have even degree. If every vertex has even degree, than all Eulerian walks are closed; if two vertices have odd degree, then every Eulerian walk starts at one of these vertices and ends at the other.

Proof. We have verified already that an Eulerian multigraph is connected and that either zero or two of its vertices have odd degree. Further, we have observed that the Eulerian walk is closed if and only if zero vertices have odd degree. Thus we now prove that if a multigraph is connected and has zero or two vertices of odd degree, then it is Eulerian. Our proof is by induction. We let S be the set of all integers n so that if a connected multigraph with n edges has zero or two vertices of odd degree, then it has an Eulerian walk. (Note: This proof has been written out in great detail because it is one of our first examples of induction which does not involve formulas. Its length should not be regarded as a measure of difficulty.)

A connected multigraph with one edge has either two vertices of degree 1 or one vertex of degree 2. Thus it has either the Eulerian walk $v_1 e v_2$ or the Eulerian walk $v_1 e v_1$. Thus 1 is in S.

Now suppose that all nonnegative integers less than n are in S. We assume the multigraph G has n edges, is connected, and has zero or two vertices of odd degree.

Let x be a vertex of G, a vertex of odd degree if G has any such vertices. Construct a walk as follows. Let $v_1 = x$, let $e_1 = \{x, y\}$ be an edge containing x and let $v_2 = y$. We continue on in this way. If v_i touches an edge yet to be used in the walk, denote the edge by e_i and let $v_{i+1} = y$ if $e_i = \{v_i, y\}$. (Note: If an edge is in E more than once, then it may be used as often as it is in E.) Continue this process until we reach a vertex v_r that does not touch any unused edges. If $v_r = x = v_1$, then the vertex x has even degree and the walk is closed. If $v_r \neq v_1$, then v_r and v_1 both have odd degree. All other vertices v_i have even degree *and* are incident with—i.e., precede or follow—an even number of edges in the walk. If the walk constructed is an Eulerian walk, we are done. If the walk is not Eulerian, we let E' consist of the edges in E but not in the walk. Then we let $G' = (V, E')$. In G', *all* vertices have even degree. (Thus, if G' were connected, we would have by induction a closed Eulerian walk in G' that could be "spliced" with the walk we constructed to get an Eulerian walk in G.) (See Figure 4.7.)

Let $C_1, C_2, \ldots, C_k$ be the connected components of G'. Each of these components contains a vertex of the walk we constructed (why?); say v_{j_i} is in C_i. Now each edge in E' connects two vertices in some C_i. Thus, $E' = E_1 \cup E_2 \cup \ldots \cup E_k$, where E_i only connects edges in C_i. The multigraph (C_i, E_i) is connected and each vertex in it has even degree. Further, (C_i, E_i) has fewer edges than G, and so by the induction hypothesis, there is a closed Eulerian walk in (C_i, E_i) starting and ending at v_{j_i}.

Now we can construct an Eulerian walk in G. Follow the original walk of v_i's until the first v_{j_i} is reached. Then follow the Eulerian walk in

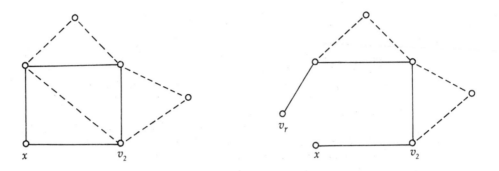

Figure 4.7. Two connected possibilities for G'. The edges shown as solid lines are in the path beginning at x; the other edges are dotted. To splice "paths" together, go from x to V_2; then follow the dotted lines until you return to v_2; finally, follow the solid lines again.

C_i, returning to v_{j_i}. Continue along the original walk to the next v_{j_i} and repeat the process until you have reached v_r. Since every edge is either in an E_i or in the original walk, this walk is Eulerian, so G is Eulerian and $n \in S$. Thus, by the principle of mathematical induction, S contains all positive integers, and this proves the theorem. (See Figure 4.8) ∎

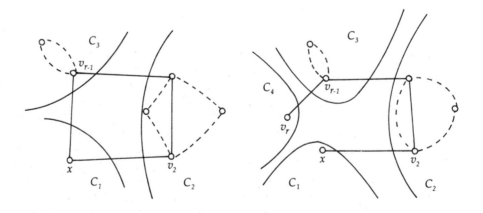

Figure 4.8. Disconnected possibilities for G'. Solid and dotted lines are as in Figure 4.7. Go from x to v_2; follow the dotted lines back to v_2; follow the solid lines to v_{r-1}; follow the dotted lines back to v_{r-1}; and then go on the solid lines again. The curved lines separate the connected components of G'.

From the theorem, we conclude there is no Eulerian walk in the multigraph of Figure 4.3, therefore, the answer to the Königsberg bridge question is "No". The only Eulerian walks in the multigraph of Figure 4.4 start and end at either vertex 3 or vertex 4. Thus the multigraph of Figure 4.4 has no closed Eulerian walks. In Figure 4.1, only *AL* and *CL* have odd degree, therefore, there is an Eulerian walk starting at *CL* and ending at *AL*. Although this means we couldn't solve the airplane shuttle problem or the mail delivery problem, some of the structure of the graph means we can still efficiently test the communications links. Namely, send the test messages from Cleveland to Albany and record what arrives in Albany. Now there is an Eulerian walk on the remainder of the graph from Albany to Albany. When the message last leaves Cleveland, check it against the original message. When the message arrives in Albany for the last time, check it against the recorded message there. If neither check shows an error, then presumably the communications links are functioning properly.

EXERCISES

1. Draw a picture of all the graphs which have the vertex set {1, 2, 3}.

2. Draw a picture of all the connected graphs which have the vertex set {1, 2, 3, 4}.

3. Explain why the relation "is connected to" is an equivalence relation.

4. If two bridges in Königsberg collapse, does Euler's question have an affirmative answer? Does it matter which two collapse? What if you don't have to start and end at the same place?

5. If one bridge in Königsberg collapses, does Euler's question have an affirmative answer? Does it matter which one collapses? What if you don't have to start and end at the same place?

6. Find an Eulerian walk in graph (a) of Figure 4.6.

7. Explain how the mail truck could enter the multigraph of Figure 4.4 at vertex 3 and traverse each street *once in each direction*.

8. Explain why the sum of the degrees of the vertices in a multigraph is an even number. (Hint: How does the number of edges relate to the sum of the degrees?)

9. Prove that in a multigraph, the number of vertices of odd degree is an even number. (Hint: Exercise 8 makes this easier.)

10. Prove that in a graph with n vertices, if two vertices are connected by a walk, then they are connected by a walk with $n - 1$ or fewer edges.

11. If a graph consists of a cycle on n vertices (with no extra edges), then how many edges does it have?

12. Prove that a connected multigraph with n vertices has at least $n - 1$ edges.

13. Prove that in a multigraph (with V finite) with exactly two vertices of odd degree, these vertices are connected.

14. Can a graph with 10 vertices have 50 edges? What is the maximum number of edges it can have?

15. A graph is *complete* if there is an edge between every pair of distinct vertices. How many edges does a complete graph with n vertices have?

16. A path in a multigraph is said to be *Hamiltonian* if it includes each vertex ex-

actly once. No easy general condition for checking whether a graph has a Hamiltonian path is known.

(a) Does a complete graph (Exercise 15) always have a Hamiltonian path?

(b) Show that if a graph G has n vertices and if the sum of the degrees of any two vertices is n or more, then G has a Hamiltonian path. (The result holds if the sum of the degrees is $n - 1$ rather than n. You might try to prove this stronger result.)

(c) Give an example that shows (b) is false for multigraphs.

17. How many graphs are there on the n-element vertex set $\{v_1, v_2, \ldots, v_n\}$?

18. Table 4.1 is called an *adjacency* table for a graph. This is one way a graph can be represented for a computer. Suppose we are given a graph which we know has an Eulerian walk. The method of proof of Theorem 1.1 can be used to find an Eulerian walk. Namely, find and remove a closed walk, determine the connected components and repeat the process on each connected component. When each connected component has size 1, no more work is necessary. There is a brute force method of finding connected components. First pick a vertex. List it and all vertices adjacent to it. For each vertex in the list, add in to the list all adjacent vertices not yet in the list. Repeat this process until the list gets no longer; then you have one connected component. Write a program which finds (by recursion if you have available a computer language which allows it) an Eulerian walk.

SECTION 2 TREES

The Chemical Origin of Trees

Another origin of graph theory was the study of molecules of hydrocarbons. A hydrocarbon is a compound formed from hydrogen atoms and carbon atoms. A molecule of a compound consists of atoms of the constituent parts held together by chemical bonds. A carbon atom can form four such bonds to hydrogen atoms—or other carbon atoms—and a hydrogen atom can form exactly one bond. This leads to a graphical representation of hydrocarbons as multigraphs in which each vertex has degree 4 or 1. Typically, a vertex of degree 4 is labelled with a C and a vertex of degree 1 is labelled with an H. Several graphs of hydrocarbons are shown in Figure 4.9. These hydrocarbons are called saturated because they contain the maximum amount of hydrogen possible for the amount of carbon they contain. Figure 4.10 shows examples of unsaturated hydrocarbons. Note that the multigraphs in Figure 4.9 are in fact graphs, while the ones in Figure 4.10 have multiple edges. Also, the second hydrocarbon in Figure 4.10 has a closed path in its multigraph. Further, the graph of each compound is connected, because all the atoms of a molecule are bound into the molecule in some way.

A connected graph with no cycles (closed paths) is called a *tree* (for the geometric reason that any path that can be traced out among the branches of a tree cannot close back upon itself). A graph with no cycles is called a *forest*. (A forest is thus a disjoint union of trees.) Figure 4.11 shows a tree and a forest.

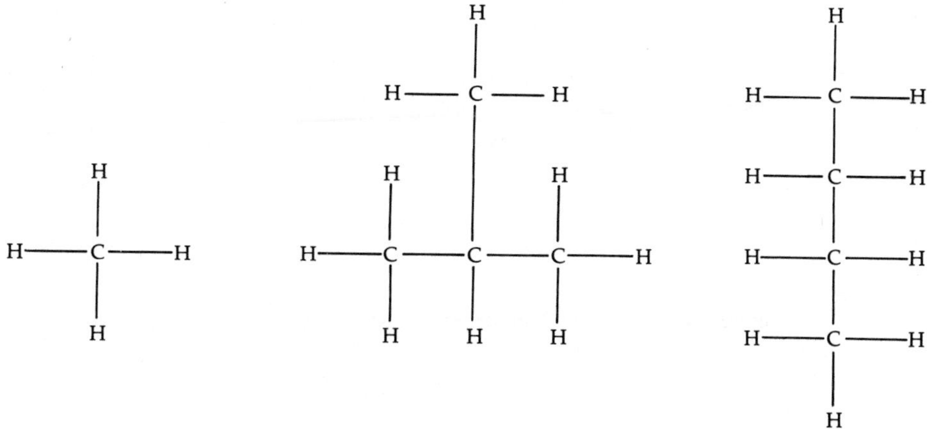

Figure 4.9 Several examples of hydrocarbons

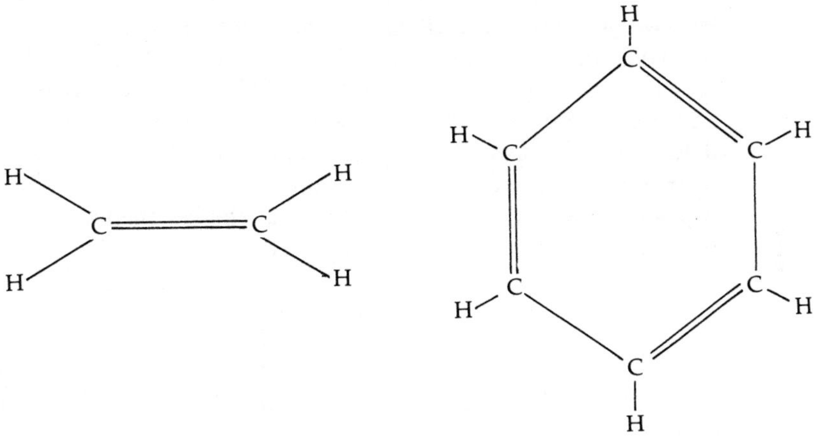

Figure 4.10

Basic Facts About Trees

There are quite a few alternate descriptions of trees. Here are two relatively simple descriptions.

Theorem 2.1. The following statements about a graph G are equivalent.

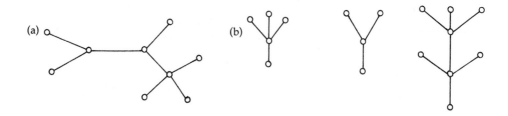

Figure 4.11 The graph labelled (a) is a tree; the graph labelled (b) is a forest, but not a tree.

(1) G is a tree.

(2) There is a unique path between any two vertices of G.

(3) G is connected, but removing any edge from the edge set E of G leaves a disconnected graph.

Proof. We show that statement 1 implies statement 2, and that statement 2 implies statement 3, finally that statement 3 implies statement 1; in this way we prove they are all equivalent. Suppose G is a tree and let x and y be two distinct vertices of G. If there are two different paths from x to y, they must have different edge sets, for the edge set of a path from x to y (in any graph) completely determines the path. (There is one edge leaving x; it determines the vertex following x and so on.) Thus x and y lie on a closed walk in which at least one edge is used only once. Now among all closed walks using at least one edge only once, choose one having no more edges than any other. Let the chosen closed walk be

$$x_0 e_1 x_1 \ \cdot \ \cdot \ \cdot \ e_n x_n$$

with $x_0 = x_n$. Now suppose $x_i = x_j$ with $i \neq j$. If $j - i < n$, there are two closed walks

$$x_0 e_1 \ \cdot \ \cdot \ \cdot \ x_i e_{j+1} \ \cdot \ \cdot \ \cdot \ x_n \text{ and } x_i e_{i+1} x_{i+1} \ \cdot \ \cdot \ \cdot \ x_{j-1} e_j x_j$$

one of which uses some edge only once. Thus the chosen closed walk has $x_i = x_j$ only if $i = 0$ and $j = n$. No edge is used twice, for then a vertex other than x_0 would appear twice. Therefore the chosen walk is a cycle, which is impossible in a tree. For this reason, there cannot be two different paths from x to y.

To show statement 2 implies statement 3, suppose G is a graph in which each pair of points is connected by a unique path. Then G is connected, and if $e = \{x, y\}$ is an edge, $x\,e\,y$ must be the unique path from x to y, so deleting e from the edge set of G leaves x and y disconnected.

Now suppose G is a connected graph such that the removal of any edge yields a disconnected graph. Then G can have no cycles, since deleting an edge of a cycle cannot disconnect a graph. Therefore G is a tree.

Thus each statement implies the other statements and so all three are equivalent. ■

The number of edges of a tree is always one less than number of vertices, as the following induction proof shows.

Theorem 2.2.   A tree with n vertices has $n - 1$ edges.

Proof.   A tree with 1 vertex has no edges and a tree with 2 vertices has one edge. Let S be the set of all integers n such that a tree with n vertices has $n - 1$ edges. Suppose all integers smaller than k are in S, and let G be a tree with k vertices. Let x and y be two vertices of G such that $\{x, y\}$ is an edge. Deleting $\{x, y\}$ from the edge set of G leaves a graph with two connected components, V_1 and V_2. Any other edge of G connects two vertices of V_1 or two vertices of V_2. Thus, using i to stand for 1 or 2, V_i together with the edges of G connecting vertices in V_i is a connected graph G_i without cycles. If V_1 has k_1 vertices and V_2 has k_2 vertices, then k_1 and k_2 are in S, so G_1 has $k_1 - 1$ edges and G_2 has $k_2 - 1$ edges. Thus since $k = k_1 + k_2$, G has $1 + k_1 - 1 + k_2 - 1 = k - 1$ edges. Thus k is in S. By the principle of mathematical induction, all positive integers are in S. ■

This theorem may be modified to give two more descriptions of a tree. A graph with n vertices and $n - 1$ edges is a tree if it is connected or if it has no cycles. Proofs that these are descriptions of a tree are asked for in the exercises.

Example 2.1.   Show that a saturated hydrocarbon with k carbon atoms has $2k + 2$ hydrogen atoms.

The graph of a saturated hydrocarbon is a tree, so if it has k carbon atoms and m hydrogen atoms, it has $k + m - 1$ edges.

An exercise in the last set asked for a proof that the sum of the degrees of the vertices of a graph is twice the number of edges of the graph, so that

$$4k + m = 2(k + m - 1),$$

or

$$4k - 2k = m - 2,$$

so that

$$m = 2k + 2. \quad \blacksquare$$

This example shows the value of the fact that

Theorem 2.3.   The sum of the degrees of the vertices of a multigraph is twice the number of edges.

Proof.   An exercise.   ■

Corollary 2.4.   A tree has at least two vertices of degree 1.

An important chemical question is whether there are two saturated hydro-carbon isomers with the same number of carbon atoms but different chemical properties.   At first we might think that since they would have the same number of hydrogen atoms, two saturated hydrocarbons with the same number of carbons would have the same properties.   Figure 4.12 shows pictures of butane and iso-butane, two saturated hydrocarbons with four carbon atoms.   As you might ex-pect, butane and isobutane are *not* exactly the same in terms of chemical proper-ties.   Different saturated hydrocarbons with the same number of carbon atoms are called isomers of one another.   What, though, do we mean when we say two saturated hydrocarbons are different?   From the discussion of butane and isobu-tane we would expect them to be different unless they have essentially the same

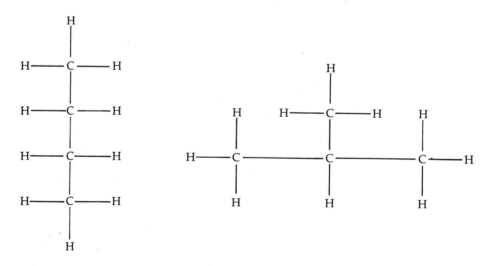

Figure 4.12

graphs. The question "when are two graphs the same?" leads us to the notion of isomorphism that is taken up in the next section. At that time, we will at least be able to give a precise meaning to the question "How many different isomers of a saturated hydrocarbon with k carbon atoms are there?" Tools to answer the question have been developed but are somewhat too complex to present here.

Spanning Trees

Trees play an important role in graph theory for many other reasons. Suppose we have a graph whose vertices represent cities and whose edges represent video communications links. A TV program originating in one city is to be transmitted to all other cities along links of the net. The links used in the process define an "edge–subgraph" of the original graph, a new graph with the same vertex set but whose edge set is a subset of the original one. There is no point in sending the program to a city in two different ways. Thus every pair of vertices should be connected by a unique path in the new graph; i.e., the new graph should be a tree.

If $G = (V, E)$, a *spanning tree* of G is a tree whose vertex set is V and whose edge set is a subset of E. A *spanning forest* of G is a forest whose vertex set is V such that two vertices connected by a path in G are also connected by a path in the forest. All graphs have spanning forests (this is an exercise), but only connected graphs have spanning trees.

Theorem 2.5.   A connected (multi)graph has a spanning tree.

Proof.   Let $G = (V, E)$ be connected. If G has no cycles, it is its own spanning tree. Let S be the set of all integers n such that a graph with n cycles has a spanning tree. Zero is in S; assume now that all integers less than k are in S, and that G has $k > 0$ cycles. Let $\{x, y\}$ be an edge in a cycle of G. Then $G' = (V, E - \{x, y\})$ is connected (because there is a path from x to y not including $\{x, y\}$). G' has $k - 1$ or fewer cycles. Thus G' has a spanning tree, which is automatically a spanning tree of G. Thus k is in S, and so by version 2 of the principle of mathematical induction, S contains all positive integers. ■

Breadth–First Search Trees

The proof of the theorem tells us that to find a spanning tree, we locate a cycle and break it; then we iterate the process. As we shall see, there are better methods. If we are going to use a spanning tree as a network for broadcasting television programs, we will have one vertex that is the source of all (or most of) the shows—the network headquarters. Since there will be delay and possibility for error each time we transmit some information, we would like to choose a special kind of tree—one that minimizes the number of retransmissions of some program required to send the program to any given vertex in the network. The *length* of a path is its number of edges. The *distance* between two vertices is the length of the shortest path connecting them. In graph–theoretic terms, we desire a tree such that the distance in the tree between each vertex and the source is as small as possible. The technique of "breadth–first search" provides us with just such a tree. The technique is illustrated in Figure 4.13 and described below.

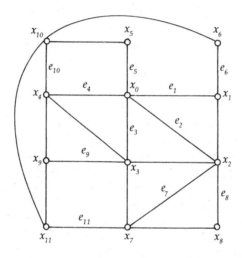

Figure 4.13 In this graph x_0 is the center x of the spanning tree; the vertex labelled x_i is the vertex added in the process of constructing set V_i and the edge labelled e_i is the edge added in the process of constructing set E_i.

A *breadth–first search tree centered at* vertex x in a connected graph $G = (V, E)$ is a spanning tree T which can be constructed by the following rules. (The rules formalize the following description. Connect x to all vertices at distance 1 from it. Once all vertices at distance i from x have been connected into the tree, use these vertices one at a time to connect into the tree all adjacent vertices at distance $i + 1$ from x that *don't* give a cycle. Continue until you run out of vertices.)

(1) Let $x_0 = x$, $V_0 = \{x_0\}$ and $E_0 = \emptyset$.

(2) Choose a y such that $\{x_0, y\}$ is in E. Let $x_1 = y$, $V_1 = \{x_0, x_1\}$ and $E_1 = \{\{x_0, x_1\}\}$.

(3) If x_0 is adjacent to another $z \neq x_1$, let $x_2 = z$, $V_2 = V_1 \cup \{x_2\}$ and $E_2 = E_1 \cup \{\{x_0, x_2\}\}$. Otherwise if x_1 is adjacent to a $v \neq x_0$, let $x_2 = v$, $V_2 = V_1 \cup \{x_2\}$ and $E_2 = E_1 \cup \{\{x_1, x_2\}\}$.

(4) Given pairs (V_i, E_i) $i = 0, 1, 2, \ldots, k$, let j be the smallest number such that

x_j is adjacent to some vertex in V but not V_k. Let $x_{k+1} = z$, $V_{k+1} = V_k \cup \{x_{k+1}\}$ and $E_{k+1} = E_k \cup \{\{x_j, x_{k+1}\}\}$.

(5) Repeat Step 4 until there is no x_j of the type described.

(6) Let $T = (V_k, E_k)$, where k has the final value encountered in Step 4.

This construction gives us a list $x_0, x_1, x_2, \ldots$ of vertices as they first appear in $V_0, V_1, V_2, \ldots$. Vertex x_0 on the list is x, followed by all vertices adjacent to x_0. Next come all vertices adjacent to v_1 not already in the list, then all vertices adjacent to v_2 not already in the list, and so on. This way of numbering the vertices is called a "breadth–first numbering". The number i is called the "breadth–first number" of x_i.

Theorem 2.6. A breadth–first search tree centered at x in a connected graph G is a spanning tree.

Proof. The graph constructed is connected, has n vertices and $n - 1$ edges; therefore, it is a tree. Since G is connected, we repeat Step 4 until V_k becomes the vertex set of G. ∎

Note both Step 2 and Step 3 of the construction are unnecessary; they are included as a helpful mental transition from Step 1 to Step 4. Each is a special case of Step 4.

Recall that we constructed breadth–first search trees in hopes of keeping distances in the tree as short as possible. We use $d(u, v)$ to stand for the distance from u to v.

Theorem 2.7. If T is a breadth–first search tree centered at the vertex x in G and m is the distance from x to v in G, then the unique path from x to v in T has length m.

Proof. Note that by construction if $d(x, x_i) < d(x, x_k)$, then $i < k$. The proof will be by induction on the breadth–first number of v. Let S be the set of all n such that if v has breadth–first number n, $d(x, v)$ is the length of the path in T from x to V. Zero is in S since x is connected to itself by a path of length 0. Suppose all nonnegative integers less than k are in S and let v have breadth–first number k. Thus $v = x_k$. There is one number $j < k$ such that $\{x_j, x_k\}$ is an edge of T. Since $j < k$, $d(x, x_j)$ is the length m of the path in T from x to x_j. Since x_k is not on the path in T from x to x_j, adding edge $\{x_j, x_k\}$ and vertex x_k to this path gives the path in T from x to x_k. Because this is also a path of G, $d(x, x_k) \leq m + 1$. We wish to prove that $d(x, x_k) = m + 1$. Thus we suppose $d(x, x_k) \leq m$ so that there is a path $x_0(x_0, v_1)v_1 \ldots v_h(v_h, x_k)$ of length $h + 1 < m + 1$ in G from x_0 to x_k. Thus $d(x_0, v_h) \leq h < m$. Now v_h has some breadth–first number i so $v_h = x_i$. Thus $d(x, x_i) < d(x, x_j)$ and $\{x_i, x_k\}$ is an edge. However this contradicts the description of j in Step 4 of the construction process, so our supposition that $d(x, x_k) \leq m$ is incorrect. Therefore $d(x, x_k) = m + 1$, so

$k + 1$ is in S. Therefore by the second version of the principle of mathematical induction, S contains all nonnegative integers; this proves the theorem. ∎

A breadth–first search tree centered at x will always give us the shortest possible path from x to any other vertex; in communications networks, such a tree will minimize the time needed to send a message from x to all other vertices. Notice that doing a breadth–first search from x gives us all vertices connected to x, so it gives a method to determine connected components. Breadth–first search is the basis of many algorithms in computer science.

The words "maximum" and "minimum," or "largest" and "smallest," or some similar pair arise naturally in the communications problem we have described. For each spanning tree, we have a maximum length path—say of length m—and we wish to choose a spanning tree that minimizes m. Problems asking for an arrangement that minimizes, over all relevant arrangements, the maximum value of some parameter are called "minimax problems." Their solution may involve a search technique such as breadth–first search, but often requires more than just a search. Problems of this nature abound in applications of combinatorial mathematics. In the communications examples we used to illustrate the spanning tree idea, different communications links might cost different amounts of money to use. In this case, we would not want just *any* spanning tree, but a minimum–cost spanning tree. A method—called the "greedy algorithm"—for finding such a minimum–cost spanning tree will be found in our next chapter. Similarly, rather than the shortest path from u to v, we might want the cheapest path. The "potential algorithm" given in Chapter 5 provides such a path.

The Number of Trees

A particularly difficult question is "How many spanning trees does a graph have?" We shall be able to answer this question later by means of some powerful techniques. If we ask the question for the complete graph with n vertices, though, it reduces to the question "How many trees are there on n vertices?" Cayley was the first person to pose and answer this question. Our solution to the problem is a modern version of Cayley's solution; it gives a sophisticated kind of generating function for trees on n vertices.

We will consider trees on the n–element vertex set $\{1, 2, \ldots, n\}$. Our generating function will involve the variables $x_1, x_2, \ldots, x_n$. We associate a monomial in these variables with each tree by using the degrees $d(1), d(2), \ldots, d(n)$ of the vertices, namely

$$M(T) = x_1^{d(1)} x_2^{d(2)} \ldots x_n^{d(n)}.$$

Another way of constructing this monomial is

$$M(T) = \prod_{\substack{\text{edges} \\ \{u,v\} \text{ of } T}} x_u x_v.$$

The *enumerator–by–degree sequence* for trees on N is given by

$$E_N(\underline{x}) = \sum_{\substack{\text{trees} \\ T \text{ on } N}} M(T).$$

The symbol $\underline{x}$ stands for the list $(x_1, x_2, \ldots, x_n)$. The coefficient of the monomial

$$M = x_1^{d(1)} x_2^{d(2)} \ldots x_n^{d(n)}$$

in the enumerator $E_N(\underline{x})$ is the number of trees on N in which vertex 1 has degree $d(1)$, vertex 2 has degree $d(2)$, . . . and vertex n has degree $d(n)$. Before considering general cases, let us examine the cases $N = \{1, 2, 3\}$ and $N = \{1, 2, 3, 4, 5\}$. The three trees on $\{1, 2, 3\}$ are shown in Figure 4.14 with their associated monomials.

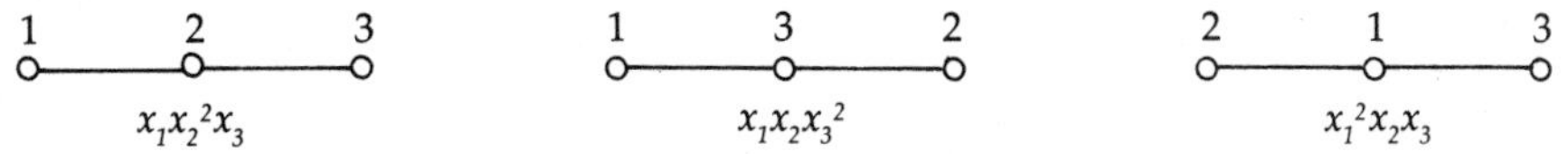

Figure 4.14

Then by adding these monomials we get

$$\begin{aligned} E_{\{1,2,3\}}(x_1,x_2,x_3) &= x_1 x_2^2 x_3 + x_1 x_2 x_3^2 + x_1^2 x_2 x_3 \\ &= x_1 x_2 x_3 (x_1 + x_2 + x_3). \end{aligned}$$

In a tree with five vertices, there will be four edges, so the sum of the degrees of the five vertices will be eight. The degree sequence will contain five positive integers that add up to eight; in other words, it will be an ordered partition of eight into five parts. The only sequences (in decreasing order) that add to up eight are (4, 1, 1, 1, 1) (3, 2, 1, 1, 1) (2, 2, 2, 1, 1), and thus, up to reordering, trees on five vertices will have one of these degree sequences. Now there is only one way to have one vertex of degree 4 and four vertices of degree 1 in a connected graph—all the degree 1 vertices must be connected to the degree 4 vertex. Thus the tree is completely determined by the choice of the degree 4 vertex. If, as in the second sequence, we have a vertex of degree 3, say a, then three vertices must connect to a, and if all three had degree 1, there would be no way to connect the vertex of degree 2 into the tree.

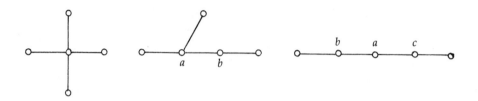

Figure 4.15

Thus one vertex, say b, must be the degree 2 vertex, and the third degree 1 vertex must connect to b.

If, as in the third sequence, we have three vertices of degree 2 and only two of degree 1, one and only one vertex, of degree 2, say a, is required to be connected to the other two degree 2 vertices, say b and c. However, since b and c each have degree 2, the degree 1 vertices must be connected to them. Thus the three trees shown in Figure 4.15 are, up to the assignment of the vertex numbers 1, 2, 3, 4 and 5, the only trees we will see on the vertex set $\{1, 2, 3, 4, 5\}$. Note that for a given assignment of degrees to vertices, there will be one tree in which vertex 1 has degree 4 and vertices 2 through 5 have degree 1, while there will be three trees in which vertex 1 has degree 3 (i.e., like a in Figure 4.15), vertex 2 has degree 2 and vertex 3, 4 and 5 all have degree 1. (See Figure 4.16.) Finally, there will be six trees in which vertices 1, 2 and 3 have degree 2 and vertices 4 and 5 have degree 1. (See Figure 4.17.) The other assignments of degrees to vertices will give either one, three or six trees for each degree sequence, just as in one of the three cases we have worked out. Thus

$$
\begin{aligned}
E_{\{1,2,3,4,5\}}&(x_1, x_2, x_3, x_4, x_5) \\
&= x_1^4 x_2 x_3 x_4 x_5 + x_1 x_2^4 x_3 x_4 x_5 + \ldots + x_1 x_2 x_3 x_4 x_5^4 \\
&\quad + 3x_1^3 x_2^2 x_3 x_4 x_5 + 3x_1^3 x_2 x_3^2 x_3 x_4 x_5 + \ldots + 3x_1 x_2 x_3 x_4^2 x_5^3 \\
&\quad + 6x_1^2 x_2^2 x_3^2 x_4 x_5 + 6x_1^2 x_2^2 x_3 x_4^2 x_5 + \ldots + 6x_1 x_2 x_3^2 x_4^2 x_5^2 \\
&= x_1 x_2 x_3 x_4 x_5 (x_1^3 + x_2^3 + \ldots + x_5^3 + 3x_1^2 x_2 + 3x_1^2 x_3 + \ldots + 3x_4 x_5^2 \\
&\quad + 6x_1 x_2 x_3 + 6x_1 x_2 x_4 + \ldots + 6x_3 x_4 x_5).
\end{aligned}
$$

Once we factor out the $x_1 x_2 x_3 x_4 x_5$; the term in parentheses will contain each product $x_{i_1} x_{i_2} x_{i_3}$ in which i_1, i_2 and i_3 are between 1 and 5. The coefficient of x_1^3 is

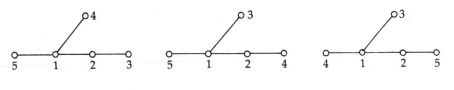

Figure 4.16

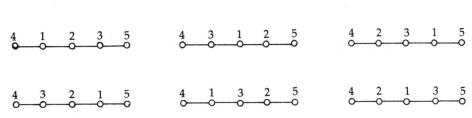

Figure 4.17

$\begin{pmatrix} 3 \\ 3, 0, 0 \end{pmatrix}$, the coefficient of $x_1^2 x_2$ is $\begin{pmatrix} 3 \\ 2, 1, 0 \end{pmatrix} = \dfrac{3!}{2! \cdot 1! \cdot 0!}$ and the coefficient of $x_1 x_2 x_3$ is $\begin{pmatrix} 3 \\ 1, 1, 1 \end{pmatrix} = \dfrac{3!}{1! \cdot 1! \cdot 1!}$. In other words, by using the multinomial theorem in reverse, we see that

$$E_{\{1,2,3,4,5\}}(x_1 x_2 x_3 x_4 x_5) = x_1 x_2 x_3 x_4 x_5 (x_1 + x_2 + x_3 + x_4 + x_5)^3.$$

This example leads us to conjecture Theorem 2.8.

Theorem 2.8. The enumerator–by–degree sequence for trees on the vertex set N is given by

$$E_N(\underline{x}) = x_1 x_2 \ldots x_n (x_1 + x_2 + \ldots + x_n)^{n-2}.$$

Proof. We note that a tree is completely determined by its edge set. We will show a one–to–one correspondence between edge sets of trees and terms with coefficient 1 in the expansion of $(x_1 + x_2 + \ldots + x_n)^{n-2}$. It will be clear from the correspondence that the degree sequence of the corresponding tree is obtained by adding 1 to the degree of *each* x_i in this term, i.e., by multiplying the term by $x_1 x_2 x_3 \ldots x_n$. This will prove the theorem.

First, list all edges containing a vertex of degree 1 in order of their degree 1 vertices (thus, for the first tree in Figure 4.16, we would have

(3, 2), (4, 1), (5, 1)). Now delete these vertices of degree 1 from T to get a new tree T_1. Continue now by listing the edges containing vertices of degree 1 in T_1. (Now for the first tree in Figure 4.16, we have (3, 2), (4, 1), (5, 1), (1, 2).) Unless T_i has only one edge, each edge listed has one vertex of degree 1, and we always list the degree 1 vertex first. We repeat the process, increasing our list each time until we have listed all the edges of T. Now the first elements of all the first $n - 2$ edges and both elements of the last edge will consist of the elements of N, each appearing exactly once.

It turns out that *any* sequence of $n - 2$ members of N can be the second elements of the first $n - 2$ edges, and further that this sequence of $n - 2$ members of N completely determines all the other entries in all the ordered pairs in the list. Note that each sequence $(k_1, k_2, \ldots, k_{n-2})$ corresponds to the term $x_{k_1} x_{k_2} \ldots x_{k_{n-2}}$ in the product $(x_1 + x_2 + \ldots + x_n)^{n-2}$, and that the number of times x_i appears in one of these terms is one less than the number of times x_i appears in an edge. (Why?) Thus, our proof will be complete when we show why any sequence of second entries can appear from one and only one tree.

Suppose we are given a sequence of $n - 2$ elements (so that our list looks like (, i_1), (, i_2), $\ldots$ (, i_{n-2}), (,), including the empty ordered pair at the end so far). Note that vertices of degree 1 in T will have to be those that do not appear in the sequence. At least two, and at most $n - 3$, numbers do not occur; suppose there are j_1 such numbers, in increasing order, $k_1, k_2, \ldots, k_{j_1}$. Thus our list looks like

$$(k_1, i_1), (k_2, i_2) \ldots (k_{j_1}, i_{j_1}), (, i_{j_1+1}), \ldots (, i_{n-2}), (,).$$

Now the edges that appear in positions $j_i + 1$ to $n - 1$ will be the edges of T_1; the numbers $k_{j_1+1}, \ldots, k_{j_2}$ not appearing in our sequence between positions $j_1 + 1$ and $n - 2$, but appearing between positions 1 and j, will be the vertices of degree 1 in T_1, and in increasing order these will be the next j_2 numbers that occur as left–hand sides of edges. Thus our list is now

$$(k_1, i_1), \ldots (k_{j_1}, i_{j_1}), (k_{j_1+1}, i_{j_1+1}), \ldots (k_{j_2}, i_{j_2}), (, i_{j_2+1}), \ldots (, i_{n-2}), (,).)$$

In the same way, the numbers that must be the left–hand sides of each of the first $n - 2$ ordered pairs in our list are uniquely determined. We never repeat a number in this list of left–hand sides, so we have $n - 2$ numbers. This determines the first $n - 2$ edges in the list of edges; the last edge contains the remaining two elements of N. Now we have exactly one graph with $n - 1$ edges and n vertices determined by each sequence of

$n - 2$ numbers from N. To prove our theorem, we must prove this graph is a tree. First note that i_{n-2} cannot be one of the numbers we listed in the first $n - 2$ left–hand sides, because we only list numbers in position $j_r + 1$ and beyond that *do not* appear in the list of right–hand sides in position $j_r + 1$ and beyond. Thus i_{n-2} is one member of the last and next–to–last edge in the list, so these two edges form a tree on the three vertices they include.

In general, each number in the right–hand side of an edge will appear eventually as a left–hand side of a later edge unless, perhaps, it is one of the two vertices we already know are in the last edge. Thus if we add the edges to the edge set of a graph in reverse order from the order in which we listed them, we will always have the edge set of a connected graph, so our sequence of $n - 1$ edges is the edge set of a tree. Thus each sequence of $n - 2$ numbers chosen from N corresponds to exactly one tree. ∎

Corollary 2.9.   There are n^{n-2} trees on n vertices.

Proof.   Set $x_1 = x_2 = \ldots = x_n = 1$ in $E_N(\underline{x})$ and you get the total number of trees on N. ∎

EXERCISES

1. We showed that all trees on three vertices have essentially the same geometric picture. In Figure 4.7 we show there are only three geometrically distinct trees on five vertices. How many geometrically *distinct* trees are there on four vertices (show their pictures)?

2. Repeat Exercise 1 for six vertices.

3. Show that there are essentially only two different pictures of four–vertex forests with two connectivity classes. What about three connectivity classes?

4. Show that every tree has at least two vertices of degree 1.

5. How many connectivity classes are there in a forest with seven vertices and four edges?

6. Draw all pictures of saturated hydrocarbons with five carbon atoms.

7. Octane is a "generic" name for saturated hydrocarbons with eight carbon atoms. Draw an example of an octane molecule in which the longest path of C's has length 7. Among all imaginable octane molecules, what is the smallest possible value for the length of the longest chain of C's? Draw such a molecule.

8. Show that a connected graph on n vertices with $n - 1$ edges is a tree.

9. Show that a graph on n vertices with no cycles is a tree if it has $n - 1$ edges.

10. How many spanning trees does a cycle (closed path) on n vertices have? There may be more than one.

11. Describe a breadth–first search tree for a cycle with an odd number of vertices; for a cycle with an even number of vertices.

12. Let G be the graph in Figure 4.18. Find a spanning tree of G such that no path has more than three vertices. Find a spanning tree of G with the longest possible path.

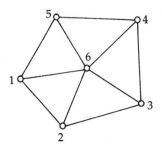

Figure 4.18

13. Find a breadth–first search tree centered at 1 for the tree of Figure 4.18. How many answers are possible?

14. True or False. Any two spanning trees of a graph have an edge in common. (Prove or give a counter-example.)

15. Let T_1 and T_2 be spanning trees of a graph G and let e be an edge of T_1 not in T_2. Show that there is a spanning tree T_3 of G containing e and all but one edge from T_2.

16. True or False. G is a connected graph. (Prove or give a counter-example.)

   (a) Given any edge of G, there is a spanning tree of G containing that edge.

   (b) Given any two edges of G, there is a spanning tree of G containing them.

   (c) Given any three edges of G, there is a spanning tree of G containing them.

17. By means of the technique used in the text to derive explicitly the enumerator–by–degree sequence for trees on $\{1, 2, 3, 4, 5\}$, derive the enumerator–by–degree sequence for trees on $\{1, 2, 3, 4\}$.

18. Show that the following three statements about a connected graph G are equivalent.

   (a) G is a tree.

   (b) Each closed walk in G uses at least one of its edges at least twice.

   (c) Each closed walk in G uses all its edges at least twice.

19. Write a computer program that takes as data the adjacency table (see Exercise 18, Section 1) of a graph and produces the connected components of the graph by means of breadth–first search.

20. Write a computer program that takes as data the adjacency table of a graph and gives the distance between each two connected vertices by means of breadth–first search.

*21. In the proof of Theorem 2.8, there is an algorithm that produces a tree from a sequence of $n - 2$ numbers between 1 and n. Write a computer program that takes a sequence as input and produces a tree as output. Modify the program to produce all trees on the set $\{1, 2, \ldots, n\}$.

SECTION 3   ISOMORPHISM AND PLANARITY

The Concept of Isomorphism

In Figure 4.19 you see pictures of three trees. The first tree is clearly a different structure from the other two trees, but though the second two trees have different pictures, each has one vertex of degree 3, surrounded by three vertices of degree 1.   Geometrically, they appear to be essentially the same.   The edge set of the second graph is $\{\{C_2, C_1\}, \{C_2, C_3\}, \{C_2, C_4\}\}$, while the edge set of the third graph is $\{\{d, a\}, \{d, b\}, \{d, c\}\}$.   Note that if we rename the vertices of the second graph by changing C_2 to d, C_1 to a, C_3 to b, and C_4 to c and carry these relabellings into the edge set, then the two edge sets of the two graphs are exactly the same.   As mathematical structures, then, the two graphs are essentially the same.   Two mathematical objects which are essentially the same are normally said to be iso-morphic (loosely translated, "isomorphic" is Greek for "same shape").   To be precise, an *isomorphism* from a multigraph $G_1 = (V_1, E_1)$ to a multigraph $G_2 = (V_2, E_2)$ is a one–to–one function f from V_1 onto V_2 such that for each edge $\{a, b\}$ in E_1, $\{f(a), f(b)\}$ is an edge in E_2 with the same multiplicity that $\{a, b\}$ has in E_1.   We say G_1 and G_2 are isomorphic if there is an isomorphism from G_1 to G_2.

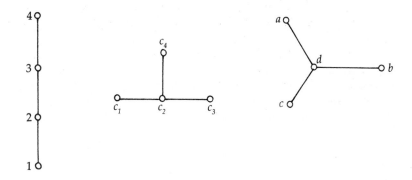

Figure 4.19

It is straightforward to prove that isomorphism between graphs is an equiv-alence relation, that is

Theorem 3.1.   Let S be a set of graphs or multigraphs.   Then the relation "is isomorphic to" is an equivalence relation on S.

Proof.   Exercise 1.   ■

Checking Whether Two Graphs are Isomorphic

It is natural to ask how many really different graphs or really different trees, etc. there are on n vertices.   We now have a concrete interpretation of this question; we are asking for the number of "isomorphism classes" of graphs for a certain set of graphs.

The two graphs in Figure 4.20 are isomorphic by the function $f(a_i) = i$.

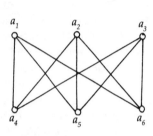
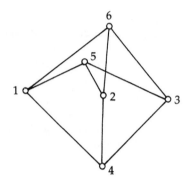

Figure 4.20

Once we are given an isomorphism f, it is easy to check that it has the desired properties.   On the other hand, given two drawings, it is very difficult to check whether they are drawings of isomorphic graphs.   For example, is either graph in Figure 4.21 isomorphic to the graphs in Figure 4.20?

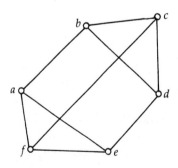
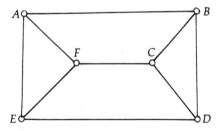

Figure 4.21

It is easy to see that the graphs in Figure 4.21 are isomorphic to each other because of the labellings; we just let the upper– and lower–case letters correspond to each other in the natural way. There is no obvious way, however, to match up the graphs in Figure 4.21 with those in Figure 4.20. *Perhaps* they are not isomorphic. Let us see how nonisomorphic graphs could be different. For example, two isomorphic graphs have the same number of edges (why?). However, all the graphs in question have nine edges. The multiset of all the degrees of vertices of one graph will be the same as the multiset of degrees of an isomorphic graph (why?). However, all the vertices of all the graphs in question have degree 3. (They are examples of so–called *regular* graphs, graphs whose vertices all have the same degree.)

One of the graphs in Figure 4.21 has been drawn without any crossings among the edges. If the graphs in Figure 4.20 and Figure 4.21 are isomorphic, we should be able to redraw the graphs in Figure 4.20 in the same way with no crossings. In this way, we would probably be able to see better what the isomorphism function is (*if* there is such a function). The second picture in Figure 4.21 divides the plane (or the flat piece of paper it is drawn on) into five regions, the two triangles and two trapezoids inside the rectangle and the region outside the rectangle. If we could draw the graph of Figure 4.20 on a plane without any crossing, the picture would also divide the plane into regions. If the graphs in question are isomorphic, there should be a redrawing of Figure 4.20 with five regions, two of them triangles, two of them quadrilaterals and one the outside of a quadrilateral. The boundary (outside or inside, depending on the region) of each region is a cycle in the graph. From this we see that the graph of Figure 4.20 and the graph of Figure 4.21 *are not* isomorphic, because all cycles in Figure 4.20 are quadrilaterals.

Planarity

This leaves us with a thorny question, though: Can the graph of Figure 4.20 be drawn on a plane surface without crossings? A graph that has such a drawing is called a *planar graph*. Thus we know the graphs of Figure 4.21 are planar, but we don't know about the graph of Figure 4.20. Euler studied this concept of planarity, especially in the case where the planar graphs are pictures in the plane of three–dimensional geometric figures called polytopes. (See Figure 4.22 for the relationship between the usual picture of a cube with its faces labelled and a planar picture.)

The second picture in Figure 4.22 can be visualized by imagining we are looking at an elastic cube from above, with the base stretched out to allow us to see all the side faces. The one face we couldn't otherwise see is punctured and folded out around the figure. Because of the relationship between graphs and polytopes the regions of a planar drawing of the graph are also called faces.

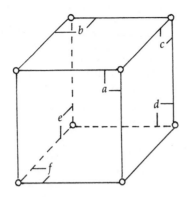

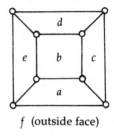

f (outside face)

Figure 4.22

Euler's Formula

For a planar drawing of a graph G, let v be the number of vertices, e be the number of edges and f be the number of faces (including the face consisting of the region outside the drawing). Euler noticed that $v - e + f = 2$ for any graph with a planar drawing. In Figure 4.22, we get $8 - 12 + 6 = 2$ and for the planar graph in Figure 4.21, we get $6 - 9 + 5 = 2$.

> *Theorem 3.2.* For any planar drawing (without crossings) of a connected planar graph, $v - e + f = 2$.
>
> *Proof.* We prove the theorem by induction on e. Let S be the set of integers n such that Euler's formula holds for all planar graphs with $e = n$ edges. If $e = 1$, the graph has two vertices, one edge and bounds only one region—the region outside the graph. Thus $v - e + f = 2 - 1 + 1 = 2$. Now suppose the theorem is true for graphs with $e - 1$ edges (that is, suppose $e - 1$ is in S), and let G be a planar graph with e edges. We consider two cases. First if G is a tree, then $f = 1$ and $e = v - 1$, so $v - e + f = 2$. Otherwise, G has at least one cycle, so that a planar drawing of G has at least two regions. Pick an edge $\{x, y\}$ of G which is between two regions in a planar drawing of G. Form G' and a planar drawing of G' by removing $\{x, y\}$ from the edge set of G and the picture of G. Then G' has $e - 1$ edges, v vertices and, because $\{x, y\}$ separates two regions from each other, G' has $f - 1$ faces. But $e - 1$ is in S, so that $2 = v - (e - 1) + f - 1 = v - e + f$. Thus e is in S. Then by the principle of mathematical induction, S is the set of all positive integers and the theorem is proved. ∎

At times, we may apply Euler's formula to determine the number of faces a planar drawing of a graph must have and from this see that a certain graph does not have a planar drawing. In a graph which is not a tree, a face has a cycle on its boundary and so must have at least three edges.   Using this, we can come up with a test for planarity that is much easier to apply in practice.

An Inequality to Check for Non–Planarity

Theorem 3.3.   In a connected planar graph with three or more vertices

$$e \leq 3v - 6.$$

Proof.   We've noted that each face has at least three edges, and each edge separates two faces (or doesn't separate faces).   Thus the total number of ordered pairs (edge, face) with the edge touching the face is *at least* $3f$, but is *no more than* $2e$.   Thus $3f \leq 2e$, but $f = 2 + e - v$, so that

$$6 + 3e - 3v \leq 2e$$

or

$$e \leq 3v - 6. \ \blacksquare$$

Example 3.1.   The *complete graph* on five vertices K_5 is the graph with five vertices each pair of which is an edge.   Show that K_5 is not planar.
Since K_5 has five vertices and $C(5; 2) = 10$ edges, we would have to have

$$10 \leq 3 \cdot 5 - 6 = 9$$

if the graph were planar, so it is not. ■

The graph of Figure 4.20 is called $K_{3,3}$—the ''complete bipartite graph on two parts of size 3.''   It is another famous example of a nonplanar graph.   We note that if it had a planar drawing, then all its faces would have to have at least 4 edges, giving us $4f \leq 2e$ rather than $3f \leq 2e$ as in proof of the main theorem. Since the graph has nine edges, we get $4f \leq 18$ or $f \leq 4.5$; i.e. $f < 5$.   But Euler's formula tells us that $f = 5$ for a planar graph with six vertices and nine edges, so $K_{3,3}$ is *not planar*.

The graphs $K_{3,3}$ and K_5 play a central role in the theory of planarity of graphs.   Kuratowski showed that at least one of these graphs occurs naturally in any nonplanar graph.   Elementary transformations are used to describe this natural occurrence.   Insertion or deletion of a vertex of degree 2 in an edge of a graph is said to be an elementary transformation of degree 2.   In particular, if $\{x, y\}$ is an edge of G, we can add z to the vertex set and replace the edge $\{x, y\}$ by $\{x, z\}$ and

$\{z, y\}$ to insert a vertex z of degree 2 in the edge $\{x, y\}$. On the other hand, given a vertex z of degree 2 connected to x and y, we can remove z from the vertex set and replace $\{x, z\}$ and $\{z, y\}$ by $\{x, y\}$ to remove the vertex z of degree 2. (See Figure 4.23.) Two graphs are 2–equivalent if a series of elementary transformations can carry one graph onto the other. By a *subgraph* of $G = (V, E)$, we mean a graph whose vertex set V' is a subset of V and whose edge set E' consists of *all* edges of E joining two vertices in V'. Kuratowski proved that

> ***Theorem 3.4.*** A graph is not planar if and only if it contains a subgraph 2–equivalent to K_5 or a subgraph 2–equivalent to $K_{3,3}$.

Proof. The proof is more properly part of a course in topology and goes beyond the scope of the present book. ■

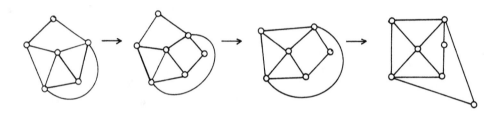

Figure 4.23 A sequence of elementary transformations between 2–equivalent graphs.

EXERCISES

1. Prove Theorem 3.1.

2. Why do two isomorphic graphs have the same number of edges?

3. Prove that if f is an isomorphism from G_1 to G_2, then $f(x)$ has the same degree in G_2 that x has in G_1.

4. Let V_1 be the positive factors of 30. Define two integers to be compatible if one is a factor of the other, and let $\{m, n\}$ be in E_1 if m and n are different but compatible. Let V_2 be the subsets of $\{x, y, z\}$ and define two sets to be comparable if one is a subset of the other. Let $\{S, T\}$ be in E_2 if the S and T are comparable but different subsets of V_2. Show that the graphs (V_1, E_1) and (V_2, E_2) are isomorphic.

5. Three houses are to be built along a line parallel to and set back 30 feet from a road. Next to the road are three utility holes, one for sewer, one for water, and one for electricity. Is it possible to connect the houses to the utility holes in such a way that no lines cross? (Lines may go behind houses.)

6. If in Exercise 5 there is a fourth utility hole for telephone wires, what is the minimum number of crossings if each house is connected to each hole?

7. Show that if a planar graph has no cycles of length 3, then $e \leq 2v - 4$.

8. Give as many examples as you can of trees on six vertices, no two of which are isomorphic.

9. Give as many examples as you can for forests on five vertices, no two of which are isomorphic.

10. Give as many examples as you can of graphs on four vertices with two or more nonisomorphic spanning trees.

11. Show that a planar graph has at least one vertex of degree 5 or less.

12. The graph K_6 has 15 edges. Show that it is impossible to remove two edges to obtain a planar graph. What about three edges?

13. Show that if every vertex of a planar connected graph has degree at least 3, then there is a face with five or fewer edges on its boundary.

14. The graph $K_{1,2,3}$ has vertices a, b_1, b_2, c_1, c_2, c_3 such that the b's are not connected to each other and the c's are not connected to each other, but a connects to all b's and c's, and all b's connect to all c's. Is $K_{1,2,3}$ planar?

15. The "integer–interval graph" $I(m, n)$ has as its vertices all closed intervals with two distinct integer endpoints between m and n (inclusive). Two vertices are connected by an edge if their intervals have at least one point in common. For what values of n is $I(0, n)$ planar?

16. Following the notation of Exercise 15, prove that $I(m, n)$ is isomorphic to $I(j, k)$ if and only if $n - m = k - j$.

SECTION 4 COLORING

The Four Color Theorem

The four color theorem has a long and colorful history. Originally a statement about maps drawn on a plane surface or a sphere, it is now usually stated as a theorem about planar graphs. On a map, it is desirable to color different political entities which share a common boundary with different colors. As long as each entity is connected (you can move between any two points without leaving it), it is possible to get such a proper coloring of a planar map with four colors. This led mathematicians to ask whether this observation was in fact a theorem. The mathematician Kempe devised what appeared to be a proof of the theorem, but a number of years after his proof, Heawood showed that though Kempe's method provided a valid proof that a planar map could be colored with five colors, there was an oversight that left the four color problem unsolved. This open problem led to a great deal of new mathematics as people tried to solve the problem.

In the early 1970's, Apel and Haken developed a strategy of classifying the kinds of problems that could occur in Kempe's approach into "configurations" and showed that these configurations could be reduced to situations where the four color problem had already been solved. They developed methods of checking for reducibility that could be programmed into a computer and set a computer to work. In 1976 they announced that any planar map can be colored with four colors; their computer work had isolated a complete set of reducible configurations. We shall study some of the ideas that developed from the four color problem over the years.

Chromatic Number

A proper coloring of a graph is an assignment of "colors" to the vertices in such a way that vertices connected by an edge receive different colors. More precisely, it is a function f from the vertex set V to a set C so that if $\{x, y\}$ *is an edge*, $f(x) \neq f(y)$. The smallest size of a set C for which there is a proper coloring is called the *chromatic number* of G.

> **Theorem 4.1.** If the largest degree of a vertex in a graph G is k, then the chromatic number of G is no more than $k + 1$.
>
> *Proof.* Pick an arbitrary vertex, and assign it any color from a set C of size $k + 1$. Now, once j vertices are colored, pick vertex $j + 1$ and assign to it any color not already assigned to any of its neighbors. (Why is this possible?) Continue until colors are assigned to all vertices. ∎

Maps and Duals

The relation between map coloring and graphs is based on the adjacency graph of a map. On a piece of paper, we use a point to represent each region of the map and connect points representing two regions if they share a boundary line. We may draw an edge for each shared boundary line, in which case we get a multi-graph. (Regions that touch only at a point are not connected by an edge in the graph.) Clearly, a planar map yields a planar graph. A proper coloring of the graph corresponds to a map coloring in which regions that share boundary lines get different colors. If it happens that the map's regions are the faces of a drawing of a multigraph G, then the new multigraph we have constructed is called a *dual* to G. Note that any planar drawing of a graph may be regarded as a map, so that each planar drawing of G yields a dual to G.

On the other hand, coloring planar graphs is no more general a topic than coloring planar maps. To see this, we show that for each planar graph G, we can construct a map whose adjacency graph is G. Note this is different from regarding the faces of the planar drawing of G as geographic regions, as in our construction of dual graphs. The construction of maps from graphs is intuitive; a glance at Figure 4.24 will suggest the method. A detailed description of the process follows.

We begin with a planar graph drawn in the plane without crossings. We place a point inside each face of the graph and mark a point on each edge of the drawing between the endpoints. Now from the chosen point inside each face, draw line segments (curved if necessary) to the marked point on each edge of the face. (Note that, since each face is a polygon these lines can be drawn so as not to cross.)   Now each vertex of the graph is surrounded by a polygon whose corners are at the previously chosen points in the faces and edges.   These polygons form the regions of a planar map, and the adjacency graph of the planar map is the original graph.   The process is illustrated in Figure 4.24.

Note that one of the points in Figure 4.24 corresponds to the infinite region in the picture.   There are two ways to reconcile this with our usual idea of a map. One is to recognize that this region is finite once the picture is drawn, because it is drawn on a finite piece of paper.   The other is to realize that in addition to all the regions in the map to be colored, there is a boundary area enclosing the regions, and we wish to treat that boundary exactly as a region, i.e., color it a different color from any region it touches.   In fact, drawings in a plane and on the surface of a sphere are equivalent, and on a sphere, the "outside" region of the planar drawing is just another region.

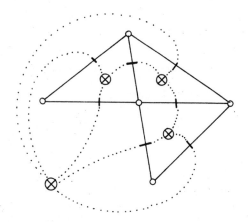

Figure 4.24

The Five Color Theorem

We shall prove that every planar graph (and thus every planar map) can be colored with five colors.   We shall translate the essence of Kempe's attempt at proving the four color theorem into the language of graph theory because it is somewhat easier to be precise with graph–theoretic terminology.   We begin with a result that says we don't have to examine all graphs.

Theorem 4.2.   If each n–vertex planar graph all of whose faces are triangles has a proper coloring with $c > 1$ colors, then each n–vertex planar graph has a proper coloring with c colors.

Proof.   Removing or adding a vertex of degree 1 doesn't change the fact that the graph can be colored with c colors. Thus we assume our graphs have no vertices of degree 1.

Let S be the set of all integers k such that if an n–vertex planar graph has k nontriangular faces, then it has a proper coloring with c colors. According to the hypothesis of the theorem $k = 0$ is in S. Now suppose $k \neq 0$ and $k - 1$ is in S. Let G be a graph with k nontriangular faces and let v_1, v_2, . . . , v_r be the vertices on the edges of one nontriangular face. Since we have no vertices of degree 1, the boundary of the face is a polygon. Then within that face, it is possible to draw (possibly curved) lines from v_1 to $v_3, v_4, \ldots, v_{r-1}$ so that none of the lines cross. This gives a planar drawing of the graph G' whose edge set is the edge set of G together with the edges $\{v_1, v_i\}$. G' has $k-1$ nontriangular faces and so has a proper coloring with c colors. However, this same assignment of colors provides G with a proper coloring. By version 1 of the principle of mathematical induction, S contains all positive integers; this proves the theorem. ∎

A second helpful result is an exercise in Section 3.

Theorem 4.3.   A planar graph has at least one vertex of degree less than 6.

Proof.   Exercise in applying $e \leq 3v - 6$. ∎

Theorem 4.4.   A planar graph has a proper coloring with five or fewer colors.

Proof.   Let S be the set of all integers n such that planar graphs with n vertices have proper colorings with five colors. Clearly, $n \in S$ if $n \leq 5$. Suppose now that $n > 5$ and $n - 1$ is in S. Suppose G is an n–vertex planar graph whose faces are all triangles and let x be a vertex of G of degree less than 6. If x has degree 4 or less, let G' be the $(n - 1)$–vertex graph obtained from G by deleting x and the edges including it. Then G' has a proper coloring with five colors. However, only four of these colors are used on vertices that share an edge with x in G. Thus we may properly color G by assigning the fifth color to x and coloring the remaining vertices as in G'.

Now if x has degree 5, suppose $\{x, v_1\}$ is one of the edges containing x, and let $\{x, v_2\}, \{x, v_3\}, \{x, v_4\}, \{x, v_5\}$ be the remaining edges arranged in clockwise fashion around x in a (fixed) planar drawing. (See Figure 4.25.) Thus $\{x, v_i\}$ and $\{x, v_{i+1}\}$ both lie on the boundary of a common face as do $\{x, v_1\}$ and $\{x, v_5\}$. Since these faces are triangles, $\{v_i, v_{i+1}\}$ (for $i \leq 4$) and $\{v_1, v_5\}$ are all edges of G. Now let G' be the graph obtained from G by deleting the vertex x and the edges $\{x, v_i\}$. Since $n - 1$ is in S, Theorem 4.2 tells us that G' has a proper coloring with five colors. If not all five of these colors are used up by the vertices $\{v_1, v_2, v_3, v_4, v_5\}$, then as before we may color G by assigning the unused color to x and coloring the remainder of G

as G' is colored. Suppose now the five colors R, B, G, Y, and W are as-signed to v_1 through v_5, respectively.

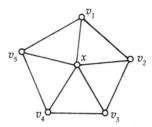

Figure 4.25

We now define two equivalence relations on the vertex set of G. They will have the property that an appropriate pair of inequivalent ver-tices may be colored the same color, giving a free color to use on x. We say y and z are $(R - G)$–equivalent if there is a path connecting y and z in G' all of whose vertices are colored R or G. Let V_1 be the set of all vertices $(R - G)$–equivalent to v_1. Then any vertex in V_1 is colored R or G, and if a vertex is in V_1, the vertices it shares edges with *outside* V_1 are colored either B, Y or W. Thus if v_3 is not in V_1, we may interchange the assign-ments of the colors R and G to the vertices in V_1. Thus now we have a proper coloring of G' using only four colors on the v_i's, and so we can five color G as above.

Now define two vertices y and z to be $(B - Y)$–equivalent if there is a path between them in G' all of whose vertices are colored B or Y. Define V_2 to the $B - Y$ class containing v_2. If v_4 is not in V_2, we may interchange the colors in V_2 to obtain a proper coloring of G' in which only four colors are used on the v_i's. Thus we can five color G as above.

Now suppose v_3 is in V_1, so that there is a path from v_1 to v_3 con-sisting of vertices colored only R or G. (See Figure 4.26.) We have two polygons in G' that, taken together, surround v_2, namely the pentagon $v_1v_2v_3v_4v_5$ and the closed path formed by v_1, v_2, v_3 and the other vertices of the "$R - G$ path" connecting v_1 and v_3; we call these other vertices u_1, u_2, . . . , u_t. Thus any line segment emanating from v_2 must either go into or along an edge of one of these polygons. Since the pentagon of v's is a face of G', no edge can go inside it. Thus any path from v_2 to v_4 must contain at least one vertex of the $R - G$ path. Then any path from v_2 to v_4 must contain a red or green vertex. Therefore there is no $B - Y$ path from v_2 to v_4, i.e., v_2 and v_4 are $(B - Y)$–inequivalent and thus the graph G may be properly colored as above. ∎

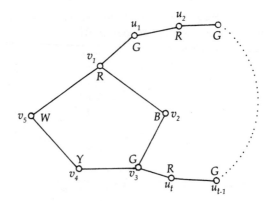

Figure 4.26

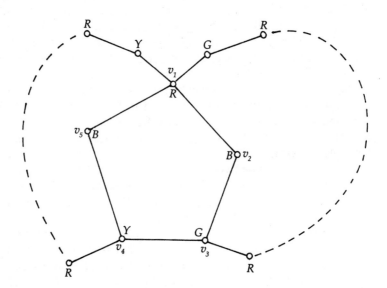

Figure 4.27

Kempe's Attempted Proof

The argument here is essentially the one used by Kempe in his attempt at proving the four color theorem. It should be clear that in the case of four colors, we would have to examine a vertex x of degree 4 and a vertex x of degree 5 separately. If there is a vertex of degree 4 it is surrounded by a quadrilateral $\{v_1, v_2, v_3, v_4\}$. If four colors are used on the quadrilateral we show we may use only three of them instead by using $R - G$ and $B - Y$ equivalence relations as above. In the degree 5 case, if all four colors appear among the vertices v_i, we can reduce the problem to the case where the W in Figure 4.26 is a B. Kempe then argued that either v_1 and v_4 were in the same $R - Y$ class, or else we interchange the R's and Y's in the class containing v_1 and extend as above to a proper coloring of G with four colors. (Similarly, we could make one R $-$ G switch unless v_1 and v_3 are in the same R $-$ G class.) In the case that v_1 and v_4 *are* connected by a path of vertices colored R or Y, and similarly v_1 and v_3 are connected by a path of vertices colored R or G (see Figure 4.27 for *one* possible picture of this situation), Kempe said we should change colors in both the $B - G$ class containing v_5 and $B - Y$ class containing v_2. In this way we get three colors on the pentagon and can proceed as above. It is instructive to find the error here.

EXERCISES

1. What is the chromatic number of the graph $K_{3,3}$? $K_{2,4}$?

2. Describe all graphs on n vertices with chromatic number n.

3. What is the chromatic number of a cycle with an even number of vertices? With an odd number of vertices?

4. What is the chromatic number of a tree?

5. A "wheel" consists of a cycle together with one more vertex connected to all vertices of the cycle. Analyze the possibilities for the chromatic number of a wheel.

6. What is the dual of a cycle?

7. Show that a wheel (see Exercise 5) is isomorphic to a graph which is its dual.

8. Find the error in Kempe's proof of the four color theorem.

9. Prove the fact that a planar graph has a proper coloring with six colors *without* appealing to the four color theorem or five color theorem.

10. A set of vertices in a graph G is called *independent* if no two of these vertices are joined by an edge. The *independence number* of a graph is the maximum of the sizes of its independent sets. Show that the product of the chromatic number and independence number of a graph is at least the number of edges.

11. Show that a graph can be properly colored with two colors if and only if it contains no cycles of odd length. What does this have to do with graphs whose vertex set is a union of two independent sets (these are called *bipartite graphs*)?

*12. (Brooks) Prove that a graph which is connected but not complete and whose vertices of largest degree have degree $c > 2$ can be properly colored with c colors.

13. Either prove or give a counter–example: Connected nonplanar graphs on six vertices are not four colorable.

14. In Exercise 15 of the previous section, we introduced the integer–interval graph $I(0, n)$. For what values of n is $I(0, n)$ four colorable?

15. What is the chromatic number of the subgraph of $I(0, n)$ whose vertices are intervals of length 2?

16. What is the chromatic number of the subgraph of $I(0, n)$ whose vertices are intervals of length 1 or 2?

17. Show that if a planar graph has fewer than 30 edges, then it has a vertex of degree 4 or less. Prove from this that a planar graph with 30 or fewer edges has a proper coloring with four colors.

18. A well–known and often discovered result in graph theory is that there is, for any n, a graph with no triangles and chromatic number n. (For $n = 3$, a cycle on five vertices is an example.) Find a proof of this result.

19. A graph is n–color critical if it has chromatic number n, while the deletion of any vertex produces a graph of chromatic number $n - 1$. Give two examples of four–color critical graphs. Show that a graph of chromatic number n contains an n–color critical subgraph.

SECTION 5   DIGRAPHS

Directed Graphs

Suppose we are given a map of a downtown area where all streets are one–way and we are asked to find a path for a shuttle bus in which each street is traversed exactly once. This seems to be a request for an Eulerian path; but if we simply constructed an Eulerian path on the obvious graph, we might have the bus going the wrong way on some one–way street! To deal with such a problem, we introduce the notion of a directed graph. A *directed graph* (digraph for short) consists of a set V of vertices and a set D of ordered pairs (called edges) of elements of D. A *directed multigraph* (multidigraph for short) consists of a set V of vertices and a multiset D of ordered pairs of vertices.

Thus a digraph consists of a set V and some relation on V. Digraphs are frequently useful in visualizing relations. We represent a digraph using a picture in which a point is drawn for each vertex and an arrow is drawn from a to b if the ordered pair (a, b) is in the edge set. (In particular, a "loop"—an arrow from a to itself—is drawn if (a, a) is in the edge set.) We introduced such pictures in Chapter 1. The first picture in Figure 4.28 is a picture of a digraph.

Given a digraph, there is an obvious multigraph underlying it, the multigraph whose edges are the two–element multisets consisting of the two vertices in each ordered pair in the edge set. We also have an underlying graph whose edges are the two–element sets $\{a, b\}$ with (a, b) or (b, a) in the digraph edge set D. The underlying multigraph and graph are shown with the digraph in Figure 4.28. We can define connectedness, paths, walks, etc. for a digraph in terms of its underlying graph. In the case of a one–way street graph, a "walk" is allowed to traverse an edge in either direction; in cases such as a digraph of the descendants of a group of people, two people are related if they are connected in the underlying graph. Thus notions about underlying graphs are useful in analyzing directed graphs. However, they cannot be used to analyze our shuttle–bus problem. In-

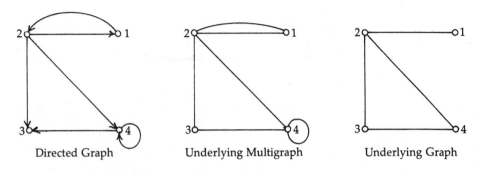

Figure 4.28

stead, we need concepts that relate to the directions as well as the connections in the digraph.

Walks and Connectivity

A *directed walk* or *strong walk* from a to b in a digraph is a sequence of vertices and edges

$$v_1(v_1, v_2)v_2 \ . \ . \ . \ (v_{n-1}, v_n)v_n$$

with $v_1 = a$ and $v_n = b$. *Strong* or *directed paths, closed walks,* and *cycles* are defined similarly. We say a is *strongly connnected* to b if there is a strong walk from a to b *and* a strong walk from b to a. It is easy to see that the relationship of being strongly connected is an equivalence relation. The equivalence classes of this relation are the "strong connectivity" classes of the directed graph.

The *indegree* of a vertex v is the number of edges in which v is the second member, i.e., the number of arrows that enter v. The *outdegree* of a vertex is the number of edges in which it is the first member. A strong Eulerian walk (which might be called an Eulerian drive) is a strong walk that includes each edge of the digraph. (A strong Eulerian walk in a multidigraph has the obvious similar definition.)

The following results are obtained by reworking the proofs of our earlier theorems about Eulerian multigraphs; thus we do not give the proofs here.

Theorem 5.1. A (multi)digraph has a closed Eulerian walk if and only if it is connected and the indegree and outdegree of each vertex are equal.

Theorem 5.2. A (multi)digraph has a nonclosed Eulerian walk if and only if it is connected, all but two vertices have equal indegree and outdegree and of these two vertices, one has outdegree 1 more than its indegree and the other has outdegree 1 less than its indegree.

Tournament digraphs

In a round-robin tournament, each player plays every other player exactly once. We may construct a directed graph on the set V of players by placing (x, y) in D if x beats y. The underlying graph (which is also the underlying multigraph) is complete. Thus we define a *tournament digraph*, or simply *tournament*, as an "oriented" complete graph; that is, a directed graph on the vertex set V which contains either (x, y) or (y, x) (but not both) for every two–element subset $\{x, y\}$ of V.

In a tournament in sporting events, we wish to find a winner, or perhaps rank the players according to ability. In a three–player tournament where x beats y, y beats z, and z beats x, there is no way to decide on a winner or rank the players. A ranking of the players would be a list of all the players such that each player beats every succeeding player on the list. Although we cannot create a ranking, we can always write down a list so that each player beats the next one on the list. This corresponds to a directed path through a graph that includes every player exactly once.

Hamiltonian Paths

A path through a graph that includes every vertex exactly once is called a Hamiltonian path. In general, it is very hard to find a Hamiltonian path in a graph and even harder to find directed Hamiltonian paths in directed graphs. In a tournament, however, it is not difficult.

Theorem 5.3. If G is a tournament digraph, then it has a directed Hamiltonian path.

Proof. Let v_1 be any vertex such that for some vertex v_2, (v_1, v_2) is an edge. This gives us two vertices and a path joining them. Suppose now we are given a directed path

$$v_1(v_1, v_2,)v_2, \; \ldots \; , (v_{i-1}, v_i)v_i.$$

Let v_0 be a vertex not on this path. If there is an edge (v_0, v_1) in the graph, we can extend the original path to the path

$$v_0(v_0, v_1)v_1 \; \ldots \; v_i.$$

If (v_0, v_1) is not an edge, then (v_1, v_0) is an edge. If (v_j, v_0) is an edge for all j with $1 \leq j \leq i$, then we can extend the original path to the path

$$v_1(v_1, v_2)v_2 \ . \ . \ . \ v_i(v_i, v_0)v_0.$$

If (v_j, v_0) is not an edge for all j, let k be the smallest value of j for which (v_j, v_0) is not an edge. Then (v_0, v_k) is an edge and so is (v_{k-1}, v_0). Thus we can extend the path to

$$v_1 \ . \ . \ . \ v_{k-1}(v_{k-1}, v_0)v_0(v_0, v_k)v_k \ . \ . \ . \ v_i.$$

In this way we can continue to extend the path until it includes all vertices of G. ∎

A tournament is said to be *transitive* if whenever (x, y) and (y, z) are edges, then so is (x, z). Note that this is equivalent to saying the relation D is transitive. If the tournament *is* transitive and $v_1, v_2, \ . \ . \ . \ , v_n$ are the vertices listed in the order they appear in a Hamiltonian path, (v_i, v_j) is an edge for each $j > i$. Thus transitive tournaments have rankings of the type we asked for earlier; in fact they have only one such ranking.

Clearly a tournament with a directed cycle (with more than one vertex) is not transitive.

Theorem 5.4.   A tournament with no cycles is transitive.

Proof.   Suppose $G = (V, D)$ is a tournament with no cycles, and let (x, y) and (y, z) be in D. Either (x, z) or (z, x) is in D, but if (z, x) is in D, then

$$x(x, y)y(y, z)z(z, x)x$$

is a nontrivial cycle. Thus G would have a cycle, so (z, x) cannot be in D. Therefore (x, z) is in D. Therefore G is transitive. ∎

Transitive Closure

In a computer program organized into related procedures, the flow of control from one procedure to another can be quite complicated. Also, data used in one procedure may or may not be passed along to a subsequent procedure when the former procedure stops and the latter starts. Other data computed in one procedure may be stored in such a way as to be accessible to some or all other procedures.

This flow of control and data gives us two digraphs associated with our program. Both have the set of procedure names of the program for their vertex set. In the "flow of control" digraph, we have an arrow from procedure P to procedure Q if P can call Q. Similarly, in the "data flow" digraph we have an arrow from P to Q if Q can access data previously accessed by P. These are special cases of communications digraphs in which vertices represent objects that can communicate with each other and arrows represent communications links along which communicators can send messages directly to other communicators. In such a situation, we are often interested in knowing who can send a message to whom along a sequence of communicators each of whom "relay" the message on. For example, if we want to modify a procedure in a way that might influence its interaction with other procedures, we need to know all the procedures that may be affected by the change. If A calls B and B calls C, a change in A may (or may not) influence C.

This leads us to the idea of the "transitive closure" of a relation or digraph. Intuitively, the transitive closure $T(R)$ of a relation R is the result of applying the transitive law (which says that if a is related to b and b is related to c, then a is related to c) again and again to the relation R. In this way, we expect to get the "smallest" transitive relation consistent with R. We can also define the transitive closure without reference to repeated applications of the transitive law. First, since a relation is just a set of ordered pairs, every relation R on a set V is a subset of at least one transitive relation, namely the set $V \times V$ of all ordered pairs of elements of V. We define the *transitive closure $T(R)$* to be the intersection of all transitive relations containing R as a subset.

Lemma 5.5. For any relation R, $T(R)$ is transitive.

Proof. Suppose (a, b) and (b, c) are in $T(R)$. Then for each transitive relation Q containing R, (a, b) and (b, c) are in Q. Thus (a, c) is in Q for each transitive relation Q containing R. Then (a, c) is in the intersection of all these relations Q, so (a, c) is in $T(R)$. ∎

The *transitive closure* of a digraph (V, D) is the digraph $(V, T(D))$. In our section on matrix representations of graphs, we will describe simple procedures that can be used for computing the transitive closure of a relation.

Reachability

A straightforward procedure can also be based on the idea of reachability. A vertex v is *reachable* from x if there is a strong walk from x to v; v is said to be *strictly reachable* from x if there is a strong walk of length 1 or more from x to v. The only difference with strict reachability is that while x is always reachable from x, x is strictly reachable from itself only if there is a strong cycle that contains x.

Theorem 5.6.   The edge (x, y) is in the transitive closure of the digraph $G = (V, D)$ if and only if y is strictly reachable from x in G.

Proof.   If y is strictly reachable from x, there is a sequence $v_0 = x$, v_1, $v_2, \ldots, v_n = y$ in which (v_i, v_{i+1}) is an edge and $n > 0$. Each (v_i, v_{i+1}) is in each transitive relation containing D, so (x, y) is in each transitive relation containing D; therefore, (x, y) is in the transitive closure $T(D)$.

To show that for each (x, y) in $T(D)$, the vertex y is reachable from x, we study the set of all vertices reachable in i or fewer steps from x. We let $D_i = \{y \mid y \text{ is strictly reachable from } x \text{ by a walk with } i \text{ or fewer edges}\}$. (Thus D_0 is empty and D_1 is D, for example.) We prove that if (x,y) is in $T(D)$, then (x, y) is in some D_i. That is, if (x, y) is in $T(D)$, then y is strictly reachable from x in i or fewer steps for some i.

In particular we will show that

$$T(D) = \bigcup_{i=1}^{\infty} D_i = \cup \{D_i \mid i \geq 1\}.$$

(The second equality is the definition of $\displaystyle\bigcup_{i=1}^{\infty} D_i$.) In the first paragraph we showed that each pair in D_i is in $T(D)$, so that $\displaystyle\bigcup_{i=1}^{\infty} D_i \subseteq T(D)$. However, $\displaystyle\bigcup_{i=1}^{\infty} D_i$ is itself transitive, for if (x, y) and (y, z) are in $\displaystyle\bigcup_{i=1}^{\infty} D_i$, then

$$(x, y) \in D_i \text{ for some } i \text{ and}$$
$$(y, z) \in D_j \text{ for some } j.$$

Now if k is the larger of i and j, we know that $D_i \subseteq D_k$ and $D_j \subseteq D_k$, so $(x, y) \in D_k$ and $(y, z) \in D_k$. Thus z is reachable from x in $2k$ or fewer steps, so $(x, z) \in \displaystyle\bigcup_{i=1}^{\infty} D_i$. Since $T(D)$ is a subset of any transitive relation containing D, we have

$$T(D) \subseteq \bigcup_{i=1}^{\infty} D_i.$$

Since we've already observed that the union is a subset of $T(D)$, it must be the case that

$$T(D) = \bigcup_{i=1}^{\infty} D_i.$$

As remarked above, this proves the theorem. ∎

Modifying Breadth–First Search for Strict Reachability

In graphs or multigraphs, the ideas of reachability and connectivity are equivalent. Recall that breadth–first search is an efficient method of finding all vertices connected to a vertex x. The same sort of search can be used in directed graphs; a slight modification ensures that we are searching for vertices strictly reachable from x. We say that x is *adjacent to* y and y is *adjacent from* x if (x, y) is in D. A strict breadth–first search from x consists of the following steps:

(1) Let x_1 be a vertex adjacent from x, $V_1 = \{x, x_1\}$ and $E_1 = \{(x, x_1)\}$.
(2) For all y adjacent from x, if $y \notin \{x_1, x_2, \ldots, x_i\}$, let $x_{i+1} = y$, $V_{i+1} = V_i \cup \{x_{i+1}\}$ and $E_{i+1} = E_i \cup \{(x, x_{i+1})\}$.
(3) Given V_i and E_i, let j be the smallest integer such that x_j is adjacent to some z not in $\{x_1, x_2, \ldots, x_i\}$. Let $x_{i+1} = z$, $V_{i+1} = V_i \cup \{z\}$ and $E_{i+1} = E_i \cup \{(x_j, x_{i+1})\}$.
(4) Repeat Step 3 until there is no j of the type described (in Step 3). Let $G = (V_n, E_n)$ be the final digraph obtained.

The final graph we get may or may not be a tree. The vertices x_1, $x_2, \ldots, x_n$ listed are all the vertices strictly reachable from x. Thus (x, x_i) is in $T(D)$ for each i. Therefore, we can construct $T(D)$ by doing a strict breadth–first search from x for each x in V and placing all the edges we get in the process into the edge set of $T(D)$. Each application of the breadth–first search might mean examining virtually all the edges in D, and $|D|$ could be as large as $|V|^2$, so in computing the transitive closure of (V, D) we might require $|V| \cdot |V|^2 = |V|^3$ steps.

EXERCISES

1. How many directed graphs are there with the three–element vertex set $\{a, b, c\}$? A ten–element vertex set? (How does the presence or lack of loops affect your answer?)
2. How many tournaments are there on the vertex set $\{1, 2, \ldots, n\}$?
3. Show that if a tournament has a ranking $v_1 v_2 \ldots v_n$ so that (v_i, v_j) is an edge whenever $i > j$, then the tournament is transitive.
4. How many transitive tournaments are there on the set $\{1, 2, \ldots, n\}$?
5. Which digraphs in Figure 4.29 have Eulerian walks? Closed Eulerian walks?
6. Show that the sum of the indegrees of the vertices in a directed graph equals the sum of the outdegrees of the vertices.

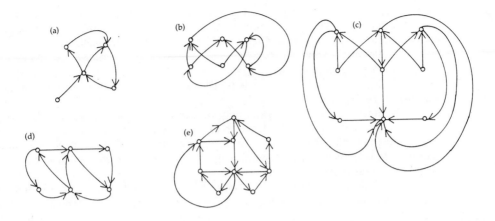

Figure 4.29

7. Around the circle in Figure 4.30 there are eight 0's and 1's. By taking each of the eight intervals of three consecutive places around the circle in turn, we encounter each list of three 0's and 1's exactly once. Show that 2^n zeros and ones can be arranged around a circle so that each possible list of n 0's and 1's occurs as a list of n consecutive digits in the sequence. (In fact, DeBruijn showed that the number of such arrangements is $2^{2^{n-1}-n}$. This is difficult to prove.) (Hint: Start with a digraph whose vertices are lists of $n - 1$

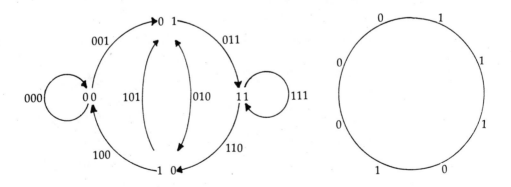

Figure 4.30

zeros and ones. Draw an arrow from vertex 1 to vertex 2 if the last $n - 2$ digits of the first equal the first $n - 2$ digits of the second. Now find an Eulerian path. Try to relate the Eulerian path to the list of numbers around the circle as in Figure 4.30.)

8. Prove Theorem 5.1.

9. Prove Theorem 5.2.

10. Prove that a tournament is transitive if and only if there is one and only one ranking of the vertices.

11. Heavy trucks are allowed only on certain roads in a city, and many of these roads are one–way. A heavily–laden truck must make stops at n stores. Show that if it is possible to go either from store X to store Y or from store Y to store X for each pair of stores (without passing other stores), then there is a route which goes by each store only once and passes none by along the way.

12. (Due to C.L. Liu) n books are to be printed and bound. There is one printing machine and one binding machine. Let p_i and b_i denote the printing and binding time for the i–th book. If it is known that either $p_i \le b_j$ or $p_j \le b_i$ for all i and j, show that there is an order in which each book may be first printed and then bound in such a way that the binding operation is kept busy from the time the first book is printed until the last book is bound.

13. Show that a tournament is transitive if and only if it has no directed cycles of length 3.

14. Show that the transitive closure of a digraph $G = (V, D)$ contains no cycles if and only if G contains no cycles.

15. Find the transitive closure of the digraphs in Figure 4.31.

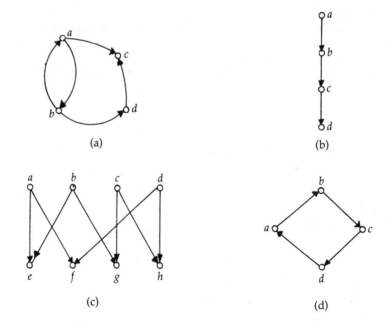

Figure 4.31

16. What is the transitive closure of a strongly–connected graph?

17. What is the transitive closure of a directed cycle on n vertices?

18. Prove that in a tournament there is a vertex v such that for *each* x in V, either (v, x) is an edge or, for some y, (v, y) and (y, x) are edges.

19. There is a number k such that if a digraph on n vertices has k or more edges then it has a directed cycle. What is k when $n = 4$? When $n = 10$?

20. Show that a digraph is strongly connected if and only if it has a strong closed walk that contains all the vertices.

21. Show that a digraph with no directed cycles has at least one vertex of outdegree zero.

22. Show that if v is reachable from u, then there is a strong path from u to v.

23. What is the maximum number of directed cycles in a strict breadth–first search graph? When will we have this number of cycles and why?

SECTION 6.   ORIENTATION AND BACKTRACKING.

Orientable Graphs

Suppose we are given a street map of a city. Is there a way to assign one–way streets so it is possible to get from any one point to any other point? If not, how do we find the streets which must be two–way? Can we perhaps assign directions so that we can drive around town, traversing each street exactly once? This last question is the easy one, for it asks for nothing more than an Eulerian walk. The first two questions are more like questions that could come up in planning a city or a communications or railroad network. (Questions of *efficient* assignments of one–way streets have partial answers in terms of the "diameter" of a digraph; we shall not investigate these questions in the text.)

An assignment of directions to the edges of a graph so that the resulting graph is strongly connected is called an *orientation* of the graph. Our first question asked how we can tell if a graph has an orientation. There is one obvious obstacle to orienting a graph. An edge of G is called a *bridge* or "isthmus" if its removal from the graph disconnects the graph. If a graph has a bridge $\{x, y\}$, then once we have chosen a direction—say (x, y)—for this edge, its endpoints will no longer be strongly connected. Thus a graph with a bridge has no orientation.

It is surprising that if a graph has no bridge, then it has a strong orientation. This means we can answer the second question at the beginning of this section as well—if we let all bridges be two–way streets, then the graph has a strong orientation. If we have a careful drawing of a graph, then we can locate all its bridges by inspection. However, if we don't already know where the bridges of a large graph are when we start drawing the graph using data of some sort, we may not draw it in such a way that the bridges are obvious. Further if we are using a computer to analyze the graph, we cannot tell it to stare at the graph. Thus we want a technique which may be used to search a graph for a bridge.

Our strategy will be to build a search tree and try to get as long a path as possible rather than keeping the paths short as in breadth–first search. If as we

search, there is a bridge which we cross then, we would be required to cross back
eventually in order to finish building the spanning tree. Thus if we can identify
when backtracking along an already used edge is necessary, we will be able to
identify our bridge.

Depth–First Search

In particular, a *depth–first search* tree from the vertex x in a graph $G = (V, E)$ is
constructed as follows:

(1) Let $v_0 = x$, $V_0 = \{v_0\}$ and $E_0 = \emptyset$.
(2) Given V_i and E_i, let j be the largest number such that v_j is adjacent to a vertex
$\quad y \notin V_i$. Let $v_{i+1} = y$, $V_{i+1} = V_i \cup \{v_{i+1}\}$ and $E_{i+1} = E_i \cup \{v_j, v_{i+1}\}$.
(3) Repeat Step 2 until there is no j of the type described.

Let $T = (V_{n,}, T_n)$, where n has the final value of i. A selection of a
depth–first search spanning tree is shown in Figure 4.32. The edges with arrows
are the edges of the tree.

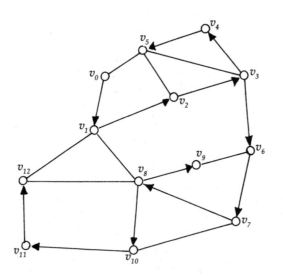

Figure 4.32

Theorem 6.1. If G is a connected graph, a depth–first search tree is a
spanning tree.

Proof. The graph has n vertices and $n - 1$ edges, is connected and since
G is connected, it must include each vertex of G. ∎

Depth–first Numbering

A numbering of v_0, v_1, v_2, . . . , v_n given by depth first search is called a *depth–first numbering* of G. Figure 4.32 suggests a natural way to regard a depth–first search tree as a directed graph, namely direct each edge from lower depth–first number to higher depth–first number; we say this *directs* the tree by increasing depth–first number. The depth–first reach of a vertex V_i is the largest j such that there is a strong (directed) path from v_i to v_j in the oriented tree.

> **Lemma 6.2.** If a depth–first search tree T is directed by increasing depth–first number and the depth–first reach of v_i is j, then the vertices reachable from v_i in T are v_i, v_{i+1}, . . . , v_j.

> *Proof.* If $i < m \le j$, we wish to show that v_m is reachable from v_i in T. Suppose the contrary, i.e., that there is a smallest m such that $i < m \le j$ and v_m is not reachable from v_i in T. This means that when v_m is added to T, it is connected to a v_k with $k < i$. This means by step 2 that *no* vertices in G are connected to v_i, v_{i+1}, . . . , v_{m-1} except for vertices v_h with $h < i$. Thus no vertex listed after v_m is connected to v_i, v_{i+1}, . . . , v_{m-1} in G, so no vertex listed after v_m is reachable in T from v_i, contradicting the definition of j. By construction, no other vertices are reachable from v_i. ∎

As Figure 4.33 shows, each vertex in T has indegree 1 when T is directed by increasing depth–first number. (By Step 2, each vertex is connected to only one vertex with a smaller number.)

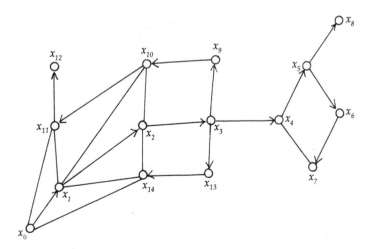

Figure 4.33

Figure 4.33 shows the result of applying depth–first search to a graph G with a bridge. The bridge is $\{x_3, x_4\}$ and there are no edges of G other than $\{x_3, x_4\}$ between vertices reachable from x_4 in T and the remainder of G. By the definition of reach and by Step 2 of the search procedure, no vertices numbered greater than the reach of x_4 are reachable from x_4 in T or even adjacent in G to a vertex reachable from x_4 in T. This gives the following theorem.

Finding Bridges

Theorem 6.3   If T is a depth–first search tree of G directed by increasing depth–first number, then the edge $\{v_h, v_i\}$ with $h < i$ is a bridge if and only if no vertex reachable from v_i in T is adjacent in G to a vertex v_k with $k < i$.

Proof.   By Step 2 of the depth–first search algorithm and Lemma 6.2, no vertex reachable from x_i is adjacent to a vertex whose depth–first number is greater than the reach of x_i. However, if $\{x_h, x_i\}$ is *not* a bridge, some vertex reachable from x_i must be adjacent in G to some vertex not reachable from x_i (otherwise, by deleting the edge $\{x_h, x_i\}$ we would separate vertices reachable from x_i from the remainder of the graph). Thus if $\{x_h, x_i\}$ is *not* a bridge, a vertex reachable from x_i is connected in G to a vertex with depth–first number less than i.

If $\{x_h, x_i\}$ is a bridge, then by the description of depth–first search, x_h and vertices with depth first number less than i are on one side of the bridge and x_i and vertices reachable from it are on the other side. Thus no vertex reachable from x_i is adjacent in G to a vertex with depth–first number less than i. ∎

A Bridge–Finding Algorithm for Computers

To use Theorem 6.3 as a basis for an algorithm for finding bridges, we introduce the vertex $L(v_i)$, which we shall call the "least vertex accessible from v_i". We *define* $L(v_i)$ to be the vertex with lowest depth–first number among all vertices adjacent in G to vertices reachable in T from v_i. In this terminology, Theorem 6.4 states that $\{v_h, v_i\}$ is a bridge if and only if the depth–first number of $L(v_i)$ is greater than or equal to i.

The depth–first number of $L(v_i)$ can be computed in the course of Step 2 of the depth–first search. However a conceptually easier way to compute $L(v_i)$ is to complete the depth–first search tree, assigning a depth–first number to each vertex. Then we review the vertices v_i in depth–first order, examine edges leaving v_i and record the smallest depth–first number of a vertex adjacent to v_i. Now working backwards, we can compute the vertex $L(v_i)$ for each vertex from the recorded information.

Example 6.1.   Which are the least vertices accessible from x_{10}, x_9, x_8 and

x_4. I.e. what are $L(x_{10})$, $L(x_9)$, $L(x_8)$, and $L(x_4)$ in Figure 4.33? What can you conclude about bridges? Since x_{11} is reachable from x_{10} in T and x_0 is adjacent to x_{11} in
G, $L(x_{10}) = x_0$.

   Since x_{10} is reachable from x_9, and $L(x_{10}) = x_0$, then $L(x_9) = x_0$. Since
nothing is reachable from x_8 and no vertices of G not in the tree are adjacent to
x_8, $L(x_8) = x_8$. Since no edge of G is an arrow from x_4, x_5, x_6, x_7 or x_8 back to
and earlier x in depth–first number, $L(x_4) = x_4$. Thus $\{x_3, x_4\}$ and $\{x_5, x_8\}$ are
bridges. ■

   We can also use depth–first search to orient graphs without bridges.

Graphs without Bridges are Orientable

   Theorem 6.4   A graph without bridges has an orientation.

   Proof.   Construct a depth–first search tree T in G, and let $x_1, x_2, \ldots, x_n$
   be the vertices of G in depth–first order. Direct T by increasing depth–
   first number and direct every other edge of G from higher to lower
   depth–first number. Then by Theorem 6.3, from each vertex (except x_1)
   we can reach a vertex with lower depth–first number. Therefore from
   each vertex in G, we can now reach x_1. But from x_1 we can reach any
   vertex in T. Thus from any one vertex we can reach any other vertex. ■

Backtracking

Depth–first search is also known as backtracking, as we can visualize the process
as follows: starting at vertex x construct a path, not stopping until you have run out
of edges. Now backtrack to the last place where you had a choice from which to
select an edge and repeat the process. Keep up the repetition until you run out of
vertices. It is possible to use depth–first search to determine the vertices connected to a vertex x because by Theorem 6.1, the search gives us a spanning tree
of the component containing x. Thus we can use depth–first search to find connected components of G. Other special features of graphs (e.g. cycles, vertices
whose removal disconnects G) can be discovered with depth–first search as well.
Backtracking or depth–first search is also a convenient way of listing all potential
solutions to a wide variety of problems. In this way it is possible to determine
whether at least one potential solution has certain desired properties. It is also
possible to determine all potential solutions that have certain desired properties.

   Problems that may be solved by backtracking involve lists of decisions.
As an example, suppose we are searching for a way out of the maze in Figure 4.34.
We are at the spot marked x, and we have a list of four choices of directions to go:

{up, right, down, left}

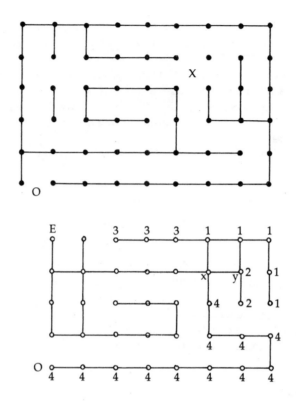

Figure 4.34

Once we decide to try one of these choices, we start to move through the maze. Our possible tracks through the maze are shown in a graph whose vertices represent "squares" inside the maze and whose edges represent openings between the squares. An attempted solution to the maze is a path in the graph. For example, we may decide to explore the maze by going up if possible, right if up is impossible, down if right is impossible and left if down is impossible. To make sure we get a path and don't cycle through the graph, we agree not to return to a vertex already covered. Our first attempt is marked with the number 1 in the graph, and leads to a dead–end. Now we back up along our path until we come to a place where another direction, other than the one chosen, is possible. Our next dead–end approach is marked with 2's. While it might seem reasonable to go left from vertex y, this would lead us back to vertex x and result in a cycle, so we do not explore this edge. On our third attempt, we get the dead–end path marked with 3's. Now we must start out again from position x. We don't want to go to vertex y, since that would duplicate an earlier position. Thus we explore the path marked with 4's and get out of the maze.

As you see, we did not need to explore all the paths leaving x. Part of this was just luck. If our exit had been at the vertex of G marked E, then we would have had more paths to explore. Note, however, that we don't explore any path which begins $x\{x, y\}y$ . . . because we have already seen that all paths leading out of y either are dead–end or lead us back to x. In other words, just by blindly applying the rules for depth–first search, we would never explore any paths starting $x\{x, y\}y$ . . . , so the use of depth–first search to explore the maze will save us a good deal of effort when there are many interconnected paths in the maze.

Decision Graphs

In a problem whose solution involves a sequence of decisions, we draw a *decision graph* by the following rules:

(1) Draw a vertex to represent the current situation.
(2) Draw an edge departing the initial vertex and entering a new vertex for each decision which changes the situation.
(3) For each vertex you add, determine which are the possible decisions that will change the situation. For those changes that lead to a situation already associated with a vertex, connect that vertex to your current situation. For each change which leads to a new situation, draw an edge to a new vertex representing the new situation.
(4) Repeat Step 3 as long as new situations are possible.

For example, this sequence of steps leads to the graph in Figure 4.34, which represents the situation in the maze. Here the situations are the "squares" of the maze and the decisions are always those that can be legally chosen from the set {up, right, down, left}.

Depth–First Listing of Colorings

The maze example is slightly misleading in that its decision graph is the same size as the maze. It is much more typical for the decision graph to be huge even if we start with what appears to be a simple or trivial problem.

Example 6.2. How many vertices are in the decision graph for deciding, one vertex at a time, on a proper coloring of a cycle on the four vertices {1, 2, 3, 4} with two colors, R and B?

Our initial vertex (the current situation) represents the graph with no colors assigned. It is adjacent to eight vertices representing the assignment of one color to one vertex. There are $2 \cdot 4 = 8$ vertices representing the assignment of two colors to two adjacent vertices and another $2 \cdot 4 = 8$ vertices representing the assignment of one or two colors to two opposite vertices. There are $4 \cdot 2 = 8$

vertices representing assignments of two colors to three of the four vertices, and two vertices representing the two final proper colorings. Thus we have 1 + 8 + 16 + 8 + 2 = 35 vertices in our decision graph. ■

A depth–first search quickly produces both colorings, but examines all 35 vertices to produce them.

Example 6.3.   Using the decision rule, "choose red first" and examining vertices in increasing numerical order, carry out a depth–first search on the decision graph to color a four–cycle with two colors.

To describe the situation in which vertex 1 is colored red and no other vertex is colored we use the notation $1R$. To describe the situation in which vertex 1 is colored red and vertex 2 is colored blue and no other vertex is colored, we write $1R2B$. To compactly say the second situation follows the first we write $1R/1R2B/$. We add more situations to this list in the order they arise from our decision rules and backtracks. We will observe the colorings in the following order ($\emptyset$ here means nothing is colored). You may wish to draw the colored graphs that correspond to a few of these sequences.

$\emptyset/1R/1R2B/1R2B3R/1R2B3R4B/1R2B4B/1R3R/1R3R4B/1R3B/1R4B/$
$1B/1B2R/1B2R3B/1B2R3B4R/1B2R4R/1B3R/1B3B/1B3B4R/$
$1B4R/2R/2R3B/2R3B4R/2R4R/2R4B/2B/2B3R/2B3R4B/$
$2B4R/2B4B/3R/3R4B/3B/3B4R/4R/4B$ ■

Notice how early in the search the two proper colorings are reached. One of the important topics discussed in a course in the design and analysis of algorithms is pruning of search trees to avoid irrelevant information. Notice also that if we want just one sequence of legal decisions rather than all the legal sequences, depth–first search should—with some exceptions—yield at least one legal sequence fairly quickly. This happens because our first priority in building a search tree is making a new decision whenever we legally can (rather than exploring all legal decisions simultaneously). Thus we get fairly long sequences of legal decisions quickly.

EXERCISES

1. Compute a depth–first numbering of the vertices in the graph of Figure 4.35, starting at x.

2. Compute the least vertex $L(v_i)$ for each vertex in your depth–first numbering of Figure 4.35.

3. Given an orientation to the portion of Figure 4.35 to the left of the bridge. Show that it is an orientation.

4. Compute the least vertex $L(v_i)$ for each vertex in Figure 4.33.

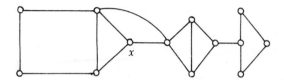

Figure 4.35

5. Using the decision sequence {up, right, down, left}, search to find the vertex E in the graph of Figure 4.34.

6. Using the decision sequence {up, left, down, right}, search for a path out of the maze in Figure 4.34.

7. Draw the decision graph corresponding to the maze in Figure 4.36.

8. Find a path through the maze in Figure 4.36 by means of depth–first search.

9. Prove that if T is a depth–first search tree for a connected graph G directed by increasing depth–first number, then each edge of G not in T connects a vertex reachable from some vertex v to a vertex from which v may be reached.

10. Prove that the edge $\{x, y\}$ is a bridge in G if every walk from x to y includes $\{x, y\}$.

11. Prove that an edge of a connected graph is a bridge if and only if it is not in any cycle.

12. Give an example of a connected bridgeless graph and a labelling of a spanning tree of G so that directing the tree edges from the smaller label to the larger label and the remaining edges from the larger numbered vertex to the smaller does not give an orientation. Why could your example not have come from a depth–first search spanning tree?

13. Each vertex in the table below is adjacent to the vertices in the table listed below it.

1	2	3	4	5	6	7	8	9	10	11	12	13
2	1	2	2	4	7	6	3	4	8	7	11	1
13	3	4	7	6	5	13	10	11	13	9	13	7
	6	8	9		2					12		12
			12							2		

Table 4.2

Use a modification of depth–first search to find a cycle in the graph.

14. Discuss how to use depth–first search to find a cycle.

15. Discuss how to use depth–first search to search for a Hamiltonian path or cycle (one that includes each vertex exactly once). Apply the technique to the graph in Exercise 13.

Enter Here

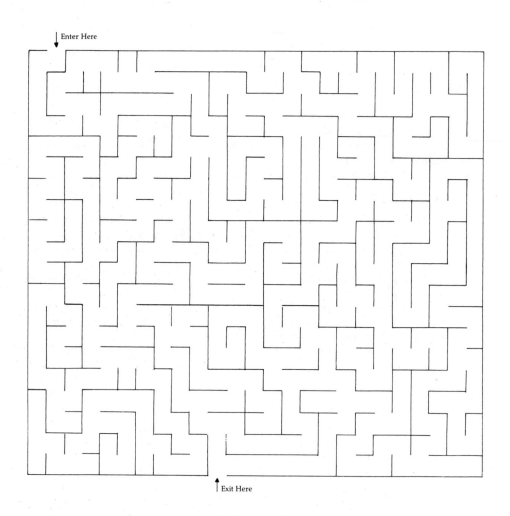

Exit Here

Figure 4.36

16. Write a computer program to determine whether a graph has a bridge and apply it to the graph in Exercise 13.

17. A magic square is a square matrix with entries 1 through n^2 such that the sum of each row, each column and the sum of the two diagonals are all the same. Discuss how to use depth–first search to find magic squares and apply your results to 3 by 3 squares.

18. The well–known eight queens problem is to enumerate all ways of placing eight nontaking queens on a chessboard. (A queen can take a piece in the same row, column or diagonal.) Describe how to use depth–first search to list the solutions.

*19. An articulation point is a vertex whose removal disconnects the graph. Develop a method of detecting articulation points by depth–first search; apply your method to the graph in Figure 4.35.

20. Write a computer program that applies depth–first search to determine a proper coloring of a graph with a minimum number of colors. Apply the program to the graph of Exercise 13.

SECTION 7   GRAPHS AND MATRICES

Adjacency Matrix of a Graph

We assume the reader is familiar with the concepts of matrices, matrix addition and matrix multiplication. If a digraph G has vertices $v_1, v_2, \ldots, v_k$, then it can be represented by an *adjacency matrix* $A(G)$ whose entry in row i and column j is 1 if (v_i, v_j) is an edge and whose (i, j)–th entry is 0 if (v_i, v_j) is not an edge. Since most computer languages allow for the storage and manipulation of matrices, an adjacency matrix provides a convenient way of storing information about a digraph in a computer. In Figure 4.37, we show a digraph and its adjacency matrix. Note that if we reassign the symbols v_1 through v_n to the graph in

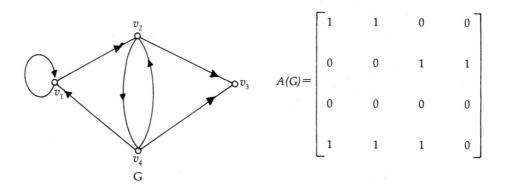

Figure 4.37

a different way, we could get a different matrix. Thus each numbering of the vertices gives a potentially different matrix as the adjacency matrix.

A multidigraph with vertices v_i can be represented by an adjacency matrix whose (i, j)–th entry is the number of times (v_i, v_j) appears in the edge multiset.

The adjacency matrix of a graph is defined similarly; we let $A(G)_{ij}$ be 1 if $\{v_i, v_j\}$ is an edge and 0 otherwise. Since saying that $\{v_i, v_j\}$ is an edge is the same as saying that $\{v_j, v_i\}$ is an edge, this means that $A(G)$ will be symmetric. The fact that a graph has no loops means $A(G)$ will have 0's along the main diagonal. For multigraphs, we let $A(G)_{ij}$ be the multiplicity of edge $\{i, j\}$. A graph and its matrix are shown in Figure 4.38.

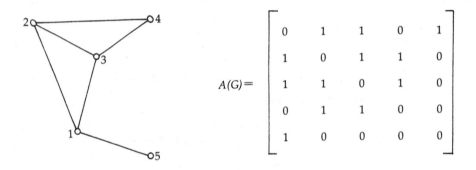

Figure 4.38

Matrix Powers and Walks

Matrix operations help us determine certain natural properties of graphs. The *length* of a walk in the theorem below refers to its number of edges.

> **Theorem 7.1.** If D is an n–vertex (multi)(di)graph with adjacency matrix A, then the (i, j)–th entry of A^k is the number of (directed) walks of length k from vertex i to vertex j in D.

> *Proof.* Clearly, A_{ij} is the number of walks of length 1 from v_i to v_j. Let S be the set of all integers m such that if m is in S, then A^m_{ij} is the number of (directed) walks of length m from v_i to v_j. We have seen 1 is in S; suppose all positive integers smaller than k are in S. Since

$$A^k = A^{k-1}A,$$

$$(A^k)_{ij} = \sum_{h=1}^{n} (A^{k-1})_{ih}A_{hj}.$$

However, we can partition the set of (directed) walks of length k from v_i to v_j into at most n classes; class h consists of all walks whose next–to–last vertex is v_h. By the multiplication principle, the number of walks in class h is the number of walks of length $k - 1$ from v_i to v_h times the number of walks of length 1 from v_h to v_j. This product is, by assumption

$$(A^{k-1})_{ih}A_{hj}.$$

Thus by the sum principle, the number of walks of length k from v_i to v_j is

$$\sum_{h=1}^{n} (A^{k-1})_{ih}A_{hj} = (A^k)_{ij},$$

and so k is in S. Thus, by version 1 of the principle of mathematical induction, S contains all positive integers and the theorem is proved. ∎

The number of (directed) *paths* from vertex i to vertex j is much more difficult to compute. For graphs, the diagonal entries of the k–th power contain interesting information. For example,

Theorem 7.2.   If A is the adjacency matrix of a graph, then the degree of vertex i is $(A^2)_{ii}$.

Proof.

$$(A^2)_{ii} = \sum_{j=1}^{n} A_{ij}A_{ji}$$

Since $A_{ij}A_{ji}$ is 0 unless $\{v_i, v_j\}$ is an edge, and is 1 if $\{v_i, v_j\}$ is an edge, $(A^2)_{ii}$ is the number of edges containing vertex i. ∎

Connectivity and Transitive Closure

Theorem 7.3.   An n–vertex (multi)graph (digraph) is connected (strongly connected) if and only if every entry of $\sum_{i=1}^{n} A^i$ is nonzero.

Proof.   Since any pair of connected vertices can be connected by a walk of length n or less, the theorem follows from Theorem 7.1. ∎

In fact, we need not compute each individual power and add them.

Theorem 7.4.   An n–vertex (multi-)graph (digraph) is connected (strongly connected) if and only if every entry of $(I + A)^n$ is nonzero.

Proof.   Since

$$(I + A)^n = \sum_{j=0}^{n} \binom{n}{j} I^j A^{n-j} = \sum_{k=0}^{n} \binom{n}{k} A^k$$

and since the entries of A^k are all nonnegative, the (i, j)–th entry of $(I + A)^n$ will be nonzero if and only if the (i, j)–th entry of some A^k is nonzero.   Thus Theorem 7.4 follows from Theorem 7.3.   ■

The matrix $(I + A)^n$ gives other information about a graph or digraph. We can define the transitive closure of a graph in a way similar to the transitive closure of a digraph.   (Just regard the edge set of the graph as a symmetric relation.)   Then we get

Theorem 7.5.   The adjacency matrix of the transitive closure of an n–vertex (multi) (di)graph has a 1 in position i, j if and only if $(A + I)^n - I$ is nonzero in position i, j.

Proof.   There is a (strong) walk from vertex i to vertex j of length k if and only if the (i, j)–th position of A^k is nonzero, and $\binom{n}{k} A^k$ is a summand of $(A + I)^n$.   However the term $\binom{n}{0} A^0 = I$ corresponds to walks of length 0 which are not used in the transitive closure.   ■

In a graph, two points are connected in the transitive closure if and only if they are in the same connected component.   This gives a conceptually simple way to determine the connected components of a graph.   The vertices in the same connected component as v_i correspond to the nonzero entries of row i of $(A + I)^n$. Since matrix arithmetic is not currently a basic electronic operation, the apparently more complicated methods based on spanning trees are more efficient for actual large–scale computation.   Even ordinary addition requires more effort than the simplest computer operations.

Boolean Operations

Another kind of addition, Boolean addition, defined for just 0 and 1 by the rules

$$1 \vee 1 = 1$$
$$1 \vee 0 = 0 \vee 1 = 1$$
$$0 \vee 0 = 0$$

where the symbol $\vee$ is read as "or," is electronically more straightforward than ordinary arithmetic. So long as we work with just 0 and 1, ordinary multiplication gives just 0's and 1's and is also electronically straightforward. All the usual rules like

$$a(b + c) = ab + ac$$

translate to rules like

$$a(b \vee c) = ab \vee ac$$

for multiplication and Boolean addition. In Boolean arithmetic, however, there is no subtraction concept, so even though

$$a + b = a + c \quad \text{implies} \quad b = c,$$

it is the case that

$$a \vee b = a \vee c \quad \text{need not imply} \quad b = c.$$

Beause only multiplication and addition are used to define matrix multiplication, we can define Boolean matrix multiplication. We use $A^{(n)}$ to denote the n-th power of a matrix of zeros and ones in Boolean arithmetic. The connectivity matrix of a (di)graph has a 1 in position (i, j) if vertex i and vertex j are (strongly) connected. In terms of Boolean matrix power and multiplication, Theorems 7.3 through 7.5 may be stated as

> **Theorem 7.6.** The connectivity matrix of a (multi)(di)graph is $(I + A)^{(n)}$ and the transitive closure has adjacency matrix $A(I + A)^{(n-1)}$.

> *Proof.* Each Boolean matrix power has a 1 in each position where the ordinary powers are nonzero. ∎

*The Matrix–Tree Theorem

More sophisticated information about a graph can be obtained from other matrix operations. The matrix–tree theorem lets us compute the number of spanning trees of a graph G from the adjacency matrix. We let $B(G)$ be the matrix whose (i, j)-th entry is 0 if $i \neq j$ and whose (ii)-th entry is the degree of vertex v_i in G.

The (i, j)–th cofactor of a square matrix M is $(-1)^{i+j}$ times the determinant of the matrix obtained by deleting row i and column j of M.

> **Theorem 7.7.** If G is a connected graph, then all cofactors of $B(G) - A(G)$ are equal to the number of spanning trees of G.

Theorem 7.7 may be proved as an easy (but clever) corollary of Theorem 7.8; this is left as an exercise. For a directed graph D, we let $C(D)$ be the matrix whose (i, i)–th entry is the outdegree of vertex i and whose (i, j)–th entry is 0 if $i \neq j$. We say a tree is directed into vertex i if its edges are directed in such a way that there is a path from any vertex to vertex i.

> **Theorem 7.8.** The (i, i)–th cofactor of the matrix $M = C(D) - A(D)$ is the number of spanning trees directed into vertex i in the connected digraph D.

Proof. (This proof may be omitted without loss of continuity.) We prove the theorem by induction on the number of edges of D. Let S be the set of all positive integers n such that if D has n edges, then the theorem holds for D. If D has one edge, say $(1, 2)$, then

$$C(D) = \begin{bmatrix} 1 & 0 \\ 0 & 0 \end{bmatrix} \qquad A(D) = \begin{bmatrix} 0 & 1 \\ 0 & 0 \end{bmatrix},$$

so that

$$M = C(D) - A(D) = \begin{bmatrix} 1 & -1 \\ 0 & 0 \end{bmatrix}.$$

The $(1, 1)$–th cofactor of M is 0, and the $(2, 2)$–th cofactor of M is 1; these are the number of spanning trees directed into, respectively, vertex 1 and vertex 2. Thus 1 is in S. Now suppose all integers less than n are in S and that D is a digraph with n edges and m vertices. We divide the proof into two cases. Case 1 is the case in which there is an edge $(1, i)$ from vertex 1 to some vertex i, so that the outdegree $od(1) > 0$. Since this edge cannot lie in any spanning tree directed into vertex 1, the number of spanning trees is the same for G and the graph G' whose edges are edges of G other than $(1, i)$. The matrix $M' = C(G') - A(G')$ will be different from M only in the first row (and there only in column 1 and column i). Thus M' and M have the same $(1, 1)$ cofactor, which, since G' has $n - 1$ edges and $n - 1$ is in S, is the number of spanning trees of G' directed into vertex 1 in G'. By the remarks above, this is also the number of spanning trees of G directed into vertex 1.

Case 2 is the case in which the outdegree of vertex 1 is 0. This case will be further divided into three steps for clarity.

Step 1: Suppose that for each vertex i such that $(i, 1)$ is an edge, vertex i has outdegree 1. Let G' be the graph obtained by choosing one such vertex, say vertex j, deleting the edge $(j, 1)$, replacing each edge $(i, 1)$ by the edge (i, j) instead and deleting vertex 1. In particular, applying this construction to a spanning tree directed into vertex 1 gives a spanning tree of G' directed into vertex j. Reversing the process on a spanning tree of G' directed into vertex j gives a spanning tree of G directed into vertex 1. These correspondences are one-to-one, so the number of spanning trees of G directed into vertex 1 equals the number of spanning trees of G' directed into vertex j. Now the matrix M' of G' may be obtained from the matrix M of G by adding column 1 to column j and then deleting row and column 1. The matrix formed by deleting rows 1 and j and columns 1 and j from M is the same as the matrix formed by deleting row j and column j from M'. Call this common resulting matrix M^*. Now in M, row j has a -1 in position 1, a 1 in position j and 0's elsewhere. Thus the $(1, 1)$–th cofactor of M is the determinant of M^*. However, the determinant of M^* is the (j, j)–th cofactor of M'; by our inductive hypothesis, this is the number of spanning trees directed into vertex j in G'. Hence the number of spanning trees of G directed into vertex 1 is the $(1, 1)$–th cofactor of M.

Steps 2 and 3 analyze the situation in which $(i, 1)$ is an edge and i has outdegree greater than 1. We compute first the number of spanning trees that use $(i, 1)$ in this situation (that is Step 2) and then the number of spanning trees that don't use $(i, 1)$ (that is Step 3) and add these two numbers.

We assume $(i, 1)$ is an edge and that vertex i has outdegree greater than 1. If a spanning tree uses $(i, 1)$, it cannot also use another edge (i, j), for the directed path from vertex j to vertex 1 would give us an (undirected) cycle in our spanning tree. Thus the number of spanning trees using edge $(i, 1)$ is equal to the number of spanning trees in the graph G'' obtained from G by deleting all edges (i, j) with $j \neq 1$. We let $M'' = C(G'') - A(G'')$; by our inductive hypothesis, we know that the number of spanning trees is the cofactor of the $(1, 1)$–th entry of M''. In this case, M'' differs from M only in row i; in row i and column 1 it has a -1; in row i and column i it has a $+1$ and all other entries in row i are 0.

Now we compute the number of spanning trees that do not use $(i, 1)$. This number is equal to the number of spanning trees directed into vertex 1 of the graph G^{**} obtained by deleting the edge $(i, 1)$ from G. Since G^{**} has $n - 1$ edges, our inductive assumption tells us that the number of spanning trees is the $(1, 1)$–th cofactor of the matrix $M^{**} = C(G^{**}) - A(G^{**})$.

Now let N, N'' and N^{**} represent, respectively, the matrices obtained from M, M'' and M^{**} by deleting the first row and column. Except for row i, these three matrices are identical, and row i of N is the sum of row i of N'' and row i of N^{**}. Thus by the addition theorem for determinants, the determinant of N is the sum of the determinants of N'' and

N^{**}. This sum is also the number of spanning trees directed into vertex 1. This proves that n is in S, and thus, by version one of the principle of mathematical induction, proves our theorem. ■

*The Number of Eulerian Walks in a Digraph

Knowing how many spanning trees a digraph has can help us determine how many Eulerian walks it has. Given a spanning tree directed into vertex i in an Eulerian digraph with no vertices of odd degree, we may construct a directed Eulerian walk as follows. Beginning at vertex i, follow any edge (i, j_1) leading out. From vertex j_1 choose any edge—say (j_1, j_2) not in the spanning tree. Repeat this process each time you arrive at a vertex j_k; follow an edge (j_k, j_{k+1}) leading out of vertex j_{k+1} but not in the spanning tree if there is such an edge; otherwise, follow the unique edge of the spanning tree that leaves vertex k. In this way, each outgoing edge from each vertex will be used, and since every vertex, except for perhaps vertex i and the final vertex, will have been in an even number of edges of the walk, vertex i must be the final vertex chosen. In Theorem 7.9, $od(v_i)$ stands for the outdegree of vertex i.

> **Theorem 7.9.**   The number of directed Eulerian walks from vertex i to vertex i in an n–vertex digraph D each vertex of which has equal indegree and outdegree is the product of the (i, i)–th cofactor of $C(D) - A(D)$ with
>
> $$od(v_i)\prod_{j=1}^{n}(od(v_j) - 1)!.$$

Proof.   Any Eulerian walk from vertex v_i to vertex v_i yields a tree directed into vertex v_i as follows: For each vertex v_j not equal to v_i, there is a last edge (v_j, v_k) in the walk in which j appears. The vertex set V with these last edges forms a tree directed into vertex v_i. We call this tree the tree of last passage of the walk. The number of walks having a given tree of last passage may be computed by observing that each such walk can be constructed by the method described before the statement of this theorem. The first time vertex v_i is used, there will be $od(v_i)$ ways of choosing an edge leaving v_i; the first time any vertex v_j for $j \neq i$ is used there will be $od(v_j) - 1$ edges leaving v_j to choose. On the k–th use of v_i or v_j (for $j \neq i$), there are respectively $od(v_i) - k + 1$ or $od(v_j) - k$ edges leaving the vertex from which to choose. By the multiplication principle, the number of walks with a given tree of last passage is

$$od(v_i)\prod_{j=1}^{n}(od(v_j) - 1)!.$$

One more application of the multiplication principle proves the theorem. ■

Unfortunately, there does not seem to be a natural parallel to Theorem 7.9 for undirected Eulerian graphs; we explore the reasons why the methods of Theorem 7.9 can't be applied to undirected graphs in the exercises.

EXERCISES

1. Draw pictures of the graphs with adjacency matrices

$$\begin{bmatrix} 0 & 1 & 1 & 0 & 0 & 1 \\ 1 & 0 & 1 & 1 & 1 & 0 \\ 1 & 1 & 0 & 1 & 0 & 1 \\ 0 & 1 & 1 & 0 & 1 & 1 \\ 0 & 1 & 0 & 1 & 0 & 0 \\ 1 & 0 & 1 & 1 & 0 & 0 \end{bmatrix} \quad \begin{bmatrix} 0 & 1 & 0 & 0 & 0 \\ 1 & 0 & 1 & 0 & 0 \\ 0 & 1 & 0 & 1 & 0 \\ 0 & 0 & 1 & 0 & 1 \\ 0 & 0 & 0 & 1 & 0 \end{bmatrix}$$

2. Write down the adjacency matrix for the graph in Figure 4.39.

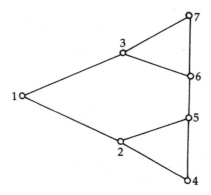

Figure 4.39

3. Write down the adjacency matrix for the multigraph in Figure 4.40.

4. Find the number of walks of length 2 and the number of walks of length 3 between vertex 1 and 5 in the graph of Figure 4.39. Do the same for paths. What about length 4?

5. Find the number of walks of length 4 between vertex 1 and vertex 4 in the multigraph of Figure 4.40.

6. Write down the adjacency matrix of the digraph in Figure 4.41.

7. Find the number of walks of length 3 and 4 from vertex 1 to vertex 3 in the digraph of Figure 4.41.

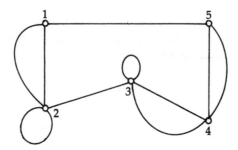

Figure 4.40

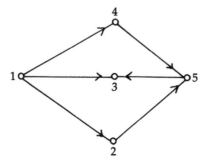

Figure 4.41

8. (a) Under what circumstances will the (i, i)–th entry of $A(G)^3$ be nonzero? (i.e., what property must a graph G have in this case?)

(b) How is the number of triangles in a graph G related to the trace (sum of the main diagonal entries) of $A(G)^3$?

9. Show that a tournament D is transitive if and only if the trace (see Exercise 8) of $A(D)^3$ is zero.

10. (a) Compute the number of spanning trees directed into vertex 3 of the digraph of Figure 4.41.

(b) Compute the number of spanning trees directed into any other vertex of the digraph of Figure 4.41.

11. (a) Compute the number of Eulerian walks starting and ending at vertex 1 in Figure 4.42.

(b) Compute the number of Eulerian walks starting and ending at vertex 3 in Figure 4.42.

12. Discuss why we need at most $\log_2(n)$ (or the next largest integer) matrix multiplications to apply Theorem 4.40 and 4.41.   What about Theorem 4.42?

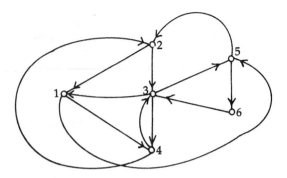

Figure 4.42

*13. Explain why any spanning tree directed into vertex 3 in Figure 4.42 must contain the edge (6, 3).

*14. Explain why in an Eulerian digraph there is at least one spanning tree directed into each vertex.

*15. Given a graph G, its double DG is a digraph that has edges (i, j) and (j, i) for each edge $\{i, j\}$ of G.

        (a) Show that there is a one–to–one correspondence between the spanning trees of G and the spanning trees of DG directed into vertex i.

        (b) Show that the diagonal cofactors of $B(G) - A(G)$ (the (i, i)–th cofactors, that is) are each equal to the number of spanning trees of G.

        (c) Show that the rows and columns of $B(G) - A(G)$ all add up to zero. It is a theorem of linear algebra that all cofactors of a matrix with zero row and column sums are equal. Thus you have proved Theorem 7.7.

*16. An Eulerian walk starting and ending at vertex i provides two orientations that turn G into a digraph. There is a tree directed into vertex i associated with such a walk. Also, *any* spanning tree of G (as a graph, not a digraph) can be oriented in exactly one way to become a spanning tree directed into vertex i. Why can't these facts be used as in Theorem 7.9 to compute the number of Eulerian walks in G?

17. Find the adjacency matrix of the digraph shown in Figure 4.42.

18. Find the Boolean squares and 4–th powers and 8–th powers of the matrices of Exercise 1.

19. The Boolean sum of two row vectors is the vector whose components are the Boolean sums of the components of the two vectors. Use A_{i-} to stand for row i of the matrix A. The following algorithm, called Warshall's algorithm (but attributed to Kleene by Reingold, Nievergelt and Deo), is very effective for computing the transitive closure because the $\bigvee$ operation is especially easy to implement. Show that this algorithm computes the transitive closure of a digraph with $n \times n$ adjacency matrix A and stores the matrix of the transitive closure in the matrix A.

$$\text{For } j = 1 \text{ to } n.$$
$$\text{For } i = 1 \text{ to } n.$$
$$\text{If } A_{i_j} = 1, \text{ then let } A_{i-} = A_{i-} \bigvee A_{j-}.$$

20. In the theorems of this section about n-th powers of matrices, the n may often be replaced by $n - 1$, but under certain situations may not. Discuss this. (Hint: There is a difference between graphs and digraphs.)

SUGGESTED READING

Behzad, M., Chartrand, G. and Lesniak-Foster, L.: *Graphs and Digraphs*, Wadsworth 1979.

Biggs, N.L., Lloyd, E.K., and Wilson, R.J.: *Graph Theory 1736–1936*, Oxford University Press 1976, Clarendon Press 1977.

Harary, F.: *Graph Theory*, Addison-Wesley 1969.

Harary, F., Norman, R.Z., and Cartwright, D.: *Structural Models: An Introduction to the Theory of Directed Graphs*, Wiley 1965.

Korfhage, R.: *Discrete Computational Structures*, Academic Press 1974.

Liu, C.L.: *Introduction to Combinatorial Mathematics*, McGraw-Hill 1968.

Nijenhuis, A. and Wilf, H.S.: *Combinatorial Algorithms*, Academic Press 1974.

Reingold, E.M., Nievergelt, Deo, N.: *Combinatorial Algorithms*, Prentice Hall 1977. '

Roberts, F.S.: *Discrete Mathematical Models*, Prentice Hall 1976.

Tucker, Alan: *Applied Combinatorics*, Wiley 1980.

Tutte, W.T.: "*Chromials*" in *Studies in Graph Theory II*, MAA Studies in Mathematics, Volume 12, Mathematical Association of America 1975.

Whitney, Hassler, and Tutte, W.T.: "*Kempe Chains and the Four Color Problem*" in *Studies in Graph Theory II, MAA Studies in Mathematics*, Volume 12, G.S. Rota, ed. Mathematical Association of America 1975.

5 *Matching and Optimization*

SECTION 1   MATCHING THEORY

The Idea of Matching

A school district has begun to advertise for teachers and plans to continue advertising until it has found enough qualified applicants for all the positions. Since a number of the applicants are qualified to teach more than one subject, it may not be entirely clear when enough qualified candidates to fill all the positions have been found. It is natural to ask whether there is a systematic way of determining when there are enough candidates. The problem has a natural graph–theoretic representation. We construct a graph whose vertices are the positions and the applicants. In this graph, we draw an edge joining a position to an applicant if the applicant is qualified for the position. If we can find a set of edges such that no two of them have a vertex in common and if every position is touched by one of these edges, then we can use the edges to match positions to people who qualify for them.

A *matching* of size m in a graph G is a set of m edges no two of which have a vertex in common. A vertex is said to be *matched* (to another vertex) by M if it lies in an edge of M. We defined a *bipartite graph* G with parts V_1 and V_2 to be a graph whose vertex set is the union of the two disjoint sets V_1 and V_2 and whose edges all connect a vertex in V_1 with a vertex in V_2. The graph for the school district's problem is bipartite. A *complete matching* of V_1 into V_2 is a matching of G with $|V_1|$ edges. A complete matching of positions to candidates is what the school district wants. Examples of bipartite graphs, first with and then without a complete matching, are shown in Figure 5.1.

If there is no complete matching of positions to candidates, the school district would still want a matching as large as possible or a matching that fills a certain critical set of positions. Thus it will be useful to have methods to build matchings of maximum size and tests to determine either the size of a maximum–sized matching or whether one part of a bipartite graph has a complete matching into the other part. Our approach to these problems follows that of

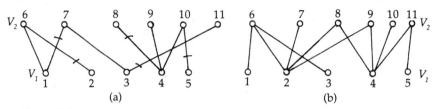

Figure 5.1

Berge, which parallels work of Norman and Rabin. Although our emphasis is on bipartite graphs, some of the fundamental theorems hold for arbitrary graphs as well. We begin with a result describing the interaction of two matchings. When we say that edges of a walk are alternately in M_1 and M_2, we mean that if e_1 and e_2 are edges with a vertex in common, then one is in M_1 and the other is in M_2. This seemingly technical lemma will have surprising applications.

Lemma 1.1. Let M_1 and M_2 be matchings of the graph $G = (V, E)$, and let $E' = M_1 \cup M_2 - (M_1 \cap M_2) = (M_1 - M_2) \cup (M_2 - M_1)$. Then the connected components of $G' = (V, E')$ fall under one of the following three types:

(1) a single vertex.
(2) a cycle with an even number of edges whose edges are alternately in M_1 and M_2.
(3) a path whose edges are alternately in M_1 and M_2 and whose two end vertices are each matched by one of M_1 or M_2 but not both.

Proof. A connected component could be of type 1. We now assume we have a connected component which is not a single vertex. Since a vertex is in at most one edge of M_1 and at most one edge of M_2, it has degree at most 2. Thus the sum of the degrees of a connected component with n vertices is at most $2n$. Since the component is connected, it has at least $n - 1$ edges so the sum of the degrees of its vertices is either $2(n - 1)$ or $2n$. (Why can't it be $2n - 1$?)

If the sum of the degrees is $2n$, then each vertex has degree 2, so the component must be a cycle. Since a vertex cannot be in two edges of an M_i, the edges must be alternately in M_1 and M_2. In particular, we have an even number of edges.

If the sum of the degrees is $2n - 2$, the connected component is a tree, but since each vertex has degree 2 or less, it must be a path. Since the end vertices have degree 1, each of them is in an edge of M_1 or an edge of M_2 but not both. As in case 2, the edges must be alternately in M_1 and M_2. ∎

From Lemma 1.1 we can derive the Berge criterion for a maximum–sized matching in a graph.   The criterion uses Peterson's idea of an alternating path for a matching.   If M is a matching, a path $v_0 e_1 v_1 e_2 \ldots e_n v_n$ is an *alternating path* for M if whenever e_i is in M, e_{i+1} is not and whenever e_i is not in M, e_{i+1} is in M. Our first step in developing the criterion shows how alternating paths may be used to enlarge matchings into bigger matchings.

Making a Bigger Matching

Theorem 1.2.   Let M be a matching in a graph G and let P be an alternating path with edge set E' beginning and ending at unmatched vertices.   Let $M' = M \cap E'$.   Then

$$(M - M') \cup (E' - M') = (M - E') \cup (E' - M)$$

is a matching with one more edge than M.

Proof.   Every other edge of P is in M.   However P begins and ends with edges not in M, so there is a number k such that P has k edges in M and $k + 1$ edges not in M.   Now the first and last vertices of P are unmatched and all other vertices in P are matched by M', so no edge in $M - M'$ contains any vertex in P.   Thus the edges of $M - M'$ have no vertices in common with the edges of $E' - M'$.   Further, since P is a path and $E' - M'$ consists of every other edge of the path, the edges of $E' - M'$ have no vertices in common.   Thus

$$(M - M') \cup (E' - M')$$

is a matching and by the sum principle, it has $m - k + k + 1 = m + 1$ edges.   Elementary set theory shows that

$$(M - M') \cup (E' - M') = (M - E') \cup (E' - M). \quad \blacksquare$$

The Berge criterion for a maximum–sized matching essentially states that when Theorem 1.2 does not apply, we cannot enlarge the matching.

Theorem 1.3 (Berge).   Suppose G is a graph and M is a matching.   Then M is a matching of maximum size (among all matchings) if and only if there is no alternating path connecting two unmatched vertices.

Proof.   If there is such an alternating path, then M is not maximum–sized by Theorem 1.2.

Now suppose there is no such alternating path and let N be a maximum–sized matching. We show that M and N have the same size by applying Lemma 1.1 to prove that $M - N$ and $N - M$ have the same size. (Since $M = (M - N) \cup (M \cap N)$ and $N = (N - M) \cup (M \cap N)$, this proves M and N have the same size.) Let G' be the graph on V whose edge set is $(M - N) \cup (N - M) = (M \cup N) - (M \cap N)$. We show that each connected component of G' has the same number of edges of M as of N. Type 1 components have no edges. Type 2 components have an even number of edges alternating between M and N, so exactly half the edges of a type 2 component are in M and the other half in N.

Since N is maximum–sized, by Theorem 1.2, a type 3 component of G' cannot both begin and end with an edge of M (since it would begin and end with N–unmatched vertices). By the hypothesis of this theorem, a type 3 component of G' cannot both begin and end with an edge of N. Therefore, exactly half the edges of a type 3 component are in M and exactly half are in M'.

Since $M - N$ and $N - M$ have the same number of edges in each connected component of G', $M - N$ and $N - M$ have the same size. Therefore, M is of maximum size. ∎

By Theorem 1.3, to produce a maximum–sized matching we need only repeat the procedure described in Theorem 1.2 until it is no longer possible to find an alternating path between two unmatched vertices. To turn this into an explicit procedure for finding maximum–sized matchings, we need only develop an explicit procedure for searching for alternating paths. In a small graph it is usually possible to find the desired path by inspection. However if we want a good method that could be programmed for a computer, inspection won't suffice. In Section 4 we develop a method of finding matchings in bipartite graphs by means of network flows. For those who wish to see how to construct a matching directly, we now outline a method of finding an alternating path.

A Procedure for Finding Alternating Paths

In our study of graphs we described two search procedures, depth–first search and breadth–first search, that allowed us to construct spanning trees for a graph. Either of these procedures can be modified and then used to search for an alternating path. Since breadth–first search yields minimal paths, we will describe how it may be modified. An *alternating search tree* centered at the vertex x in a graph G with matching M is a tree containing x such that if edges e_1 and e_2 of the tree share a vertex v, then one of the two edges e_i is in M and the other is not. To construct an alternating breadth–first search tree, we modify the procedure of Chapter 4, Section 2 as follows.

Begin at an unmatched vertex x for the matching M, and use this as the x in step 1. Recall that steps 2 and 3 are superfluous. Suppose in step 4 we have constructed the tree (V_k, E_k) and are examining a vertex x_j adjacent to a z in V but not in V_k. Then there is an x_i such that $\{\{x_i, x_j\}\}$ is in E_j. If $\{x_i, x_j\}$ is in M, then no matter what z we examine, $\{x_j, z\}$ will not be in M. Thus we proceed with step 4 without modification. If, however, $\{x_i, x_j\}$ is *not* in M, then we examine *only* vertices z such that $\{x_j, z\}$ *is in* M.

With this single modification, the breadth–first search procedure will produce an alternating search tree; however, it need not produce a spanning tree. If, in a bipartite graph, there are some unmatched vertices connected to x by alternating paths, this procedure will give us such an alternating path of minimum length to an unmatched vertex. (If G is not bipartite, this search procedure does not suffice.)

In particular, we get our alternating path (if there is one) by constructing an alternating breadth–first search tree starting at an unmatched vertex. If we find an unmatched vertex in this tree, we stop. If we do not find an unmatched vertex, we repeat the process starting at another unmatched vertex. Once we find an unmatched vertex in our search tree and stop, we work backwards to find the unique path in the tree from x to the unmatched vertex. In a non-bipartite graph we might use a backtracking search instead; however this could be time consuming if we had to back-track through all alternating paths before finding the desired one. Edmonds pointed out the difficulties with such standard search techniques and developed a technique without these difficulties in his article listed in the Suggested Reading.

Constructing Bigger Matchings

Example 1.1.   To see how to enlarge a matching, note that in Figure 1.1 (a) we can see by inspection that $M = \{\{1, 6\}, \{3, 7\}, \{4, 8\}, \{5, 10\}\}$ is a matching. Enlarge it.

The only unmatched vertex in V_1 is vertex 2. By inspection, we see that $\{2, 6, 1, 7, 3, 11\}$ is the vertex set of an alternating path.

If we prefer, we may form the alternating breadth–first search tree centered at vertex 2. It turns out to be the path whose vertices are $\{2, 6, 1, 7, 3, 11\}$.

Vertex 11 is unmatched, so we have our alternating path. Along this path, the edges $\{1, 6\}$ and $\{3, 7\}$ are in the matching, while edges $\{2, 6\}$, $\{1, 7\}$ and $\{3, 11\}$ are not. Throwing out $\{1, 6\}$ and $\{3, 7\}$ from M and adding in $\{2, 6\}$, $\{1, 7\}$ and $\{3, 11\}$ gives the matching shown by means of edges marked with a slash in Figure 1.1. ■

Example 1.2.   By inspection, $\{1, 6\}$, $\{2, 7\}$, $\{4, 9\}$ and $\{5, 11\}$ is a matching in Figure 5.1(b). Show it is maximum–sized.

Since 3 is the only unmatched vertex in V_1, we look for alternating paths beginning at vertex 3. By inspection, we see there is no such path.

Rather than use inspection, we may construct an alternating search tree

centered at 3.   The tree consists of the vertices {3, 6, 1} and the edges {3, 6} and
{1, 6}, because {2, 6} is not in the matching.   There are no unmatched vertices
other than 3 in this tree.

Thus, there is no alternating path from 3 to an unmatched vertex, so by the
Berge criterion this matching is maximum–sized.   (A larger matching would have
to match a currently unmatched vertex in V_1 and a currently unmatched vertex in
V_2.   Therefore, we need not examine unmatched vertices in V_2 as well.) ∎

Testing for Maximum–Sized Matchings by Means of Vertex Covers

In bipartite graphs, there are two well–known tests for determining the size of a
maximum matching or the possible existence of a complete matching.   These
tests do not lead to an algorithm as efficient as the Berge criterion, but they do
provide simple techniques that in many situations can be applied just by looking at
a graph.

Our first method, which uses the idea of a vertex cover, is usually attributed to König.   A *vertex cover* C of a graph G is a set of vertices such that each
edge of G contains at least one vertex in C.   If M is a matching and C is a vertex
cover, then $|M| \leq |C|$ because each edge of M will have to contain a different
element of C.   In Figure 5.2(a) the vertices marked 1, 2, 3 and 4 cover all the
edges.   Thus, we know there is no complete matching.   There is an obvious
matching of size 4 in Figure 5.2(a), namely the vertical edges.   On the other hand,
Figure 5.2(b) has a complete matching (can you find one?) and has a vertex covering of size 5 (the bottom row of vertices, for example).

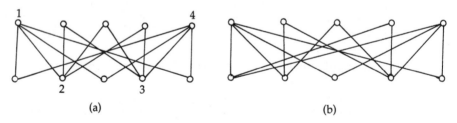

(a)   (b)

Figure 5.2

In fact, in a bipartite graph the size of a maximum–sized matching and the
size of a minimum–sized vertex cover are always equal.

In the proof of this theorem, we use the idea of the set of "relatives" $R(X)$
of a set X of vertices.   For a subset X of the vertex set of a graph with edge set E,
we let

$$R(X) = \{y \mid \{x, y\} \in E \text{ for some } x \in X\}.$$

For example, in Figure 5.1(a) $R(\{6, 7\}) = \{1, 2, 3\}$.

Theorem 1.4.   In a bipartite graph, the size of a minimum–sized vertex cover and the size of a maximum–sized matching are equal.

Proof.   We use the Berge criterion to show that given a maximum–sized matching, we can construct a vertex cover of the same size. Since we already have seen that $|M| \le |C|$ for any matching M and cover C, this will prove the theorem. Suppose G is a bipartite graph with parts V_1 and V_2, and let M be a maximum–sized matching of G. Let U_1 be the set of *unmatched vertices in* V_1. If U_1 is empty, then $|M| = |V_1|$ and we may let $C = V_1$. Otherwise, denote by A be the set of vertices connected by alternating paths to vertices in U_1 and let $A_2 = V_2 \cap A$. That is,

$$A_2 = \{v \in V_2 \mid \text{an alternating path connects } v \text{ to a vertex in } U_1\}.$$

The set

$$C = A_2 \cup (V_1 - R(A_2) - U_1)$$

is a vertex cover, for if an edge contains a member of U_1 or $R(A_2)$, it is covered by a member of A_2, while any other edge is covered by a vertex in $V - R(A_2) - U_1$. Each member of C is matched, since members of $V_1 - U_1$ must be matched and if a member of A_2 were not matched, we would contradict the Berge criterion. Since each member of A_2 is matched to something in $R(A_2)$, no two members of C lie in the same edge of M. Thus $|C| \le |M|$, so $|C| = |M|$. ∎

Hall's "Marriage" Theorem

The idea of the set $R(X)$ of relations of a set X provides us with another test for whether there is a complete matching of V_1 into V_2. It is clear that if there is a complete matching, then each subset X of V_1 has as relations at least the $|X|$ elements matched to X. In other words, to have a complete matching of V_1 into V_2, it is necessary that $|X| \le |R(X)|$ for all $X \subseteq V_1$. In fact, it is also sufficient to check this one condition.

Theorem 1.5.   (König-Hall) A bipartite graph on two sets V_1 and V_2 has a complete matching from V_1 to V_2 if and only if $|X| \le |R(X)|$ for each subset X of V_1.

Proof.   Assume $|X| \le |R(X)|$ for all $X \subseteq V_1$, since we've already seen this holds if there is a complete matching. Since each edge relates something in V_1 to something in V_2, V_1 is a vertex cover. We show V_1 is a minimum–sized vertex cover. Suppose T is a minimum–sized vertex cover and $T \neq V_1$. Suppose $S_1 = T \cap V_1$ and let $S_2 = V_1 - S_1$. By assumption, $|R(S_2)| \geq |S_2|$. However, $R(S_2) \subseteq T \cap V_2$ because edges not covered by vertices in V_1 must be covered by vertices in V_2. Thus $|S_2| \leq |T \cap V_2|$, so $|V_1| = |S_1| + |S_2| \leq |S_1| + |T \cap V_2| = |T|$. Thus V_1 is a minimum–sized vertex cover, and by Theorem 1.4 there is a matching of size $|V_1|$ which must be a complete matching from V_1 to V_2. ∎

The idea of a matching sometimes occurs in disguised forms in combinatorics.

Example 1.3. A dean is appointing a student activities committee which is to contain a representative of each student organization on campus. To avoid any conflict of interest, the dean has decided that a student can represent only one organization even if that student is a member of several organizations. Thus the question is, "Can the dean find a system of distinct representatives, one from each organization?"

If we draw a graph in which the vertices represent students and organizations and in which a line represents membership, then a matching from organizations to students will give us a system of distinct representatives, and a system of distinct representatives would give a matching. Thus the dean can find the desired system if and only if for each k, every k organizations have at least k members among them. ∎

Given a family $(S_1, S_2, \ldots, S_k)$ of sets, we call a list $(a_1, a_2, \ldots, a_k)$ a *system of distinct representatives* (SDR) if the a_i's are all different and if $a_i \in S_i$ for each i. Then using the techniques of Example 1.3, we can prove "Hall's marriage theorem":

Theorem 1.6. A finite family **S** of sets has a system of distinct representatives if and only if for each i, the union of any i sets in **S** has at least i elements.

Proof. Essentially given in Example 1.3. ∎

Term Rank and Line Covers of Matrices

The basic results of matching theory can be profitably reformulated in terms of matrices. By a *line* of a matrix, we mean a row or column. By a *line cover* of a matrix, we mean a set of lines containing all the nonzero entries of the matrix. An $m \times m$ matrix has m^2 positions, each containing an entry of the matrix. The *term rank* of a matrix is the maximum number of positions, no two in the same row or column, all containing nonzero entries. A theorem of König, which is also referred to as the König-Egerváry theorem (Egerváry's contribution is a more detailed result), relates term rank and line covers in the way matchings relate to vertex covers in bipartite graphs.

Theorem 1.7.   The term rank t of a matrix m is equal to the minimum number m of lines that contain all the nonzero entries of the matrix.

Proof.   We construct a bipartite graph G whose vertices are the lines of M. We put $\{L_1, L_2\}$ in the edge set of G if L_1 and L_2 are different and have a nonzero entry in common.   Thus our graph must be bipartite because, for L_1 and L_2 to be connected, one must be a row and the other must be a column.   The term rank of M may be interpreted as the maximum number of edges of G no two of which have a vertex in common, i.e., as the maximum size of a partial matching.   By Theorem 1.2, this is the size of a minimum vertex cover of G—and this is interpreted as a minimum–sized set of lines such that each nonzero entry in M lies in one of these lines.   Thus the term rank is the minimum number of lines in a line cover. ■

Permutation Matrices and the Birkhoff–von Neumann Theorem

As an example of the use of Theorem 1.7, we develop another celebrated result known as the Birkhoff–von Neumann Theorem.   A *permutation matrix* is a zero–one matrix with exactly one nonzero element in each row and column. Multiplying a vector by such a matrix simply permutes (interchanges) the entries of the vector.   We begin with a special case of the theorem for integral matrices.

Theorem 1.8.   If M is an $m \times m$ matrix of nonnegative integers such that all the elements of each row and each column add up to k, then M is a sum of k permutation matrices.

Proof.   To prove the theorem by induction, let S be the set of all integers n such that if the row and column sums are n, then the matrix is the sum of n permutation matrices.   Then $n = 1$ is in S because if the row and column sums are all 1 in a nonnegative integral matrix, then the matrix is a permutation matrix.   Now suppose $k - 1$ is in S.

If all nonzero entries could be covered by r lines, the sum of all these nonzero entries would be no more than rk.   However, the sum of all the entries in the matrix is mk, and $mk \leq rk$ implies $m \leq n$.   The rows form an m–line cover, so by Theorem 1.7, there are m positions, no two in the same row or column, with nonzero entries.   Let P be the zero–one matrix with 1's in exactly these positions.   Then P is a permutation matrix.

Returning to the original M, note that $M - P$ is a nonnegative integral matrix whose row and column sums are all $k - 1$.   Thus since $k - 1$ is in S, $M - P$ is a sum of $k - 1$ permutation matrices, so M is a sum of k permutation matrices.   Thus, by the first version of the principle of mathematical induction S is the set of all positive integers. ■

Theorem 1.9. (Birkhoff–von Neumann). If M is a matrix of nonnegative real numbers such that each row and column adds up to 1, then $M = c_1P_1 + c_2P_2 + \ldots + c_kP_k$ where the c_i's are nonnegative numbers that add up to 1 and the P_i's are permutation matrices.

Proof.   Outlined in the exercises.   ■

EXERCISES

1. Explain why, if at a school dance, each k boys are liked by at least k girls, then every boy can find a girl to dance with. (For reasons related to this example, Hall's matching theorem is sometimes called the marriage theorem.)

2. Apply the Berge criterion to show that the vertical edges in Figure 5.2(b) do not form a maximum–sized matching.

3. Find a complete matching for Figure 5.2(b).

4. Which of the bipartite graphs in Figure 5.3 have complete matchings? Find matchings for those that do. (Inspection should suffice for finding matchings.)

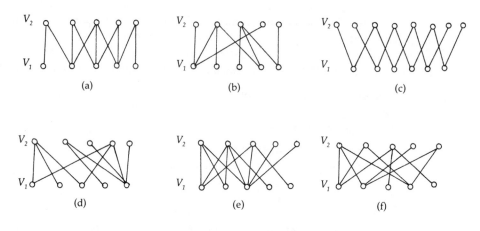

Figure 5.3

5. Find minimum–sized vertex covers and maximum–sized partial matchings for the graphs in Figure 5.3 which lack complete matchings.

6. Find a system of distinct representatives for $\{1, 2, 4, 8\}$, $\{2, 3, 5, 7\}$, $\{1, 3, 8, 6\}$, $\{2, 4, 6, 8\}$, $\{1, 3, 7, 4\}$.

7. Determine whether each family below has a system of distinct representatives. Find such a system if the family has one.

   (a) $\{1, 2, 7\}$, $\{1, 4, 8\}$, $\{2, 4, 6\}$, $\{1, 6, 8\}$, $\{2, 3, 5\}$
   (b) $\{1, 7, 9, 4\}$, $\{2, 3, 5, 8\}$, $\{3, 8, 2\}$, $\{4, 6, 7\}$, $\{5, 6, 3, 8, 2\}$

(c) {1, 2, 4}, {6, 7, 8}, {1, 3, 4}, {1, 6, 7}, {2, 7, 8}, {1, 6, 8}, {6, 7, 2}

(d) {1, 3, 6, 9}, {3, 9, 7}, {1, 8, 7}, {3, 6, 8}, {6, 7, 8}, {7, 8, 9}, {1, 5, 6}, {3, 4, 5}, {1, 3, 7, 9}

(e) {1, 5, 6}, {6, 8, 9}, {3, 5, 9}, {1, 7, 9}, {1, 6, 7, 9}, {2, 6, 8}, {2, 5, 7}, {2, 3, 6, 7, 8}

8. A Latin rectangle is an $m \times n$ matrix with $m \leq n$ whose rows are each a permutation of 1, 2, . . . , n and whose columns are lists of m distinct integers. By using an SDR for the sets of numbers not used in each column, show that if $m < n$, an $m \times n$ Latin rectangle may be extended to an $(m + 1) \times n$ Latin rectangle.

9. Show that Theorem 1.5 msy be derived as a consequence of Theorem 1.6.

10. Derive Theorem 1.5 as a consequence of Theorem 1.7.

11. Find a minimum–sized vertex cover and a maximum–sized matching in the graphs of Figure 5.4.

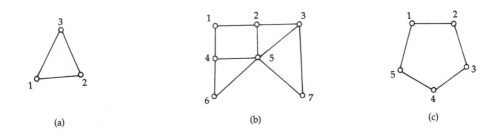

(a)   (b)   (c)

Figure 5.4

12. What are the sizes of the minimum vertex covers and maximum matchings in

   (a) a cycle on an even number of vertices?
   (b) a cycle on an odd number of vertices?
   (c) a complete bipartite graph?
   (d) a complete graph?

13. In Figure 5.1(b), construct an alternating breadth–first search tree centered at vertex 3 using the matching {{2, 6}, {4, 9}, {5, 11}}. Find an alternating path you can use to make a larger matching. Find the larger matching.

14. In the graph of Figure 5.5, the set of edges {a, c}, {d, f}, {b, e} is clearly a matching.

   (a) Construct an alternating breadth–first search tree centered at each unmatched vertex.
   (b) Determine whether or not there is a larger matching and find one if it exists.

15. Beginning with the three vertical edges in Figure 5.6, construct a maximum–sized matching using alternating paths. (You may find the alternating paths by trial and error or by constructing alternating search trees.)

*16. Prove the Birkhoff–von Neumann theorem by induction on the number of non-

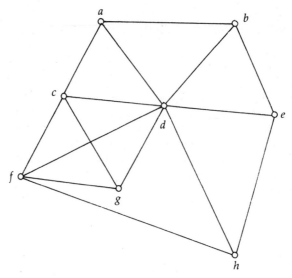

Figure 5.5

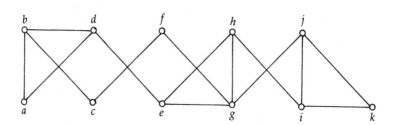

Figure 5.6

zero entries using the following outline. Argue as to why there must be a set of m nonzero entries, one in each row and column. Pick the smallest of these m nonzero entries, and using a multiple of a permutation matrix, subtract it from all m positions. This gives a matrix with one more 0.

17. Prove or disprove. There is a matching from V_1 to V_2 in a bipartite graph if there are k or more edges containing each vertex of V_1 and no more than k edges containing each vertex of V_2.

18. Suppose P and Q are partitions of N into m parts such that the union of any k classes of P is not a subset of the union of any $k - 1$ classes of Q. Prove that there is a system of common representatives of P and Q, i.e., a list of elements a_i that is an SDR for both P and Q.

19. Show there are at least $n!(n-1)! \ldots (n-r+1)!$ $r \times n$ Latin rectangles. What does this say about the number of Latin squares? (See Exercise 7.)

20. Write a computer program that finds alternating search trees. Expand it to a program that constructs maximum–sized matchings.

21. In the construction of a breadth–first search tree, each edge is examined at most once. In examining an edge, one checks its vertices to see if they are both in the tree. (This can be done quickly if one records for each vertex that it is in the tree when one adds it.) In constructing a matching, one might need to construct a breadth–first search tree for each vertex. In a graph with v vertices and e edges, explain why a matching of maximum size may be found in approximately ve^2 steps (each of these steps taking some constant amount of time).

22. Give an example of a matching non-bipartite graph in which an alternating breadth of a first search tree centered at an unmatched vertex x fails to find an alternating path from x to an unmatched vertex, even though such a path exists in the graph.

23. Show that if the unmatched vertex x in a bipartite graph is connected by alternating paths to other unmatched vertices, an alternating breadth-first search tree contains one of these paths.

SECTION 2   THE GREEDY ALGORITHM

How the SDR Problem Changes when Representatives Cost Money

We began our study of matching with a problem that has a natural interpretation in terms of either systems of distinct representatives or matchings. Recall we have a school district trying to fill positions. For each position, the school board has a set of candidates qualified to fill it, and the school board wants a system of distinct representatives which matches distinct candidates with positions they are qualified for. We can turn any such SDR problem into a matching problem in a bipartite graph. We construct the bipartite graph whose parts are the set of positions and the set of candidates; an edge from a candidate to a position means the candidate qualifies for the position. A maximum–sized matching from the candidates to the positions gives us a system of distinct representatives which fills as many positions as possible. If all the positions are matched with candidates, then we have a system of distinct representatives. We can use the matching algorithm of the last section to find such a system of distinct representatives.

In light of financial constraints, however, the school board might not find this solution particularly attractive. Each candidate may have some minimum salary he or she will accept and the school board may wish to fill as many jobs as possible at the lowest possible total cost. In other words, among all ways of choosing the set of people to fill the jobs, we wish to choose the set whose total salary is the smallest. Does this mean we must examine all sets of representatives of maximum size to choose the cheapest set? Fortunately, there is a much less time–consuming method.

The Greedy Method

The so-called "greedy method" of choosing a set of representatives is as follows. First, choose the cheapest representative possible. Next, choose the cheapest representative such that these two representatives can fill two positions. Once you have a certain set of k representatives for k positions, choose as your next representative the cheapest possible representative so that this representative and the k previously chosen ones can fill $k + 1$ positions among them. It turns out that if we keep this process up until it is no longer possible to find a new representative to add, then we will end up with a set of representatives which is both as large as possible *and* as cheap as possible among all maximum–sized sets of distinct representatives.

Notice that the method concentrates on the choice of the representatives and not on the assignment of the representatives to positions. For this reason we introduce new terminology to emphasize our concentration on representatives alone. In particular, given a bipartite graph with parts V_1 and V_2 we say that a subset I of V_1 is *independent* for matchings of V_1 into V_2 if there is a matching which matches all the elements of I to elements of V_2. In the same way, given a family $\mathbf{F}$ of subsets of a set X, we say that a subset I of X is an *independent* set of representatives for $\mathbf{F}$ if the elements of I may be matched with sets in $\mathbf{F}$ so that the elements of I (in some order) form a system of distinct representatives for the sets with which they are matched.

We want to build maximum–sized independent sets by continually adding the cheapest element possible to the set we already have. How do we know that we will in fact have a *maximum-sized* independent set when the process stops and we can no longer find representatives to add? In other words, when we can no longer add representatives to the particular independent set we have, how do we know there isn't some other independent set that is larger? Theorem 2.1 assures us that we will get a maximum–sized independent set.

> *Theorem 2.1.* The independent sets for matchings of V_1 into V_2 in bipartite graphs and the independent sets of representatives for a family $\mathbf{F}$ of subsets of a set X satisfy the rule:
>
> *Expansion Rule.* If I and J are independent sets and $|I| < |J|$, then there is an element x of J such that $I \cup \{x\}$ is independent.
>
> *Proof.* We prove the theorem for independent subsets for matchings of V_1 into V_2 in a bipartite graph G. Because of the relationship between matchings and systems of distinct representatives, the same result will hold for independent sets for a family $\mathbf{F}$ of subsets of a set X.
>
> Suppose M_1 is the matching of I into V_2 and M_2 is the matching of J into V_2. As in Theorem 1.1, let G' be the graph on $V_1 \cup V_2$ with edge set

$$E' = (M_1 \cup M_2) - (M_1 \cap M_2) = (M_1 - M_2) \cup (M_2 - M_1).$$

From Theorem 1.1 we know that the connected components of G' fall
under one of three types. Since $|M_2| > |M_1|$, at least one of these con-
nected components must be a type 3 component with more edges from M_2
than from M_1. Thus there is an alternating path (for both M_1 and M_2)
whose first and last edges are in M_2. Each vertex of this path touched by
an M_1 edge is touched by an M_2 edge. Also one V_1 vertex x (and one V_2
vertex) touched by an M_2 edge is not touched by an M_1 edge. Let E' be the
edge set of the alternating path. By Theorem 1.2, the set of edges

$$(M_1 - E') \cup (E' - M_1) = M'$$

is a matching with one more edge than M_1. Thus M' is a matching of $I \cup$
$\{x\}$ into V_2. Thus $I \cup \{x\}$ is independent and by our choice of x, we know
that $x \in J$. ∎

To use our greedy method to build up a set of independent representatives,
we need some independent set of representatives to start with. The subset rule
below simply formalizes something we already know about independent sets and
says we can start with the empty set.
   Subset Rule.   Every subset of an independent set (including the empty set)
is independent.
   In our problem, we have a cost C_x associated with each element x of our set
X of candidates, and we wish to find a set of distinct representatives such that the
sum of the costs of these representatives is a minimum. In the language we have
introduced, we can give the greedy method the following precise description.

The Greedy Algorithm

Greedy Algorithm

Step 1. Let $I = \emptyset$.
Step 2. From the set X, pick an element x with minimum cost.
Step 3. If $I \cup \{x\}$ is independent, replace I by $I \cup \{x\}$.
Step 4. Delete x from X.
Step 5. If X is not empty, return to Step 2.

   The greedy algorithm works to select a maximum sized independent set of
least cost because of the expansion and subset properties. To see why, we study
these properties in isolation by introducing the concept of the matroid.

Matroids Make the Greedy Algorithm Work

A *matroid* on a set X is a collection **C** of subsets of X (called independent subsets
of X) satisfying the subset rule and expansion rule stated previously. (Note that

we are ignoring the family **F** and concentrating instead on the collection **C** of independent sets. Though matching problems provide one example of a matroid, we shall see other examples later.) A set of maximum size in the collection **C** is called a *basis*. A *cost function* f is simply a numerical function whose domain consists of all subsets of **X**, whose value on the empty set is zero and whose value on $\{x_1, x_2, \ldots, x_k\}$ is $\sum_{i=1}^{k} f(\{x_i\})$. In other words, f is defined on subsets by summing its values on single elements.

Theorem 2.1 may be reformulated to state that the independent subsets for matchings of a bipartite graph (with $X = V_1$) and the independent sets of representatives of a family **F** of subsets of X are the independent sets of a matroid on X.

Theorem 2.2. Suppose we are given a matroid on X and a cost function on the subsets of X. The greedy algorithm selects a basis of minimum cost (a basis such that the cost function evaluated on this basis is no more than the cost function evaluated on any other basis).

Proof. One consequence of the expansion rule is that the greedy algorithm will be able to continue adding to I until it reaches a maximum–sized independent set. Thus it selects a basis. Suppose it selects the basis B and B' is another basis. List the elements of B as $(b_1, b_2, \ldots, b_k)$ and the elements of B' as $(a_1, a_2, \ldots, a_k)$ in order of increasing cost, i.e., so that if $i < j$, then the cost of b_i is less than or equal to the cost of b_j and the cost of a_i is less than or equal to the cost of a_j. By Step 2 of the greedy algorithm, cost$(b_1) \le$ cost(a_1). If cost$(b_i) \le$ cost(a_i) for all i, then the cost of B is no more than that of B', so assume that for some i, but for no previous i, cost$(a_i) <$ cost(b_i). Then the two sets

$$\{a_1, a_2, \ldots, a_i\}$$

and

$$\{b_1, b_2, \ldots, b_{i-1}\}$$

are both independent. Then, by the expansion property, for some a_j with $j \le i$,

$$\{b_1, b_2, \ldots, b_{i-1}, a_j\}$$

is independent. Since cost$(a_j) \le$ cost$(a_i) <$ cost(b_i), a_j would have been selected by the greedy algorithm. Thus the cost of b_i is no more than the cost of a_i and therefore B is a minimum–cost basis. ∎

If we get a k–element set when we apply the algorithm to find a minimum–cost basis B for the matroid of sets independent relative to a family $\mathbf{F}$ of k subsets of X, then B will be a complete set of independent representatives. Thus we will know our family has an SDR. If B has fewer than k elements, then there is no independent set of representatives with k elements and so there is no SDR.

Frequently we will know the size of a basis in our application; in this case, we can change Step 5 of the greedy algorithm to

Step 5'. If I is not a basis, return to Step 2.

How Much Time Does the Algorithm Take?

The algorithm can be easily programmed for a computer once we have a representation of the elements of X in a form appropriate for our computer language and once we have a test for independence. When we study algorithms, we are faced with the practical question, "How many steps will our algorithm use?" Conceivably, even with the new Step 5', we would have to examine each element of X in Steps 2 and 3 before getting a maximum–sized I. Further we will need to pick the lowest cost element of X each time, and so will either have to arrange the elements of X in increasing order of cost before starting or search for the lowest cost x each time. Finally, we won't know how many steps are involved until we know how many steps it takes to test $I \cup \{x\}$ for independence. This will vary from application to application.

From Theorems 1.6 and 2.1, it is clear that a set $I \cup \{x\}$ will be independent if and only if each j–element subset lies in at least j sets in the family F. Further since we already know that I is independent, we need only check subsets that contain x. This is a straightforward way to proceed, but since our eventual SDR (if there is one) contains k elements, we will be making 2^k checks, one for each subset of the set of distinct representatives.

Since the time needed to use the algorithm is proportional (more or less) to the number of steps, and we have 2^k checks to make, we say this implementation of the algorithm is exponential in k. Suppose we also search for the lowest cost x each time; then potentially we must examine

$$n + (n - 1) + (n - 2) + \ldots + 1 = \frac{(n^2 - n)}{2}$$

possibilities for x. Thus we say the algorithm is quadratic in n. Since an algorithm which requires 2^k steps would be quite time–consuming, we would naturally like to improve on the time needed for the independence test.

By thinking in terms of matchings instead of representatives of a family of sets, we can develop a considerably improved implementation of the greedy algorithm for this application. Recall that we say a subset I of V_1 in a bipartite graph with parts V_1 and V_2 is independent if and only if there is a matching of I into V_2. Suppose we not only keep track of I as we apply the greedy algorithm but also, at each step, record a matching $M(I)$ of I into V_2.

We can modify the proof of Theorem 1.2 to show that given an element x not matched by $M(I)$, there is a matching of $I \cup \{x\}$ into V_2 if and only if there is an alternating path joining x to another vertex unmatched by $M(I)$. We can use inspection to find such a path or we can search for such a path by constructing an alternating search tree centered at x. From this alternating path, we write down the matching $M(I \cup x)$. Thus if we keep track of $M(I)$ as well as I, the number of steps required to check $I \cup \{x\}$ for independence is no more than the number of edges of the graph. Thus the number of steps required by the greedy algorithm is no more than the number of edges times the number of vertices in the graph. The number of edges is less than n^2 if n is the number of vertices in the graph, and so the number of steps involved in this version of the algorithm is approximately n^3. This is certainly superior to an algorithm that is exponential in k. Often we can find a matching by examining a picture; thus in most cases when we apply the greedy algorithm to construct a set of independent representatives by hand, we will not actually use a search tree.

Finding Minimum–Cost Independent Sets for a Matching Using the Greedy Algorithm

Example 2.1. Apply the greedy algorithm to find a minimum–cost independent subset of maximum size in the graph of Figure 5.7. The costs of vertices in V_1 are shown below the labels of these vertices. We begin with $I = \{g\}$, $M(I) = \{\{g, b\}\}$. Note we could let $M(I)$ be $\{\{g, c\}\}$ just as well. The next cheapest member of V_1 is i, so we let $I = \{g, i\}$, $M(I) = \{\{g, b\}, \{i, c\}\}$. Next we wish to add j to I; since we can match j with e, we let $i = \{g, i, j\}$ and $M(I) = \{\{g, b\}, \{i, c\}, \{j, e\}\}$,

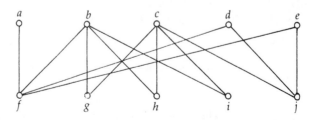

Figure 5.7

Now we ask if $I \cup \{h\}$ is independent. Since h can only be matched with b and c and they are already used, we will either have to change our matching or else we won't be able to use h. You may wish to check that there is no alternating path connecting h with an unmatched vertex. In this case, however, it is simpler to note that the three vertices g, h and i are connected to just the two vertices b and c, so there is no way to match these to three distinct vertices in V_2. (In other words, $|R(\{g, h, i\})| < |\{g, h, i\}|$.) Thus $I \cup \{h\}$ is not independent.

Now $I \cup \{f\}$ is independent because $\{f, a\}$ is an edge without vertices in common with edges of $M(I)$. Thus we may let $I = \{g, i, j, f\}$ and $M(I) = \{\{g, b\}, \{i, c\}, \{j, e\}, \{f, a\}\}$. Then I is the basis for the "matching matroid" chosen by the greedy algorithm and costs $2 + 3 + 4 + 6 = 15$. Note that there are many matchings of I into V_2; there is nothing special about the particular choices we wrote down as we went along. ■

The matroid we use in studying independent sets for a matching is called a "matching matroid" or a "transversal matroid". The matroid itself consists of the set V_1 and the independent subsets of V_1. In Example 2.1 we saw several independent sets of our matroid but not all of them. For example, $\{f, h, i, j\}$ is independent as well. Can you find any other independent sets of size 4?

The Forest Matroid of a Graph

The greedy algorithm has many applications aside from the determination of lowest–cost sets of distinct representatives. That is one reason for stating it in terms of matroids; to find whether we can use the algorithm in another situation, we analyze the situation to see if it has a matroid interpretation. In our introduction of spanning trees for graphs, one practical problem was to determine a minimum–cost spanning tree. Recall that in the problem we have a communications network connecting various cities. We know the cost of using the communication line from city i to city j (if there is such a line) and we wish to construct a system without redundant links that links together all the cities at minimum cost. Such a system is a spanning tree with the sum of the costs of its edges having minimum value.

Based on our solution to the problem of a set of representatives of lowest cost, it seems natural to try to build our spanning tree one edge at a time. If we always choose the cheapest edge, we might first take the Concord–Boston edge, then the Miami–Atlanta edge, etc. This kind of set of edges won't look like a tree while we are building it, but so long as we keep building bigger and bigger forests of edges, eventually the pieces will join together to form a tree. A set F of edges in a graph $G = (V, E)$ is called a *forest of* G if $F \subseteq E$ and (V, F) is a forest.

The cost of an edge is given, and the cost of a set of edges is the sum of the costs of its elements. Thus we have a cost function, as defined before Theorem 2.2. To apply Theorem 2.2, we need a matroid.

Theorem 2.3. The edge sets of forests of a graph $G = (V, E)$ form the independent sets of a matroid on E.

Proof. If (V, F) has no cycles, then (V, F') has no cycles for any subset F' of F, and so the forests satisfy the subset rule. Recall that a tree is a connected graph with k vertices and $k - 1$ edges. Thus a forest on v vertices with c connected components will consist of c trees and thus will have $v - c$ edges. Suppose F' and F are forests with r edges and s edges with $r < s$. If no edges of F can be added to F' to give an independent set, then adding any edge of F to F' gives a cycle. In particular, each edge of F must connect two points in the same connected component of (V, F'). Thus each connected component of (V, F) is a subset of a connected component of (V, F'). Then (V, F) has no more edges than (V, F'), so that $r \geq s$, a contradiction. Therefore the forests of G satisfy the expansion rule, so the collection of edge sets of forests of G is a collection of independent sets of a matroid on E. ∎

Minimum–Cost Spanning Trees

Since a maximum–sized forest in a connected graph must be a tree (if it had two connected components, you could join them without adding any cycles), we get

Theorem 2.4. The greedy algorithm applied to cost–weighted edges of a connected graph produces a minimum–cost spanning tree.

Example 2.2. The greedy algorithm applied to the graph in Figure 5.8 yields the following sequence of results.

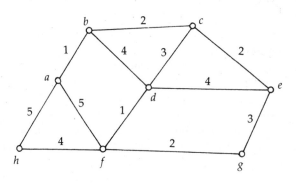

Figure 5.8

$\{\{a, b\}\}$; $\{\{a, b\}, \{d, f\}\}$; $\{\{a, b\}, \{d, f\}, \{f, g\}\}$; $\{\{a, b\}, \{d, f\}, \{f, g\}, \{b, c\}\}$; $\{\{a, b\}, \{d, f\}, \{f, g\}, \{b, c\}, \{c, e\}\}$; $\{\{a, b\}, \{d, f\}, \{f, g\}, \{b, c\}, \{c, e\}, \{c, d\}\}$; $\{e, g\}$: REJECT $\{d, e\}$: REJECT; $\{\{a, b\}, \{d, f\}, \{f, g\}, \{b, c\}, \{c, e\}, \{c, d\},$ $\{h, f\}\}$.

The second variant of the greedy algorithm stops with this set of seven edges since a spanning tree of a graph on eight vertices can have only seven edges. The first variant would have continued testing and rejecting edges until all edges had been tested. ■

In order to analyze the efficiency of the algorithm, we would have to specify how the test for independence would work. We can test a set $S \cup \{e\}$ of edges for "cyclicity" by the following method. As S grows, make a connectivity table for S by placing a 1 in position i, j of a matrix if v_i and v_j are connected by a path in S. Before adding e to S, check the table to see if e connects two vertices already connected in S, and if this is the case, reject it. Otherwise, to update the table use the fact that e connects two vertices v_j and v_k and thus anything previously connected to v_j is now connected to v_k and vice–versa. Now the table is ready to test another e.

Alternatively, we can keep a list in which we put a 1 in position i if v_i is already in the tree. Then to test edge $\{v_i, v_j\}$ we check to see if there are already 1's in positions i and j. If not we may add this edge. Is one of these two methods preferable?

The concept of a matroid is quite important in modern combinatorial mathematics. There are many examples of matroids other than the two given here; for applications, the matching matroids and forest matroids are among the most important.

EXERCISES

1. Is $\{a, b, c\}$ an independent set of representatives for $(\{a, b\}, \{a, b, c\}, \{a, c\}, \{a\})$?

2. Is $\{a, b, c\}$ an independent set of representatives for $(\{b\}, \{a\}, \{a, b, c\})$?

3. Is $\{a, b, c\}$ an independent set of representatives for $(\{a, b, c\}, \{a, b\}, \{a, e\})$?

4. Can a set of six edges be independent in the complete graph on six vertices?

5. Does $\{1, 2, 5, 8\}, \{2, 3, 4, 7\}, \{1, 5, 8, 6\}, \{2, 5, 6, 8\}, \{1, 3, 7, 4\}, \{1, 2, 5, 6\}, \{1, 2, 6, 8\}, \{2, 8, 5, 6\}, \{1, 2, 5, 8\}$ have a system of distinct representatives?

6. Find a maximum–sized independent set of representatives for the family of Exercise 5.

7. Let i be the cost of using i to represent a set in Exercise 5. Find a minimum–cost basis of representatives for the family in Exercise 5.

8. Prove that if two subsets of a set X are independent, but neither is a proper subset of an independent set, then they have the same size.

9. Find a maximum–cost spanning tree in the graph of Figure 5.7.

10. Find all independent sets of size 4 in the set V_1 of Figure 5.6.

11. The graph of Figure 5.9 shows the minimum acceptable hourly wage of applicants for various construction jobs, and lists which applicants are qualified for (and willing to accept) which jobs. Find a minimum cost set of job applicants who can fill each job.

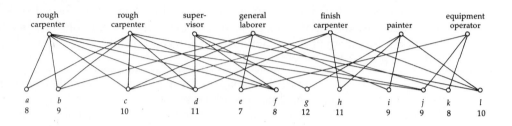

Figure 5.9

12. In how many ways may you match your set of applicants in your solution of Exercise 11 to the jobs?

13. Find a maximum–cost set of job applicants to fill each job of Exercise 11. (To see why this problem might make sense, think of the numbers as representing years of experience; then we are getting the most experienced crew.)

14. Find a minimum–cost spanning tree in the graph of Figure 5.10.

15. Find a maximum–cost spanning tree for the graph of Figure 5.10.

16. Explain why the following algorithm finds minimum–cost spanning trees. Choose a minimum–cost edge from among those which include an already chosen vertex

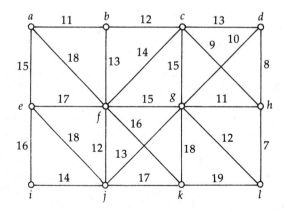

Figure 5.10

but which do *not* include two vertices connected by edges you have already chosen.   Continue until no further choices are possible.

   17. Let X consist of all points in ordinary three–dimensional space.   Define independence as follows.   A set $I \subseteq X$ is independent if it is empty, has one element, two elements, three elements which are not on the same line or has four elements which are not in the same plane.   Show that X together with these independent sets forms a matroid.

   18. Rewrite the proof of Theorem 2.2 to show that, given an element x not matched by $M(I)$, there is a matching of $I \cup \{x\}$ into V_2 if and only if there is an alternating path joining x to another vertex unmatched by $M(I)$.

   19. (For those who have had linear algebra.)   Show that the independent sets of vectors in a vector space form the independent sets of a matroid.

   20. (For those who have had linear algebra.)   Let X be the set of one–dimensional subspaces of a vector space and let a set I of subspaces be independent if the dimension of the subspace sum of I is $|I|$.   Explain why this gives a matroid.

   21. Let X be a finite set of size n.   Let $\mathbf{I}$ be the collection of subsets of X of size less than $\frac{n}{2}$.   Show that $\mathbf{I}$ is the family of independent sets of a matroid.

   22. Analyze the number of steps required in the tests for independence of edges of a graph given after Example 2.2.

   23. Write a program that takes for its data an adjacency matrix of a graph and a cost function and produces the adjacency matrix of a minimum–cost spanning tree (if such exists).

   24. Write a computer program that takes as data the incidence matrix of a family of subsets of a set and a cost function and produces a set of distinct representatives of minimum cost if one exists.   Modify the program so that it produces a minimum–cost system of distinct representatives.   (The incidence matrix of a family $\mathbf{F}$ has rows corresponding to the elements of X, columns corresponding to the sets of $\mathbf{F}$ and a 1 in row i, column j if $x_i \in F_j$.)

SECTION 3   NETWORK FLOWS

Transportation Networks

A *network* is a directed graph (or multigraph) which has numbers (usually called capacities) assigned to its edges.   Such a network might arise as a communications network in which the capacities represent the *number* of messages that can simultaneously be sent from vertex i to vertex j.   If our edges represent transportation routes (highways, railroads, etc.) between cities, the numbers on the edges might represent the capacity of the routes in terms of the number of vehicles per hour.   If the network represents a city water system with the vertices the pumping stations and the connections among them the water lines, the numbers might represent the volume of water per hour that can be moved through the pipes at standard operating pressures.   In many of these cases, as in the case of electrical networks in which the numbers represent conductances (which are the reciprocals of resistances), we might also consider networks based on graphs rather than digraphs.

A typical problem in such a network occurs when we have a *source* of material or messages destined for some destination (called a *sink* in the technical terms) and we want to know how much of this material can be sent from the source to the destination over some standard unit of time. For this reason, we call a network with two chosen vertices named the *source* and *sink* (or destination) a *transportation* network. In most practical situations, edges only leave the source and only enter the sink (because the material enters the network at the source and leaves the network at the sink); however, we need not assume this is always the case. In a transportation network, the numbers assigned to the edges are called *capacities;* $c(e)$ or $c(x, y)$ denotes the capacity of the edge $e = (x, y)$.

The Concept of a Flow

We think of material flowing along a transportation network from the source to the sink; thus what enters a vertex in a network must leave it—unless the vertex is the source or sink. For this reason, we define a *flow* in a network N with capacity function c, source s, and sink t and edge set E to be a function f defined on the edges of N that satisfies the conditions

(1) $0 \le f(e) \le c(e)$ for each edge e of n
(2) for each vertex $v \notin \{s, t\}$

$$\sum_{\substack{(x,v):\\(x,v)\in E}} f(x, v) - \sum_{\substack{(v,y):\\(v,y)\in E}} f(v, y) = 0.$$

The interpretation of (1) is clear; in words Condition 2 says that the flow into vertex v equals the flow out of vertex v. Because of our motivating problem, it is natural to ask how much material is flowing from the source to the sink. It turns out that the two expressions

$$F_s = \sum_{\substack{(s,v):\\(s,v)\in E}} f(s, v) - \sum_{\substack{(v,\ s):\\(v,s)\in E}} f(v, s)$$

and

$$F_t = \sum_{\substack{(v,t):\\(v,t)\in E}} f(v, t) - \sum_{\substack{(t,v):\\(t,v)\in E}} f(t, v)$$

are equal. The first difference is normally interpreted as the amount of material flowing from the source and the second is interpreted as the amount of material arriving at the sink. The common value of these two expressions is called the *value* of the flow and is interpreted as the amount of material flowing from the source to the sink.

Theorem 3.1.   In a network, the flow F_s leaving the source and the flow F_t arriving at the sink, are equal.

Proof.   We sum all the flow entering and leaving each vertex as below

$$\sum_{\substack{v \in V}} \sum_{\substack{(x,v): \\ (x,v) \in E}} f(x, v) - \sum_{\substack{(v,y): \\ (v,y) \in E}} f(v, y) = F_s - F_t.$$

Each term (other than F_s and F_t) cancels out because of Condition 2 in the definition of a flow. However, if we distribute the terms of the sum, we obtain

$$F_s - F_t = \sum_{v \in V} \sum_{(x,v) \in E} f(x, v) - \sum_{v \in V} \sum_{\substack{(v,y): \\ (v,y) \in E}} f(v, y)$$

$$= \sum_{(x,v) \in E} f(x, v) - \sum_{(v,y) \in E} f(v, y)$$

$$= 0$$

since both sums add up all the $f(x, y)$ for all edges (x, y) in E. Thus, $F_s - F_t = 0$ or $F_s = F_t$. ∎

A natural question to ask for a given network is, "How big can we make the flow—i.e., what is the maximum value of the flow?" Given a drawing of a network, imagine cutting the drawing with a pair of scissors without cutting through any vertex. Thus, the source ends up on one piece of paper and the sink on the other. If you further imagine a flow going along the edges of the network before it is cut, material will still be flowing out of edges directed away from the source after the cut, and in that case will leave the network. Intuitively, it seems that the amount going from the "source half" to the "sink half" is at least as much as the flow into the sink itself. This flow can be no more than the sum of the capacities of the edges leading from the "source half" to the "sink half," and so the value of the flow itself should be no more than the sum of these capacities. This intuition holds true, and we now make these ideas sufficiently precise to show why it holds true.

Cuts in Networks

A *cut* of a network with vertex set V is a partition of V into two (disjoint) sets such that the source is in one (called the "source set") and the sink is in the other (called the "sink set"). The *capacity* of the cut $\{S, T\}$, denoted by $c(S, T)$, is the sum of the capacities of the edges leading from a vertex in the source set to a vertex in the sink set. As our intuition suggests, given a flow on our network, the sum of the flow values along edges from the source set S of a cut $\mathbf{C}$ to the sink set T of $\mathbf{C}$ minus the flow in the reverse direction is the common value F of F_s and F_t.

Theorem 3.2. If $\mathbf{C}$ is a cut with source set S and sink set T, and F is the value of a flow f on the network, then

$$F = \sum_{\substack{(x,y)\in E: \\ x\in S, y\in T}} f(x, y) - \sum_{\substack{(y,x)\in E: \\ x\in S, y\in T}} f(y, x).$$

Proof. We construct a new network and flow by replacing all vertices in T with a single vertex w and construct an edge between w and a member of S (of the same capacity, in the same direction and with the same flow) for each edge between a member of T and a member of S. (This may give a multidigraph, but that is allowable in a network.) Then w is the sink in the new network, s is the source in the new network and F_s is unchanged, so $F_w = F$. However, F_w is simply the difference of the two sums above, so this difference must be F. ∎

Theorem 3.3. For every flow f with value F and every cut $\{S, T\}$ of the network, $F \le c(S, T)$.

Proof. Since $f(x, y) \le c(x, y)$ for each edge (x, y) leading from S to T, the first sum in the expression for F in Theorem 3.2 is no more than $c(S, T)$. Subtracting the second term can only make the expression smaller. But then $F \le c(S, T)$. ∎

Theorem 3.3 means that if for some cut, the value F of the flow f is the capacity $c(S, T)$, then the flow is as large as it can conceivably be. Thus the maximum flow value is no larger than the minimum of the capacities of all cuts. In fact, in each network there is a maximum flow whose value is equal to the minimum of the capacities of all cuts in the network. This result is called the "max–flow min–cut theorem."

The max–flow min–cut theorem will follow from the fact that any flow which has a value less than the minimum cut capacity can be modified to give a larger value. Thus, given a flow in one network, we may ask whether and how we can improve it to get more flow from the source to the sink. Intuitively, there are two things we could do to make more material move towards the sink.   If an edge is not being used to capacity, we could try to send more material through it; if an edge is working against us by sending material back towards the source, we could try to reduce the flow along this edge and redirect it in a more practical direction. These two ideas are combined in the concept of a *flow–augmenting path*.   To give this idea a concrete definition, we assume there is already a flow f through our network.

Flow–Augmenting Paths

An undirected path (or semipath) connecting s and t is an alternating sequence of vertices and edges that form a path from s to t in the underlying graph.   Thus an undirected path is an alternating sequence $s = v_0 e_1 v_1 e_2 v_2 \ldots \ldots v_{n-1} e_n v_n = t$ of vertices and edges such that $e_i = (v_{i-1}, v_i)$ or else $e_i = (v_i, v_{i-1})$. If $e_i = (v_{i-1}, v_i)$, we say e_i is a *forward edge* in the path; if $e_i = (v_i, v_{i-1})$, we say e_i is a *reverse edge*. In Figure 5.11, the edges (v, x) and (x, y) are forward edges but the edges (z, y) and (w, z) are reverse edges.   Edges of the graph not in our undirected path are shown as dashed lines.   They don't get either the name forward or reverse.   We say an undirected path from s to t is a *flow–augmenting path* of *worth* $w > 0$ if adding w to $f(e_i)$ for each forward edge e_i and subtracting w from $f(e_i)$ for each reverse edge e_i yields a new flow f'.   (Note that by definition the new function f' must be a flow for the path to be called flow–augmenting.)

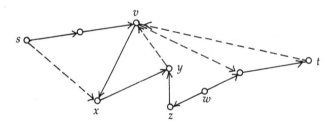

Figure 5.11

Theorem 3.4.   If f is a flow of value v in a network and if the flow f' is obtained from f by adding w to each forward edge of a flow–augmenting path and subtracting w from each reverse edge of this flow–augmenting path, then f' is a flow of value $v + w$.

Proof.   By assumption, f' is a flow.   Note that $F'_s = F_s + w$.   Thus f' has value $v + w$. ∎

The Labelling Algorithm for Finding Flow–Augmenting Paths

Our theorem tells us that the problem of improving a flow may be solved by finding a flow–augmenting path. In fact, we shall see that as big a flow as possible may be constructed by starting with the all–zero flow f_0, augmenting it with a flow–augmenting path to get a flow f_1, and repeating this process until we find a flow f_n that has no flow–augmenting path. For these reasons, we develop an algorithm to find a flow–augmenting path, if one exists, or to show that none exists otherwise. Our algorithm is a variant of the Ford–Fulkerson labelling algorithm. Vertices that might be useful in a flow–augmenting path are labelled in such a way that we can use the labels to determine the path. In essence, we are using a breadth–first search tree, placing edges in the tree only if they are useful for increasing the flow. We start at the source and search until we reach the sink or run out of edges.

The steps of the labelling algorithm are:

(1) Assign to the vertex s the label "source". Make s the first element of a list L (for labelled).
(2) Let u be the first element of the list L.
    (a) For each unlabelled vertex v such that (u, v) is an edge and $f(u, v) <$ $c(u, v)$, assign v the label (u, v) and place v at the end of the list L.
    (b) For each unlabelled vertex v such that (v, u) is an edge and $f(v, u) > 0$, assign v the label (v, u) and place v at the end of the list L.
(3) Remove u from the list L.
(4) Repeat Steps 2 and 3 until the list L is empty.

We will complete the process when (or before) all vertices of the network have been labelled. A vertex cannot have more than one label assigned to it, and different vertices are assigned different labels. Note that we have labelled vertices v with edges (u, v) if the flow from u to v can potentially be increased and have labelled vertices u with edges (u, v) if the flow from u to v can potentially be decreased. These potential increases and decreases will be important when we wish to choose the forward and reverse edges of a flow–augmenting path. Before we try to construct such a path, it would be nice to know if we already have a flow whose value equals the minimum–cut capacity. Suppose the sink is *not* labelled. Then the partition $\{P, Q\}$ of the set V in which P consists of labelled vertices and Q consists of unlabelled ones is a cut. By Step 2a of our algorithm, $f(u, v) = c(u, v)$ for each u in P and v in Q such that (u, v) is an edge. (In words, our flow uses each edge leading from P to Q to full capacity.) By Step 2b of our algorithm, $f(v, u) = 0$ for each v in Q and u in P such that (v, u) is an edge. Thus the value of the flow equals the capacity of this cut $\{P, Q\}$. We have just proved a theorem that characterizes flows of maximum value.

Theorem 3.5. A flow f has maximum possible value (i.e., has value equal to the minimum–cut capacity of the network) if, after the labelling algorithm is complete, the sink remains unlabelled.

As has already been suggested, if the sink is labelled, we may choose a flow–augmenting path. The construction works backwards through the search tree from the sink to the source as follows. (Recall that t is the sink and s is the source.)

(1) Let $u_0 = t$.
(2) Given a vertex u_i assigned the edge e as label, let u_{i+1} be the other vertex in e.
(3) If $u_{i+1} = s$, then stop; otherwise repeat Step 2.

Now suppose we have vertices $u_0, \ldots u_n$. Since they "range" from the sink to the source, these are in reverse order from the way we want to send material. Therefore, we let $v_0 = u_n$, and in general $v_i = u_{n-i}$. Let e_i be the edge e we used to choose v_{i-1} given v_i. Note that if $e_i = (v_{i-1}, v_i)$ (i.e., if e_i is a forward edge), it has been encountered in Step 2a of the labelling algorithm, and if $e_i = (v_i, v_{i-1})$ (i.e., e_i is a reverse edge), it has been encountered in Step 2b of the labelling algorithm. Thus, if $e_i = (v_{i-1}, v_i)$, then $f(e_i) < c(e_i)$ and if $e_i = (v_i, v_{i-1})$, then $f(e_i) > 0$. Let w_1 be the minimum value of $c(e_i) - f(e_i)$ for all forward edges $(v_{i-1}, v_i) = e_i$, and let w_2 be the minimum value of $f(e_i)$ for all reverse edges $(v_i, v_{i-1}) = e_i$. Now let w be the smaller of w_1 and w_2. This description of w is equivalent to the symbolic description of w in the statement of the following theorem.

Theorem 3.6. Let

$$w = \min(\min\{c(e_i) - f(e_i)|e_i \text{ is forward}\}, \min\{f(e_i)|e_i \text{ is reverse}\}).$$

Then $v_0 e_1 v_1 \ldots e_n v_n$ is a flow–augmenting path of worth w.

Proof. Note that $f(e_i) + w \leq c(e_i)$ for each forward edge e_i and that $f(e_i) - w \geq 0$ for each reverse edge e_i. To show that adding w to each flow in a forward edge and subtracting w from each flow in a reverse edge gives a flow f', we must still show that for each vertex $v \notin \{s, t\}$,

$$\sum_{\substack{(x,v): \\ (x,v) \in E}} f'(x, v) - \sum_{\substack{(v,y): \\ (v,y) \in E}} f'(v, y) = 0.$$

However, each $f'(e)$ is equal to $f(e) + w$ or else $f(e) - w$, and f is a flow, so we need to show only that the w's cancel out. In particular, we only need to check vertices v on our flow–augmenting path, because otherwise f'

equals f. Further, only along two edges e_i and e_{i+1} touching v_i do we change the value of f. Now if e_i and e_{i+1} are both forward or both reverse, one of the w's will occur in the left–hand sum and one in the right–hand sum, so they cancel out. If either e_i or e_{i+1} is forward while the other is reverse, then the two occurrences of w are in the same sum but with opposite sign, so they cancel out. Therefore f' is a flow, and our path is a flow-augmenting path of worth w. ■

The Max–Flow Min–Cut Theorem

Note that if the initial flow and capacities are integral, then the increase w in the value F will be an integer. Thus we can continue using flow–augmenting paths to increase F until its value reaches a maximum. The same conclusion follows if all the capacities are rational numbers. This method will also produce a maximum flow if the capacities are arbitrary real numbers, but we shall not prove this. Thus we have the following theorem.

Theorem 3.7. The maximum value of a flow in a network is the minimum of the capacities of any of its cuts.

This theorem is called the "max–flow min–cut" theorem.

Example 3.1. Find a maximum flow in the network of Figure 5.12. The capacity of each edge is written beside the edge. We repeat the process of labelling and constructing flow–augmenting paths.

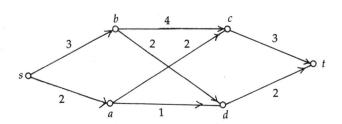

Figure 5.12

Stage 1 $f(x, y) = 0$ for all x, y.
    Label s: source, b: (s, b), a: (s, a), c: (b, c), d: (b,d), t: (c, t)
    Path t (c, t) c (b, c) b (s, b) s
    Find d   min $(3, 4, 3) = 3$
    Find f' $f'(s, b) = 3, f'(b, c) = 3, f'(c, t) = 3, f'(x, y) = 0$ for all other x, y.
Stage 2 $f(s, b) = f(b, c) = f(c, t) = 3; f(x, y) = 0$ for all other x, y.

  Label s: source, a: (s, a), d: (a, d), c: (a, c), t: (d, t)
  Path t (d, t) d (a, d) a (s, a) s
  Find d min $(2, 1, 2) = 1$
  Find f' $f'(s, a) = f'(a, d) = f'(d, t) = 1$
     $f'(s, b) = f'(b, c) = f'(c, t) = 3$
     $f'(x, y) = 0$ for other (x, y).
Stage 3 $f(s, a) = f(a, d) = f(d, t) = 1, f(s, b) = f(b, c) = f(c, t) = 3$
   $f(x, y) = 0$ for all other (x, y).
  Label s: source, a: (s, a), c: (a, c), b: (b, c), d: (b, d), t: (d, t)
  Path t (d, t) d (b, d) b (b, c) c (a, c) a (s, a) s
  Find d min $(2- 1, 2, 3, 2, 2 - 1) = 1$
  Find f' $f'(s, a) = 2, f'(a, d) = 1, f'(a, c) = 1$
     $f'(s, b) = 3, f'(b, c) = 2, f'(b, d) = 1$
     $f'(c, t) = 3, f'(d, t) = 2$.
Stage 4 $f(a, d) = f(a, c) = f(b, d) = 1, f(s, a) = f(b, c) = f(d, t) = 2,$
   $f(s, b) = f(c, t) = 3$
  Label s: source, no more labelling is possible.

  The value of the flow shown in Stage 4 is 5; the cut with $S = \{s\}$ and $T = \{a,$ $b, c, d, t\}$ has cut capacity 5. ■

EXERCISES

Exercises 1–8 deal with the digraph in Figure 5.13.
  1. If $f(e) = 1$ for all edges e of Figure 5.13, is f a flow?
  2. Suppose $f(s, c) = f(s, b) = f(s, a) = 1$ and $f(d, a) = 0$. How should $f(e)$ be defined for other edges to achieve a flow?
  3. What is the value of the flow of Exercise 2?
  4. If all the edges of Figure 5.13 have capacity 2, locate at least seven cuts of capacity 6.

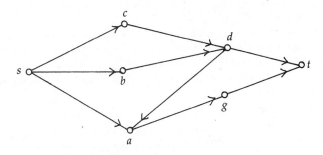

Figure 5.13

5. Using the flow of Exercise 2 and capacities of Exercise 4, apply the labelling algorithm to Figure 5.13. Find a flow–augmenting path.

6. Find a maximum flow in Figure 5.13 using the capacities of Exercise 4.

7. If all the edges of Figure 5.13 have capacity 1, what is the maximum value of a flow? Find a flow with this value.

8. Suppose $c(d, t) = c(g, t) = 3$ and all other edges have capacity 2 in Figure 5.13. Find a maximal flow and a cut whose capacity is the value of the flow.

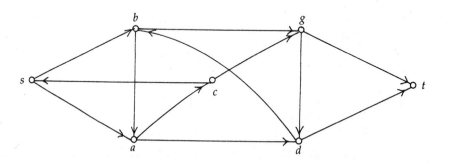

Figure 5.14

9. In Figure 5.14, suppose all edges touching s have capacity 3, all edges touching g have capacity 2 and $c(d, t) = 6$, $c(a, c) = c(a, d) = 8$, $c(d, b) = 4$ and $c(b, a) = 2$. Determine a maximum flow.

10. In Figure 5.14 with the capacities as in Exercise 9, given a flow with $f(s, a) = 3$, $f(s, b) = 3$, $f(c, s) = 3$, $f(a, c) = 3$, $f(b, g) = 2$, $f(g, d) = 2$, $f(d, t) = 3$, $f(b, a) = 1$ and $f(a, d) = 1$, find the flow in all the remaining edges. Apply the labelling algorithm to construct a maximal flow from f by means of flow–augmenting paths.

11. Is there a way to assign capacities to the edges of Figure 5.14 such that all cuts have the same capacity and the network has a flow that is positive?

12. Redirect the edge (d, b) to be (b, d) instead in Figure 5.14. Let $c(a, c) = c(s, b) = 4$, $c(b, a) = c(b, g) = 1$, $c(d, t) = c(c, s) = c(s, a) = 3$, $c(c, g) = c(g, t) = 5$ and $c(b, d) = c(a, d) = c(g, d) = 2$. Find a maximum flow.

13. Explain why the labelling algorithm assigns a label to each vertex at most once and labels different vertices differently.

14. Write a computer program which produces a maximum flow in a network given the $n \times n$ matrix C given by

$$C_{ij} = \begin{cases} 0 \text{ if there is no edge from } i \text{ to } j \\ c(e) \text{ if } e \text{ is the edge from } i \text{ to } j. \end{cases}$$

Assume vertex 1 is the source and vertex n is the sink.

15. Work out the changes needed to find a maximum flow from a source to a sink in an undirected network, that is a graph (multigraph) with capacities assigned to the edges and a chosen source and sink.

SECTION 4 APPLICATIONS OF FLOWS TO CONNECTIVITY AND MATCHING

Connectivity and Menger's Theorem

The concept of maximum flow in a network has applications in a wide variety of problems that do not seem to be transportation problems. The first application we will discuss is the concept of n–connectivity in a graph. We say u is connected to v in a digraph if there is a directed path from u to v. Often there will be many paths from u to v. We say that u is n–edge–connected to v if there is a set of n (directed) paths from u to v such that no two of these paths have any edges in common. (Also u is n-edge connected to itself.) A set of edges is called a u–v cutset (or u–v edge–cutset) if deleting these edges from the edge set of the digraph yields a digraph in which u is not connected to v. It is clear that if u is n–edge–connected to v, then a u–v cutset will have to have at least n edges. (Why?)

The idea of a u–v cutset seems analogous to the idea of a cut in a network with source u and sink v. To be precise, removing the cutset edges from the edge set D of the digraph leaves a digraph with at least two strongly connected components, one of which contains u and the other v. Thus a u–v cutset gives a cut for any network whose digraph has edge set D, source u and sink v. Now given a digraph (V, D), we define a network by letting each edge have capacity 1. Then if u and v are n–edge–connected, we can use the n "edge–disjoint" paths to send n units of flow from u to v. The max–flow min–cut theorem then tells us that if a cut has capacity n, then n is the maximum flow that can be achieved. From such a flow we shall see that we get n "edge–disjoint" paths. This leads to one form of a result known as Menger's Theorem.

Theorem 4.1. If a minimum–sized u–v cutset in a digraph (V, D) has n edges, then u is n–edge–connected to v.

Proof. As above, we define a network on (V, D) by letting each edge have capacity 1. Then if a minimum cut in the network has capacity n, there are n edges whose removal disconnects u from v. Suppose a set S of edges is a minimal u–v cutset, so that if all but one edge in S is removed from D, there is a path from u to v, but if all the edges in S are removed from D, there is no path from u to v. Suppose U is the set of all vertices x such that there is a path from u to x in $(V, D - S)$, and let **C** be the partition $\{U, V - U\}$. Then **C** is a cut in the network, and the edges in S are the only edges that lead from an element of U to an element of $V - U$, so the capacity of C is the size of S. Thus if a minimum–sized u–v cutset has n edges, a minimum cut of the network will have capacity n.

From the max–flow min–cut theorem, it follows that there is a flow of value n, and that all its edge values are integers. We now prove by induction that this means there are n edge–disjoint paths from u to v. Let S be the set of all n such that if there is an integral flow of value n, then there are n edge–disjoint paths from u to v. Clearly, 1 is in S; assume $k - 1$ is in S and that the flow has value k. There is a path from u to v each edge of which has flow and capacity 1. Thus, the remaining $k - 1$ units of flow go entirely through edges not in this path, so that deleting the edges of the path gives a flow of value $k - 1$ which, since $k - 1$ is in S, yields $k - 1$ edge–disjoint paths. These paths, together with the deleted path, form a set of k paths no two of which have an edge in common. Thus by version 1 of the principle of mathematical induction, S contains all the positive integers and the theorem is proved. ∎

The original form of Menger's Theorem uses the idea of a u–v vertex cutset. A u–v vertex cutset is a set S of vertices of (V, D) such that if S is removed from V and all edges incident with (i.e., containing) a vertex in S are removed from D, then u and v are in the resulting digraph, but there is no path from u to v. The vertex form of Menger's Theorem says that if a minimum u–v vertex cutset has n vertices, then there is a set of n paths from u to v such that no two of the paths have a vertex (other than u or v) in common. There is a way to derive this theorem from Theorem 4.1; namely, for each vertex x of (V, D) other than u and v, replace x by two vertices x_1 and x_2 such that (using $u_1 = u_2 = u$ and $v_1 = v_2 = y$)

(1) (x_1, x_2) is an edge.
(2) (w_2, x_1) is an edge if and only if (w, x) is an edge.
(3) (x_2, w_1) is an edge if and only if (x, w) is an edge.

By thus "splitting" the vertices, we are replacing each vertex–cutset by an edge–cutset of the same size; applying Theorem 4.1 to this new network provides a set of paths having none of the new edges in common; this set of paths may be translated to a set of paths in the original graph with no vertices in common. (This fact may seem clear; technically, though, it requires proof. To prove it, note that if two paths translate to paths with a vertex x in common, you can show that these paths have the edge (x_1, x_2) in common.)

There are also two versions of Menger's Theorem (a vertex form and an edge form) for ordinary (undirected) graphs. The definition of an edge–cutset in a graph is the same as the definition of an edge–cutset in a digraph. One version of Menger's Theorem for graphs is

Theorem 4.2. If a minimum–sized u–v edge–cutset in a graph (V, E) has n edges, then there are n paths from u to v such that no two of these paths have an edge in common.

Proof.   From the graph (V, E), construct a digraph (V, D) with (x, y) and (y, x) in D if and only if $\{x, y\}$ is in E.   Then a $u-v$ edge–cutset of size n in (V, E) gives a $u-v$ edge–cutset of size $2n$ in (V, D).   However, we may delete half of these edges, after which we have no directed path from u to v. (Why?).   Thus we have a $u-v$ edge–cutset in (V, D) of size n, (but there is no smaller cutset) so there are n directed paths from u to v in (V, D) such that no two of them have an edge in common.

Now suppose one path uses the edge (x, y) and another path the edge (y, x).   Thus (as shown in Figure 5.15 geometrically), one path is

$$u = x_0 e_1 x_1 \ . \ . \ . \ . \ x_{k-1}(x_{k-1}, x)x(x, y)y(y, x_{k+2})x_{k+2} \ . \ . \ . \ v$$

and the other is

$$u = y_0 e_1' y_1 \ . \ . \ . \ y_{k-1}(y_{k-1}, y)y(y, x)x(x, y_{k+2})y_{k+2} \ . \ . \ . \ v.$$

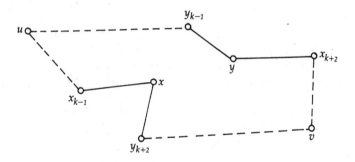

Figure 5.15

We can replace these two paths with paths that use neither (x, y) nor (y, x), namely

$$u = x_0 e_1 x_1 \ . \ . \ . \ x_{k-1}(x_{k-1}, x)x(x, y_{k+2})y_{k+2}, \ . \ . \ . \ v$$

and

$$u = y_0 e_1' y_1 \ . \ . \ . \ y_{k-1}(y_{k-1}, y)y(y, x_{k+2})x_{k+2}, \ . \ . \ . \ v.$$

Repeating this process (using induction), we can eliminate all simultaneous occurrences of (x, y) and (y, x) among the paths. Thus the corresponding paths in the original graph will have no edges in common as desired. ∎

There is a vertex form and an edge form of Menger's Theorem for multigraphs as well; we leave the exploration of these theorems to the reader.

Flows, Matchings, and Systems of Distinct Representatives

In our discussion of matchings and systems of distinct representatives, we only outlined algorithms for finding matchings or SDR's or minimum–cost matchings or SDR's. In contrast, our flow algorithm was given in a step–by–step process suitable for writing a computer program. The reason for waiting until flows to give such details is that all the algorithms we outlined are special cases of network–flow algorithms. In particular, alternating paths are special cases of flow–augmenting paths. We shall now show how flow algorithms may be applied to solve all our matching and SDR problems.

Recall that a *matching* from V_1 to V_2 in a bipartite graph on the vertex set $V_1 \cup V_2$ is a set of edges no two of which have an endpoint in common; a matching is called *complete* if each vertex of V_1 is included in it, and it is called a *maximum* (but not necessarily complete) matching if no other matching has more edges. A subset X of V_1 is *independent* if there is a matching of X into V_2 (or in other words, if there is a complete matching from X to V_2 in the bipartite subgraph whose vertex set is $X \cup V_2$). We use network flows to find matchings and test whether a matching of an independent set I into V_2 can be modified to give a matching of the set $I \cup \{x\}$ into V_2.

Given $G = (V_1 \cup V_2, E)$, we define a network N with vertex set $V_1 \cup V_2 \cup \{s\} \cup \{t\}$ (adding two new vertices s and t) by letting (s, x) be an edge for all x in V_1, (y, t) be an edge for all y in V_2 and (x, y) be an edge for all $\{x, y\}$ in E with $x \in V_1$ and $y \in V_2$. We assign capacity 1 to all the edges. Then a flow f whose value on each edge is either 0 or 1 will correspond to a matching; the edges in the matching are the edges of E whose directed counterparts have flow 1.

Example 4.2. Find a matching from the set $\{1, 2, 3\}$ into $X = \{1, 2\}$, $Y = \{1, 3\}$, $Z = \{1, 2, 3\}$. The network corresponding to this problem is shown in Figure 5.16. The process of constructing a maximal flow is outlined below.

Stage 1 $f(e) = 0$ for all e.

Vertex	1	2	3	X	Y	Z	t
Label	(s, 1)	(s, 2)	(s, 3)	(1, X)	(1, Y)	(1, Z)	(X, t)
Path	t, X, 1, s						
Flow	$f'(s, 1) = f'(1, X) = f'(X, t) = 1$; $f'(e) = 0$ all other e.						

Stage 2 $f(s, 1) = f(1, X) = f(X, t) = 1$; $f(e) = 0$ all other e.

Vertex	1	2	3	X	Y	Z	t

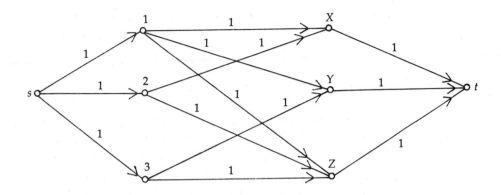

Figure 5.16

Label   (1, X) (s, 2) (s, 3) (2, X) (3, Y) (3, Z) (Z, t)
Path   t, Y, 3, s
Flow $f'(s, 1) = f'(1, X) = f'(X, t) = f'(s, 3) = f'(3, Z) = f'(Z, t) = 1$;
          $f'(e) = 0$ all other e.
Stage 3 $f(s, 1) = f(1, X) = f(X, t) = f(s, 3) = f(3, Z) = f(Z, t) = 1; f(e) = 0$ all
          other e.
Vertex   1   2   3   X   Y   Z   t
Label   (1, X) (s, 2) (3, Z) (2, X) (3, Y) (2, Z), (Y, t)
Path   t, Y, 3, Z, 2, s
Flow $f'(s, 1) = f'(s, 3) = f'(s, 2) = f'(1, X) = f'(3, Y) = f'(2, Z) =$
           $f'(X, t) = f'(Y, t) = f'(Z, t) = 1; f(e) = 0$ for all other edges e.
From this flow we see immediately that one matching is (1, X) (3, Y) (2, Z). ■

Minimum Cost SDR's

Now suppose we want to use the greedy algorithm to build a minimum–cost SDR
for the sets X and Z of Example 4.1, given the cost function (denoted by
upper–case C) $C(1) = 10$, $C(2) = 8$ and $C(3) = 6$. The first element we would
choose for our minimum–cost independent set of representatives would be 3.
Vertex 3 did not appear until Stage 2 of the flow construction process above, so
that process does not correspond to the greedy algorithm. Let us analyze how
else we might proceed. We know that $S \subseteq \{1, 2, 3\}$ is independent if and only if it
has a partial matching into $\{X, Z\}$, and we know $\{3\}$ has such a matching. We
could find one such matching by observation, or by applying the flow algorithm to
the network in Figure 5.17 to get the flow $f(s, 3) = f(3, Z) = f(Z, t) = 1; f(e) = 0$
for all other e. Now we ask whether we can add 2 to $S = \{3\}$ to get a
minimum–cost independent set of size 2. To find out, apply the flow algorithm to

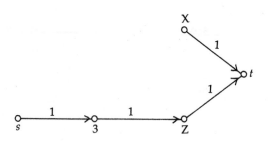

Figure 5.17

the network in Figure 5.18, starting with the flow f just constructed. In this case, we get the flow $f(s, 2) = f(s, 3) = f(2, X) = f(3, Z) = f(X, t) = f(Z, t) = 1$, $f(2, Z) = 0$, so our set $\{2, 3\}$ is an independent set of minimum cost *and* the list $(2, 3)$ is a *system* of distinct representatives for X and Z.

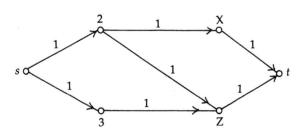

Figure 5.18

Although this example is so elementary that we could have constructed our minimum–cost SDR without the network flow algorithm, the method shown works in all situations.

Theorem 4.3.   Given a family $S_1, S_2, \ldots, S_n$ of subsets of a set X, the algorithm that follows selects a maximum–sized, minimum–cost independent set $S = \{x_1, x_2, \ldots, x_j\}$ of representatives and a matching $(x_1, S_{i_1}), (x_2, S_{i_2}), \ldots, (x_j, S_{i_j})$ with $x_k \in S_{i_k}$.

Step 1. Let $M = S = \emptyset$, and let D be the digraph with vertices s, t, S_1, $S_2, \ldots, S_n$ and edges (S_i, t). Let $f(e) = 0$ for each edge e.

Step 2. Choose a minimum–cost element x of $X - S$.

Step 3. Add the vertex x and the edges (s, x) and (x, S_i) for each S_i with

$x \in S_i$ to D. Let N be the network with digraph D and all edge capacities 1. Let $f(e) = 0$ for all new edges e just added to D.

Step 4. Apply a network–flow algorithm to the flow f on D to increase f to a maximum flow.

Step 5. If $f(x, S_i) = 1$ for some i, add x to S. Otherwise, delete x from X.

Step 6. If $|S| = k$, let $M = \{\{x, S_i\} \mid f(x, S_i) = 1\}$ and stop.

Step 7. If $|S| = |X|$, then stop, otherwise return to Step 2.

Proof. At each stage, we add x to S if and only if there is a maximum flow in the network with a flow of 1 out of each vertex in $S \cup \{x\}$; therefore we add x to S if and only if $S \cup \{x\}$ is independent. If the set M is nonempty when we stop, then it has the edges of a matching from the j–element independent set S into $S_{i_1}, S_{i_2}, \ldots, S_{i_j}$ with $x \in S_{i_k}$ for each $\{x, S_{i_k}\}$ in M. ∎

The matching M in Theorem 4.3 gives us a minimum cost *system* of distinct representatives. This would mean, for example, in the case of a school board wishing to match applicants to positions that the school board had not only a minimum–cost choice of qualified applicants but also an assignment of the chosen applicants to the positions. Note that Step 4 involves constructing one flow– augmenting path and then augmenting the flow. That process requires several steps, the number of which is proportional to the number of edges. Since we carry out the process only once for each x_i (after listing them in increasing order of cost), the number of steps needed is proportional to the product of $|X|$ and the number of edges. This product is no more than $|X|^2 m$.

Minimum–Cost Matchings and Flows with Edge Costs

A subtly different kind of problem of finding minimum–cost systems of distinct representatives or minimum–cost matchings is as follows. We have a family S_1, $S_2, \ldots, S_k$ of sets, a set $x_1, x_2, \ldots, x_m$ of representatives and the price p_{ij} it costs to use x_i to represent S_j. We want to find a system of distinct representatives of minimum total price. This problem was first considered on its own by Philip Hall who discovered the "marriage theorem"; however, the problem falls naturally into the class of network flow problems. Suppose we are given a network with not only a capacity $c(e)$ defined for each edge e, but also a price $p(e)$ defined for each edge e. Thus it costs $x p(e)$ to send x units of flow along edge e. Then the cost of a flow f on our network is the sum

$$\sum_{e \in E} f(e) p(e) = \text{cost}(f).$$

In this situation, we desire a maximum–valued flow of minimum cost–that is, a flow of maximum value whose cost is no more than that of any other flow of maximum value. To get a minimum cost matching, we construct a network from the

x_i's and S_j's as in Example 4.2.   Using C_{ij} as the cost of edge (x_i, S_j) and 0 as the cost of edges including s or t, we find a maximum flow of minimum cost.   If we had to construct all possible maximum flows and compute their costs to solve this problem, we would have a time–consuming chore.   Fortunately this is not necessary, as the following theorem shows.

> ***Theorem 4.4.***   Let f be a minimum cost flow of value V.   Augmenting f by means of a minimum–cost flow–augmenting path of worth w yields a minimum–cost flow of value $V + w$.
>
> *Proof.*   We assume that each $f(e)$ is rational and that among all flows of value $V + w$ the flow g is a rationally–valued flow of minimum cost.   We multiply by the least common multiple of all the denominators to convert f, g, w and all capacities and prices to integers.   Note now that if we can prove the theorem with this new w equal to 1, we can iterate the process to take care of arbitrary positive integral values of w.   Dividing by the least common multiple again preserves relative costs of flows and so we may assume all relevant numbers are integers and w is 1.   Now consider a new network with the same vertices, edges and prices, but with $c(v_i, v_j)$ the maximum of $f(v_i, v_j)$ and $g(v_i, v_j)$.   Then the capacity of the network is at least the value of g, so by labelling we may find a flow–augmenting path of worth 1 in this network.   Among all such paths, pick one of minimum cost. Note this is a flow–augmenting path of worth 1 in our original network.
>
> An edge e used in this flow–augmenting path must be an edge with $g(e) > f(e)$.   If we reduce the value of g by 1 in each of these edges, we get a flow g' of value V, the value of f.   The *cost* of g is the cost of g' plus the cost of the flow–augmenting path; the cost of f is no more than the cost of g', so the cost of g is at least the cost of f plus the cost of the flow–augmenting path.   ■

The Potential Algorithm for Finding Minimum–Cost Paths

This theorem reduces the problem of finding a minimum–cost flow to that of finding a minimum–cost flow–augmenting path.   To find this, we use a variant of the labelling algorithm, called a "potential" algorithm.   We start by assigning either a number or the symbol ∞ as a label $P(v)$ on each vertex v.   We update these numbers as we search through the network to give us our current "best estimate" of the cost of getting a unit of flow from the source to vertex v.   We use the following algorithm.

(1) For each vertex v other than the source, let $P(v) = \infty$.   Let $P(s) = 0$ for the source.
(2) For each edge (u, v), if $P(u) + p(u, v) < P(v)$, replace $P(v)$ by $P(u) + p(u, v)$.
(3) Repeat Step 2 until $P(u) + p(u, v) \geq P(v)$ for each edge (u, v).

In Step 2 we examine each edge of the network once.   If we examine them in a random way, we might take more steps than necessary. A good order for examining edges is to first list the vertices in the order in which they appear in a breadth–first search; then examine all edges leaving the first vertex in the list, all leaving the second vertex, and so on.

Theorem 4.5. The potential algorithm stops in a finite number of steps, and when it stops, $P(t)$ is the cost of a minimum–cost path from s to t.

Proof. A proof by induction on the distance from vertex s to v proves that there is a path of cost $P(v)$ from the source to v. (The details are left as an exercise.) ∎

Theorem 4.6. A minimum–cost flow–augmenting path for a flow f may be found by subtracting $f(e)$ from $c(e)$ for each edge e, adding the edge (v, u) of capacity $f(v, u)$ and cost $-c(u, v)$ whenever $f(u, v) > 0$, applying the potential algorithm and then tracing back from t to s along edges (u, v) satisfying $P(u) + p(u, v) = P(v)$ (starting with t and continuing until s.)

Proof. Exercise. ∎

Finding a Maximum Flow of Minimum Cost

Example 4.3. Find a maximum flow of minimum cost in the network of Figure 5.19. The circled numbers are the prices associated with the edges shown; the uncircled numbers are the capacities.

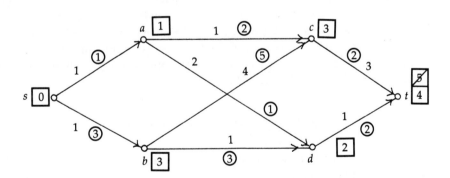

Figure 5.19

One application of the potential algorithm assigns to each vertex v the potential $P(v)$ shown in the square near the vertex. The two boxes near vertex t show that, using edge (c, t), we may first assign potential 5 to t, and then using edge

(d, t), assign potential 4 to t. This assignment now satisfies the rule $P(v) \le P(u) + p(u, v)$ for each edge (u, v). By working backwards from t, we see that the path whose vertices are t, d, a and s is the path of cost 4 leading to s from t, so $s(s, a)a(a, d)d(d, t)t$ is our flow–augmenting path. The worth of this path is 1, so our first flow is $f(s, a) = f(a, d) = f(d, t) = 1$. This flow gives us the digraph in Figure 5.20, to which we apply the potential algorithm. No capacities are shown in Figure 5.20, for once we have the new digraph the capacities are irrelevant to the potential algorithm.

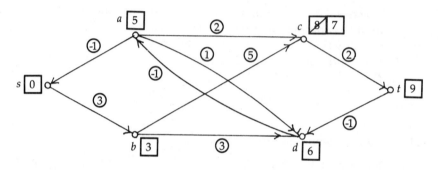

Figure 5.20

The boxes show the potentials assigned to the vertices traversed in the order $s\ b\ c\ d\ a\ c\ t$. Note how the potential assigned to c drops from 8 to 7 in the process. Working back from t along the edges and vertices with $P(v) = P(u) + P(u, v)$ gives us the sequence $t\ c\ a\ d\ b\ s$. Therefore, our minimum–cost flow–augmenting path, which we choose by examining our sequence of vertices in Figure 5.19 rather than Figure 5.20 is

$$s(s, b)b(b, d)d(a, d)a(a, c)c(c, t)t.$$

The edge (a, d) is a reverse edge on this path. This illustrates a subtle point. When we apply the potential algorithm to an auxilliary digraph (such as the digraph in Fig. 5.20) and obtain edge *not* in our network, the edge pointing the opposite direction is a reverse edge in the flow augmenting path. Our flow–augmenting path has worth 1, so our new flow is

$$f(s, a) = f(a, c) = f(c, t) = f(s, b) = f(b, d) = f(d, t) = 1$$

and $f(e) = 0$ for all other e. This flow has value 2 and the cut capacity of the network is 2, so we have a maximum flow of minimum cost. The cost of the flow is $1 + 2 + 2 + 3 + 3 + 2 = 13$. ■

EXERCISES

1. Find the largest n such that a is n–edge–connected to b in the digraph of Figure 5.21.

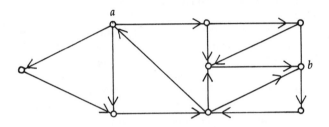

Figure 5.21

2. Use a network–flow algorithm to find a maximum–sized set of edge–disjoint paths from a to b in the digraph of Figure 5.21.

3. Find the largest n such that a is n–vertex–connected to b in the digraph of Figure 5.21.

4. Find a maximum–sized set of vertex–disjoint paths from a to b in the digraph of Figure 5.21.

5. Change the directed edges to undirected edges in Figure 5.21 and determine the largest m and n such that a is m–edge–connected and n–vertex–connected to b.

6. Change the directed edges to undirected edges and find a maximum–sized edge–disjoint set of paths in Figure 5.21 connecting a to b.

7. Find an SDR for the family $\{1, 3, 5\}, \{2, 3, 5\}, \{7, 5, 3\}, \{6, 2, 5\}, \{2, 3, 7\}$ or explain why none exists. (Use network flows.)

8. Let the cost of i be i and find a minimum–cost SDR for the family of Exercise 7, or explain why none exists.

9. Do Exercise 11 of Section 2 (finding an assignment of people to jobs).

10. Prove that the relations "a is n–edge–connected to b" and "a is m–vertex–connected to b" are equivalence relations.

11. Write out the details of the proof of the vertex form of Menger's Theorem for directed graphs.

12. State and prove the vertex form of Menger's theorem for graphs.

13. State and prove the edge form of Menger's theorem for multigraphs.

14. Write a computer program that finds a minimum–cost SDR for a family of sets.

15. Write out a complete proof of Theorem 4.5.

16. Prove Theorem 4.6.

17. In Figure 5.19, change the capacity of edge (s, a) to 2 and find a maximum–sized flow of minimum cost. What is the cost of the flow?

18. In Figure 5.19, change all capacities to 1 and the costs of edges touching s or t to 0. Why may this network be interpreted as the network of a minimum–cost matching problem? Find the minimum–cost matching using network flows.

19. In Figure 5.19, change the capacities of edges (s, b) and (s, a) to 2 and find a maximum–sized flow of minimum cost. What is the cost of the flow?

20. In the graph of Figure 5.8, use the potential algorithm to find a minimum–cost path from h to c. (Assume all edges are directed in both directions and have the same cost in each direction.)

21. In the graph of Figure 5.9, use a variant of the potential algorithm (which you must somehow change slightly) to find a maximum–cost path from h to c.

22. In Exercise 7, let the cost of representing set S_j by integer i be i plus the minimum number in S_j. Find a minimum–cost SDR.

23. In Exercise 11 of Section 2, let the jobs be numbered 1 = rough carpenter, 0 = general laborer, 3 = painter, 4 = equipment operator and 5 = supervisor. Let the cost of assigning person i to job j be the number beneath i plus the number just assigned to j. Find a minimum–cost assignment of people to jobs.

24. In Example 3.1, let the costs be $p(s, a) = c(a, d) = 3$, $p(s, b) = p(b, d) = 4$ and $p(e) = 2$ for all other e. Find a minimum–cost flow of maximum value.

25. Either give an example of a digraph with prices on the edges for which the potential algorithm does not stop after all edges have been examined once (in breadth–first ordering) or explain why there is none.

26. Write a computer program to find minimum–cost paths between two vertices in a digraph with a price function on the edges.

27. Write a computer program to find maximum matchings of minimum cost.

Suggested Reading

Berge, Claude: *Graphs and Hypergraphs,* North Holland 1973, Chapter 7.

Edmonds, J. "Paths, Trees and Flowers," *Canadian Journal of Math 17* (1963) 449–467.

Even, Shimon: *Graph Algorithms,* Computer Science Press 1979.

Fulkerson, D. Ray (Ed.): *Studies in Graph Theory I, MAA Studies in Mathematics,* Volume 11, Mathematical Association of America 1975.

Lawler, Eugene L.: *Combinatorial Optimization: Networks and Matroids,* Holt, Reinhart, Winston, New York 1976.

Welsh, D.J.A.: *Matroid Theory,* Academic Press 1976.

6 *Combinatorial Designs*

SECTION 1   LATIN SQUARES AND GRAECO–LATIN SQUARES

How Latin Squares are Used

Suppose we have five varieties of gasolines formulated in different ways, each designed to provide high mileage per gallon. We wish to test the different varieties against each other so as to choose an optimal formulation. Each variety is to be tested in a 400–mile drive. It would be nice to use the same car for each test, but then we would have to make the tests on different days and face such variable extraneous factors as wind, temperature variation, humidity, traffic, etc., any one of which might affect gasoline mileage as much as the gasoline variety itself. Thus we use five different cars so that they may all drive the same course in the same day. Now we have the problem that different cars have different gas mileage characteristics! Suppose, however, instead of running the experiment just once, we run it five times in such a way that each car uses each variety of gasoline exactly once. In this way, if a variety of gasoline is truly superior, we should be able to notice that fact because the car using that gasoline on a given day will have improved performance relative to the other cars on other days. Correspondingly, if one of the cars is a consistently superior performer, we will be able to notice that. Suppose the cars are designated by capital letters A through E. A possible schedule of cars and varieties of gasoline (numbered 1 to 5) is shown in Figure 6.1.

Gasoline Variety	Day				
	1	2	3	4	5
1	A	B	C	D	E
2	B	C	D	E	A
3	C	D	E	A	B
4	D	E	A	B	C
5	E	A	B	C	D

Figure 6.1

The array of letters in Figure 6.1 is called a Latin square. A *Latin square* is a square matrix of symbols such that each symbol occurs in each row exactly once and in each column exactly once. It is clear from the figure that given n symbols we can always construct an $n \times n$ Latin square.

Orthogonal Latin Squares

Since the five different cars are to be driven at the same time, they must have different drivers. However, the gasoline mileage that a car gets depends on who is driving it. If we could arrange things so that each driver drives each car exactly once and each driver uses each variety of gasoline exactly once, then perhaps we could sort out which effects are due to the drivers and which are due to the gasoline. Figure 6.2 is an arrangement in which the lower–case letters stand for the drivers. This arrangement is called a *Graeco–Latin* square or a pair of *orthogonal Latin squares*. More precisely, a pair of Latin squares S and s are *orthogonal* if all ordered pairs (S_{ij}, s_{ij}) are different. In our example, there are 25 ordered pairs and 25 positions, so orthogonality is equivalent to saying that each ordered pair occurs exactly once in the positions of the superimposed squares.

What if we had four, or even six, varieties of gasoline? This leads us to ask, "Is there a pair of orthogonal n by n Latin squares for each value of n? Is there a pattern in Figure 6.2 that can be duplicated for other values of n besides 5?" In Figure 6.2 we can obtain the second row from the first by shifting all entries of row one (except the first entry) to the left and moving the first entry to the end of the row. We get succeeding rows by repeating this process. If you imagine erasing the upper–case letters in Figure 6.2, you will see that for the lower–case letters, we get the second row by shifting everything but the first two entries *two* places to the left, moving the first two to the end, and then repeating the process to get the remaining rows. Is this a general method of constructing orthogonal squares? If we tried it with $n = 6$, we would not get a Latin square for the lower–case letters because the fourth row would be the same as the first. We shall soon show that whenever the construction method gives two Latin squares, they are orthogonal. It is easy to show that there is no pair of orthogonal Latin squares of side 2. (Can you explain why?)

Aa	Bb	Cc	Dd	Ee
Bc	Cd	De	Ea	Ab
Ce	Da	Eb	Ac	Bd
Db	Ec	Ad	Be	Ca
Ed	Ae	Ba	Cb	Dc

Figure 6.2

Euler's 36 Officers Problem

Euler observed that no pair of 2×2 squares existed and found he was able to construct examples of $n \times n$ orthogonal squares for n up to 5, but had trouble with 6. He proposed the following problem.

Thirty–six officers of six ranks and from six different regiments are to march in a square formation of size 6×6. Each row and each column of the formation is to contain one and only one officer of each rank and one and only one from each regiment. Is such a formation possible? The officers are the 36 places of a Graeco–Latin square in which the symbols are the six names of the regiments and the six ranks. In 1782, Euler conjectured that such an arrangement was not possible. Further, he conjectured that so long as $n = 4k + 2$ for some integer k, then no pair of $n \times n$ orthogonal Latin squares exists. Around 1900, Tarry showed that there is no pair of 6×6 orthogonal Latin squares by systematically checking all possible constructions.

Euler's more general conjecture remained unsettled until 1959 when Bose, Shrikhande and Parker constructed first a pair of 22×22, and then a pair of 10×10, orthogonal Latin squares. Currently, it is unknown whether there is a system of three Latin squares of size 10, all orthogonal to each other. It is known now that there is a pair of orthogonal $n \times n$ Latin squares if n is different from 2 or 6. We shall not prove this, but instead show that the construction used in Figure 1.2 can be generalized to yield many pairs of orthogonal squares. Our construction uses the concept of congruence modulo an integer n; we now review this concept.

Congruence Modulo an Integer n

We define an equivalence relation on the integers by saying a is congruent to b mod n, written $a \equiv b \pmod{n}$ if and only if n is a factor of $a - b$. Let us check that this relation is an equivalence relation. Since n is a factor of $a - a = 0$, our relation is reflexive. Since n is a factor of $b - a = -(a - b)$ whenever it is a factor of $a - b$, our relation is symmetric. Now suppose $a \equiv b \pmod{n}$ and $b \equiv c \pmod{n}$, so $a - b = kn$ and $b - c = hn$. Then

$$a - c = a - b + b - c = kn + hn = (k + h)n,$$

so our relation is transitive. Thus it is an equivalence relation. The equivalence class containing an integer i will also contain $i + n$ and $i - n$; in fact, it will contain $i + kn$ for each (positive or negative) integer k. Further if $x \equiv i \pmod{n}$, then for some k, $x - i = kn$ since n is a factor of $x - i$; this gives $x = i + kn$. Thus the

n equivalence classes,

$$\{0 + kn \,|k \text{ is an integer}\}, \{1 + kn \,|k \text{ is an integer}\},$$

$$\ldots \{n - 1 + kn \,|k \text{ is an integer}\}$$

are the complete set of equivalence classes for congruence modulo n. We will use the symbol $\underline{i}$ to stand for the equivalence class containing i.

We will define arithmetic operations on equivalence classes, so that we can talk about $\underline{i} + \underline{j}$ and $\underline{i} \cdot \underline{j}$. For this purpose, we prove the following theorem.

Theorem 1.1.   If $i \equiv i_1 \pmod{n}$ and $j \equiv j_1 \pmod{n}$, then

$$i + j \equiv i_1 + j_1 \pmod{n}$$

and

$$ij \equiv i_1 j_1 \pmod{n}.$$

Proof. $i_1 = i + hn$ and $j_1 = j + kn$.   Then

$$i + j - (i_1 + j_1) = -(h + k)n$$

and

$$\begin{aligned}
ij - i_1 j_1 &= ij - (i + hn)(j + kn) \\
&= ij - (ij + ikn + jhn + hkn^2) \\
&= -(ik + jh + khn)n. \quad \blacksquare
\end{aligned}$$

By Theorem 1.1, we see that whenever i is congruent to i_1, and j is congruent to j_1, then $i + j$ lies in the same class as $i_1 + j_1$ and ij lies in the same class as $i_1 j_1$. Thus if we define $\underline{i} + \underline{j}$ to be $\underline{i + j}$, then we will have

$$\underline{i} + \underline{j} = \underline{i_1} + \underline{j_1} \text{ whenever } \underline{i} = \underline{i_1} \text{ and } \underline{j} = \underline{j_1}$$

and

$$\underline{ij} = \underline{i_1 j_1} \text{ whenever } \underline{i} = \underline{i_1} \text{ and } \underline{j} = \underline{j_1}.$$

Thus defining $\underline{i} + \underline{j}$ to be $\underline{i + j}$ and $\underline{i} \cdot \underline{j}$ to be $\underline{ij}$, respectively, makes sense since the definition doesn't depend on which member of $\underline{i}$ and $\underline{j}$ we choose—that is, we could have just as well defined them to be $\underline{i_1 + j_1}$ and $\underline{i_1 j_1}$ and we would have gotten the same sets.

It is convenient to visualize $\underline{i}$ as being represented by the smallest nonnegative integer in the set $\underline{i}$. We don't introduce a new notation though, but we call the smallest nonnegative element of $\underline{i}$ the *residue* of i (mod n).

Example 1.1.   When n is 5, the classes are $\underline{0}, \underline{1}, \underline{2}, \underline{3}$, and $\underline{4}$; we often let 0 represent $\underline{0}$. Since $\underline{0} = \underline{5}$, we are letting 0 represent $\underline{5}$ as well, but we don't let 5 represent $\underline{5}$. Rather than writing $\underline{4} + \underline{3} = \underline{7}$, it is customary to write $\underline{4} + \underline{3} = \underline{2}$ or $4 + 3 \equiv 2 \pmod 5$. Note that $\underline{4} \cdot \underline{3} = \underline{2}$ as well; that is $4 \cdot 3 \equiv 2 \pmod 5$. We call 2 the residue of 7 mod 5. It is also the residue of 12 or $-3 \pmod 5$. ∎

Using Arithmetic Modulo n to Construct Latin Squares

We state the relationship between arithmetic modulo n and Latin squares as a theorem.

Theorem 1.2.   If the rows of a matrix are labelled rows 0 through $n - 1$ and the entries of row i are the residues of the numbers from i to $n + i - 1$ in the order of the numbers, then this matrix is a Latin square.

Proof.   Each of the numbers from i to $n - 1$ appears as a residue in row i—but the numbers from 0 to $i - 1$ are the residues of n through $n + i - 1$, and so each residue appears in each row exactly once. Number the columns from left to right as 0, 1, through $n - 1$. Then column k consists of the residues of $k + i$ for $i = 0, 1, \ldots, n - 1$ and thus, as with the rows, contains each residue exactly once. ∎

Now it is clear that we could have used the residues of 5 to construct the array in Figure 6.1 and the upper–case array in Figure 6.2. Of course, the lower–case letters in Figure 6.2 could be replaced by symbols denoting residues too. In fact, if we replace the lower–case letters by residues modulo 5 in the natural way, we get the array in Figure 6.3 in place of the lower–case letters. Note that we get the second row by adding 2 to everything in the first row and taking residues modulo 5. In fact, we get each row by adding 2 to everything in

```
0  1  2  3  4
2  3  4  0  1
4  0  1  2  3
1  2  3  4  0
3  4  0  1  2
```

Figure 6.3

the preceding row and taking residues modulo 5. If we think of this as a matrix B whose rows and columns are indexed by the numbers 0 through 4 (rather than 1 through 5, as we would normally index them), we can write

$$B_{i+1,j} \equiv B_{ij} + 2 \pmod 5$$

and since row 0 is given by

$$B_{0j} = j,$$

we get

$$B_1 \equiv B_{0j} + 2 = 2 + j \pmod 5,$$

and in general,

$$B_{ij} \equiv B_{0j} + 2i = 2i + j \pmod 5.$$

From this, it is clear that the residues of $2i$, $2i + 1$, . . . , $2i + 5 - 1$ all appear in row i, and so all residues mod 5 appear in row i. Since the residues of 0, 2, 4, 6 and 8 are 0, 2, 4, 1 and 3, the first column contains distinct elements. Column j is derived by adding j to column 0, so every other column contains distinct elements because adding j to distinct elements gives distinct elements. Therefore B is a Latin square.

The Latin square of Theorem 1.2 is given by $A_{ij} \equiv i + j \pmod n$. Let us consider the general matrix B we get by replacing the integers between 0 and 4 by the integers between 0 and $n - 1$ and otherwise carrying out the construction of B just as before. This will give us $B_{ij} \equiv 2i + j \pmod n$. To check that A and B are orthogonal, we must show that each pair of residues appears once and only once in the form of the pair

$$(A_{ij}, B_{ij}).$$

Thus we ask for solutions to the symbolic equation

$$(A_{ij}, B_{ij}) = (A_{kh}, B_{kh})$$

which reduces to

$$i + j \equiv k + h \pmod{n}$$
$$2i + j \equiv 2k + h \pmod{n}$$

by setting the two components of the pairs equal.   Now applying Theorem 1.1 and subtracting the first equation from the second, we get

$$2i + j - (i + j) \equiv 2k + h - (k + h) \pmod{n}$$
$$i \equiv k \pmod{n}.$$

Now subtract this equation from

$$i + j \equiv k + h \pmod{n}$$

to get

$$j \equiv h \pmod{n}.$$

Thus two pairs are equal only if they are in the same position in the paired arrays. Notice that this computation did not use the actual value of n; however, we have seen that if n were 6, then the matrix B would not be a Latin square because rows 1 and 4 would be identical.   In fact, for any even n, the matrix B would not be a Latin square for the same reason.   However, the methods above yield

> **Theorem 1.3.**   If n is odd and if the matrices (indexed by 0 through $n - 1$) A and B have $A_{ij} \equiv i + j \pmod{n}$ and $B_{ij} \equiv 2i + j \pmod{n}$, then A and B are orthogonal Latin squares.

Orthogonality and Arithmetic Modulo n

By the same sort of computations, we can prove

> **Theorem 1.4.**   Suppose that k and j are numbers such that for $i = 0$ to $n - 1$, the products $k \cdot i$ are all distinct modulo n and the products $j \cdot i$ are all distinct modulo n.   If $r - s$ has no factors in common with n, then the arrays of residues mod n A and B with $A_{ij} = ri + j$ and $B_{ij} \equiv si + j \pmod{n}$ are orthogonal Latin squares.

Proof. As above, if

$$(A_{ij}, B_{ij}) = (A_{kh}, B_{kh}),$$

then

$$ri + j \equiv rk + h \ (\text{mod } n)$$
$$si + j \equiv sk + h \ (\text{mod } n)$$

so that subtraction gives

$$(r - s)i \equiv (r - s)k \ (\text{mod } n)$$

which is the same as

$$(r - s)(i - k) \equiv 0 \ (\text{mod } n).$$

But then n is a factor of $(r - s)(i - k)$, and since n has no factors in common with $r - s$, either it must have all its factors in common with $i - k$, or else $i - k$ must be 0. Since i and k are both less than n, the only possibility is that $i - k$ is 0, so that $i = k$. Then by substituting into one of the first two equations, we get $j = h$ as well. ■

Corollary 1.5. If n is a power of a prime p, then there are $p - 1$ $n \times n$ Latin squares all orthogonal to each other.

This corollary tells us that we could, for example, also consider the effects of tires on gas mileage at the same time we are testing the five formulations of gasoline with the five different drivers in the five different cars—if we were willing to change tires from car to car following each run! In fact, if n is a power p^m of a prime, there are $p^m - 1$ mutually orthogonal Latin squares. This follows from the theory of vector spaces over finite fields and thus is beyond the scope of this book.

EXERCISES

1. Write down all the 4×4 Latin squares with first row and first column consisting of 1, 2, 3 and 4 in order. (Trial and error is the most appropriate method.)
2. Explain why there is no pair of orthogonal 2×2 Latin squares.

3. Find a 3×3 Graeco–Latin square.

4. Find a pair of 7×7 orthogonal Latin squares.

5. Give an example using the integers mod 6 to show that the rule "if $\underline{ab} = \underline{0}$, then $\underline{a} = \underline{0}$ or $\underline{b} = \underline{0}$" doesn't always hold.

6. Prove that $\underline{a} + (\underline{b} + \underline{c}) = (\underline{a} + \underline{b}) + \underline{c}$.

7. Prove that $\underline{a}(\underline{b} + \underline{c}) = \underline{ab} + \underline{ac}$.

8. (a) Show that if n is a prime number and a is a fixed number between 1 and $n - 1$, then the products $\underline{a} \cdot \underline{1}, \underline{a} \cdot \underline{2}, \dots , \underline{a} \cdot (\underline{n - 1})$ are all distinct.

   (b) Show that if n is a prime, then for each class $\underline{a}$, there is a class $\underline{b}$ such that $\underline{ab} = \underline{1}$. (Hint: Use (a) and the pigeon–hole principle.)

9. Using the integers mod 15, find a class $\underline{a}$ for which there is no class $\underline{b}$ with $\underline{a} \cdot \underline{b} = 1$.

10. Prove Corollary 1.5.

11. Find three 5×5 Latin squares all orthogonal to each other.

12. Find a pair of 4×4 orthogonal Latin squares.

13. Find three 4×4 Latin squares all orthogonal to each other.

14. Is there a Latin square orthogonal to

$$
\begin{array}{cccc}
1 & 2 & 3 & 4 \\
2 & 4 & 1 & 3 \\
3 & 1 & 4 & 2 \\
4 & 3 & 2 & 1?
\end{array}
$$

15. Under what conditions on n does the cancellation law "$\underline{a} \cdot \underline{b} = \underline{a} \cdot \underline{c}$ implies $\underline{b} = \underline{c}$" hold?

16. A Latin square which is orthogonal to its transpose is called self–orthogonal.

(a) Prove no Latin square of side 3 is self–orthogonal.

(b) Find a 4×4 self–orthogonal Latin square.

17. Show that it is impossible to find k mutually orthogonal $k \times k$ Latin squares.

18. A Latin rectangle is an $m \times n$ matrix whose rows are permutations of $1, 2, \dots , n$ and whose columns each contain m distinct elements. Using the theory of systems of distinct representatives, show that it is always possible to keep adding rows to a Latin rectangle until you get a Latin square.

19. Write a backtracking program to list $n \times n$ Latin squares whose first row and first column each consist of the symbols 1 through n in order.

20. (a) Using inclusion and exclusion, compute the number of $2 \times n$ Latin rectangles whose first row consists of the symbols 1 to 6 in their usual order.

   (b) Do the same for $3 \times n$ Latin rectangles whose first row is 1 through 6 in order.

21. Write a backtracking program that searches for one $n \times n$ Latin square orthogonal to a given $n \times n$ Latin square. Experiment with squares of size 5, 6 and 7.

22. Write a program to confirm Tarry's result that there is no pair of orthogonal 6×6 Latin squares. (Warning: There are 9408 6×6 Latin squares with first row and column in increasing order.)

SECTION 2   BLOCK DESIGNS

How Block Designs are Used

Suppose now we are interested in determining the effect of seven motor oil formulations on gas mileage.   Each variety of motor oil will have to be tested over a significant distance, say 10,000 miles.   A car will have to be broken in for 5000 or 10,000 miles before we begin testing, and if a car goes further than, say 40,000 miles, its age will have some effect on gasoline consumption.   Thus we should not run our cars too long, or else we will get bad data.   In addition, it would take some time to drive more than 40,000 miles in a car, especially if the driving is to be done to simulate normal driving habits.   For these reasons, we decide to test only three motor oils per car.   If two motor oils are never compared against each other in the same car, we will have reason to be nervous about any conclusions about these motor oils.   Thus we would like to have each pair of motor oils used in the same one of the cars at least once (at different times, of course).   We will arrange the seven motor oils into a collection of overlapping blocks of size 3, each block to be tested in a given car.   Each block is tested in exactly one car, so if each pair gets into exactly one block together, then each pair gets compared by exactly one car. How many cars do we need?   Since we have $C(7; 2) = 21$ pairs to be compared and each car is used to make $C(3; 2) = 3$ comparisons, we need seven cars.   This means our arrangements of motor oils into blocks requires seven blocks.   Does such an arrangement exist?   We shall see that it does.

A *block design* on a set V with v *vertices* or *varieties* is a multiset of $k-$ element multisets, called *blocks,* chosen from V.   Further, if each variety appears (is replicated) in exactly r blocks and each pair of varieties occurs together in exactly λ blocks, the design is called *balanced*.   The design is called *complete* if each of the blocks is V and *incomplete* if the blocks form a *set* of *proper* subsets of V.   A block design with v vertices arranged into b blocks of size k in such a way that each vertex appears in r blocks and each pair of vertices appears together in λ blocks is called a (b, v, r, k, λ) design.   Balanced incomplete block design is often abbreviated BIBD.

The designs we used in the last section were complete $(5, 5, 5, 5, 5)$ designs; the motor oil design we asked for above is a $(7, 7, r, 3, 1)$ design.   In fact, we haven't demanded that each oil be tested in exactly the same number r of cars, but this occurs nonetheless.   Our first theorem tells us how to compute r when we know b, k, and v.

Basic Relationships among the Parameters

Theorem 2.1.   In a (b, v, r, k, λ) design $bk = vr$.

Proof.   Since there are b blocks each with k elements, the number of ordered pairs consisting of a block and an element of that block is bk.   However, since each element of V must occur in r blocks, there must be vr ordered pairs consisting of a vertex and a block containing it. Since bk and vr are the number of elements in the same set of ordered pairs, they are equal. ∎

Theorem 2.2. In a (b, v, r, k, λ) design, $r(k - 1) = \lambda(v - 1)$.

Proof. For each x in V, both sides of this equation count the number of ordered pairs consisting of a block the vertex x is in and one other element of that block. ∎

Given any three parameters of a design, we can use them to compute the other parameters.

Example 2.1. In the motor oil example we had seven varieties of oil in V; further we wanted three varieties per block and wanted each pair of varieties to occur together in a single block. Thus $v = 7, k = 3$ and $\lambda = 1$. We computed b above, but will compute it here differently. Since $r(3 - 1) = 1(7 - 1), 2r = 6$, so $r = 3$ and since $bk = vr$, $b \cdot 3 = 7 \cdot 3$ and thus $b = 7$. ∎

Theorem 2.3 provides another example of how the basic equations are used.

Theorem 2.3. In a BIBD, $\lambda < r$.

Proof. We know that in an incomplete design, $k < v$ because the blocks are proper subsets. Since $\lambda(v - 1) = r(k - 1)$, we conclude that $\lambda/r = \dfrac{(k - 1)}{(v - 1)} < 1$ so $\lambda < r$. ∎

Each block design has an *incidence matrix* N whose rows are indexed (labelled) by the elements of V and whose columns are indexed by the blocks. (Think of the vertices and blocks as being numbered.) The entry N_{ij} is the number of times variety i appears in block j. For a BIBD, N_{ij} is either 0 or 1. The matrix NN^t (where N^t stands for the transpose of N) is a matrix with v rows and v columns. Its form gives us considerable insight into the parameters of the design. In Theorem 2.4, I stands for the $v \times v$ identity matrix and J stands for the $v \times v$ matrix consisting of 1's exclusively.

Theorem 2.4. If N is the incidence matrix of a BIBD on a v–element set, then

$$NN^t = (r - \lambda)I + \lambda J.$$

Proof. The ij entry of NN^t is formed by taking row i of N and column j of N^t (or in other words, row i and row j of N), multiplying corresponding entries and adding up the results. However, row i and row j will both be 1 in position k if and only if vertex i and vertex j are in block k. Thus the sum, if $i \neq j$, is the number of blocks in which i and j occur together, namely λ, and if $i = j$, the sum is the number of blocks variety i lies in, namely r. Thus $(NN^t)_{ij}$ is λ if $i \neq j$ and r if $i = j$, which are precisely the i, j entries of $(r - \lambda)I + \lambda J$. ∎

Theorem 2.5.   In a balanced incomplete block design, $b \geq v$.

Proof.   Note that adding one or more columns of 0's to a matrix A does not change the value of AA^t. For example,

$$\begin{bmatrix} a & b & 0 & 0 \\ c & d & 0 & 0 \end{bmatrix} \begin{bmatrix} a & c \\ b & d \\ 0 & 0 \\ 0 & 0 \end{bmatrix} = \begin{bmatrix} a^2 + b^2 & ac + bd \\ ca + db & c^2 + d^2 \end{bmatrix} = \begin{bmatrix} a & b \\ c & d \end{bmatrix} \begin{bmatrix} a & c \\ b & d \end{bmatrix}.$$

Assume that $b < v$. Then the $v \times b$ matrix N has more rows (v) than columns (b). We can change N into a square matrix M by adding $v - b$ columns of 0's to N. Then

$$NN^t = MM^t.$$

But since M is a square matrix with a column of 0's, det $M = 0$. Therefore det $MM^t = 0$, so det $NN^t = 0$.

Now we show that the determinant of NN^t is nonzero. This contradiction will show that the assumption that $b < v$ is invalid. Subtract column 1 from all the other columns. This gives a matrix A whose first column is the same as the first column of NN^t, whose first row is $\lambda - r$ in position 2 through v, whose diagonal entries are $r - \lambda$ in position 2 through v and whose other entries are 0. Now add rows 2 through v of A to row 1. This gives a matrix which is 0 above the main diagonal and nonzero in each diagonal position. Thus it has a nonzero determinant. Therefore, it is impossible for b to be less than v, so that $v \leq b$. ∎

Constructing a Design

It is not particularly easy to construct examples of designs. However in the case we began with, we can make a fairly straightforward construction.

Example 2.2.   Construct a (7, 7, 3, 3, 1) BIBD.
We will use the symbols 0, 1, 2, 3, 4, 5, 6 for our seven varieties. We may

assume $\{0, 1, 2\}$ is one of our three–element sets. Now 0 appears in only three sets and must appear with 3, 4, 5 and 6, so we may assume two other sets are $\{0, 3, 4\}$ and $\{0, 5, 6\}$. (If the symbols 3, 4, 5 and 6 occur differently, we can rename them.) Now 1 occurs in two other sets, but has already occurred with 0 and 2, so in these sets it must occur with 3, 4, 5 and 6. However 3 has already appeared with 4, so when it appears with 1, only 5 or 6 can occur with it. Assume 5 occurs with 1. (If not, we have the option of interchanging the labels of 5 and 6.) Then $\{1, 3, 5\}$ and $\{1, 4, 6\}$ must be the two sets containing 1. There are two sets left to construct. Now, 0 and 1 have been used three times and so only 2, 3, 4, 5 and 6 may be used. Similarly, 3, 4, 5 and 6 have been used twice, but 2 has been used once, so both sets contain 2. Now 4 and 5 have not appeared together, but 3, 4 and 5, 6 have, so one of the sets with 2 must be $\{2, 4, 5\}$. Then the other set with 2 must be $\{2, 3, 6\}$. ∎

Note that, except for our labelling of the elements of V, there really is no choice as to how we might construct the designs. Two designs on V and V' are *isomorphic* if there is a one–to–one function f from V onto V' such that for each block B of design 1,

$$f(B) = \{f(x) | x \in B\}$$

is a block of design 2 and conversely each block of design 2 is $f(B)$ for some block B of design 1. Thus, since we had no choice except for labelling in the construction, any two $(7, 7, 3, 3, 1)$ designs are isomorphic.

Symmetric Designs

In our example, we had $v = b$ rather than $v < b$; a design is called *symmetric* if $v = b$. Since $bk = vr$, then a symmetric design also has $k = r$. Such a design is sometimes called a (v, k, λ) design. In Exercise 4, you will be asked to show that det NN^t is $(r - \lambda)^{v-1} \cdot rk$; for a symmetric design, this is $(r - \lambda)^{v-1} \cdot r^2$. Also for a symmetric design, N and N^t are square and det $N = \det N^t$ gives us

$$\det(N)^2 = \det(N)\det(N^t) = \det(NN^t) = (r - \lambda)^{v-1} \cdot r^2.$$

Since $\det(N)$ is an integer, the left–hand side is a square of an integer, so the right–hand side must be also. Since r^2 is a square, the same goes for $(r - \lambda)^{v-1}$. If v is odd, this will happen automatically; if v is even, then $r - \lambda$ must be a square integer. This yields our last theorem, which is the easier part of a more general theorem sometimes called the Bruck–Ryser or Bruck–Ryser–Chowla theorem; the part we proved was also discovered by Schutzenberger and Shrikhande at about the same time.

Theorem 2.6.   In a symmetric BIBD with v even, the number $r - \lambda$ is a square.

Example 2.3.   Is there a $(46, 10, 2)$ design?

Note that this would be a $(46, 46, 10, 10, 2)$ design. The parameters satisfy $bk = vr$ and $r(k - 1) = \lambda(v - 1)$. However, $r - \lambda = 10 - 2 = 8$ which is not a square. ∎

It is tempting to assume that if the parameters b, v, r, k and λ satisfy all the tests we have devised, then there is a design with these parameters. However, the sequence $(21, 15, 7, 5, 2)$ passes our tests and there is no design with these parameters. (See the book by Hall in the suggested reading for further explanation.) When we need to know if there *is* a design with certain parameters that pass our tests, there is no substitute for constructing such a design. Generally, the construction process is one of intelligent guesswork supplemented by good luck as in Example 2.2. In the next section, we give a few general construction methods that are sometimes helpful.

EXERCISES

1. What are k and λ in a $(4, 4, 3, k, \lambda)$ design?
2. What are k and r in a $(7, 7, r, k, 2)$ design?
3. Show that if $\lambda = r$ in a balanced block design, then the design is complete.
4. Show that $C(v; 2) \cdot \lambda = C(k; 2) \cdot b$ in any block design. (Hint: The most interesting proof comes from observing that $C(v, 2) \cdot \lambda$ is the total number of pairs of vertices that appear among the blocks of the design.) Write out the resulting equation in explicit form.
5. Show that the determinant of NN^t is $(r - \lambda)^{v-1} \cdot rk$.
6. Write out the matrices N and NN^t for Example 2.2.
7. (a) Show that if N is the incidence matrix of a BIBD, then $NJ = rJ$.
    (b) What is JN for a BIBD?
8. Are there (b, v, r, k, λ) designs with the following parameters?
    (a) $b = 8$, $v = 6$, $r = 5$, $k = 3$, $\lambda = 2$
    (b) $b = 12$, $v = 8$, $r = 6$, $k = 4$, $\lambda = 2$
9. How many $(4, 4, 3, 3, 2)$ designs are there? Find them.
10. Find a $(7, 7, 4, 4, 2)$ design.
11. Are all $(7, 7, 4, 4, 2)$ designs isomorphic?
12. Find all $(6, 4, 3, 2, \lambda)$ designs.
13. Find a $(14, 7, 6, 3, 2)$ design.
*14. Find a symmetric $(11, 5, 2)$ design.
15. Find a $(10, 6, 5, 3, 2)$ design.
16. Show that the condition which states that each element appears in exactly r blocks may be removed from the definition of a BIBD and proved as a consequence of the remainder of the definition.
17. Show that given v and k, there is a design with
$$b = C(v; k), r = C(v - 1; k - 1), \lambda = C(v - 2; k - 2).$$

18. True or False. If $k = 3$, then either λ is even or v is odd. (Prove or give a counter–example.)

19. Is it possible to distribute 22 brands of detergent to different households so that each household tests seven brands over a seven–month period and each pair of brands is tested by two different households?

20. (a) Why is Exercise 19 made significantly harder if 22 is replaced by 16 and seven is replaced by six?

   *(b) Solve this more difficult problem.

*21. Show that if M_1 and M_2 are the matrices below, then there are three more matrices whose first column is the first column of M_2 such that the rows of these five matrices form the blocks of a (16, 20, 5, 4, 1) design.

$$M_1 = \begin{bmatrix} 1 & 2 & 3 & 4 \\ 5 & 6 & 7 & 8 \\ 9 & 10 & 11 & 12 \\ 13 & 14 & 15 & 16 \end{bmatrix} \quad M_2 = \begin{bmatrix} 1 & 5 & 9 & 13 \\ 2 & 6 & 10 & 14 \\ 3 & 7 & 11 & 15 \\ 4 & 8 & 12 & 16 \end{bmatrix}$$

(Hint: Put three orthogonal Latin squares with first row $ABCD$ down on M_1, in which case A, B, C and D each land on a four–element set.)

22. Discuss the potential use of backtracking for finding BIBDs. Write a back-tracking program to search for designs and apply it to various *small* parameter values. Discuss how many steps a backtracking search for a symmetric (7, 3, 1) design would take in comparison with checking all families of seven three–element sets chosen from a seven–element set.

SECTION 3   CONSTRUCTION AND RESOLVABILITY OF DESIGNS

A Problem that Requires a Big Design

The director of the Lebanon Softball League recently asked the following question. "We have fourteen teams—probably, but there may be a couple dropouts. We can use the municipal field at night just long enough to play three games before the lights go out. We would like to have three teams show up each time and play three games; that way, one team can do the umpiring, etc. while the other two play. Is there any way to schedule the season of about 12 weeks so that each team plays each other team twice and each team comes out just once a week?"

   Although the statement of the problem wasn't exactly a problem of design, it is fairly clear that the question deals with the existence of some sort of $(b, v, r, 3, 2)$ design where v would probably be 14 but might be less. In the case $v = 14$, we can write

$$b \cdot 3 = 14 \cdot r$$

and

$$r(2) = 2(13),$$

so $r = 13$ and thus $b \cdot 3 = 14 \cdot 13$.   However, since 3 is not a factor of either 13 or 14, there is no solution.   If the softball league had a dropout or consolidated two teams, then v would be 13 or 12.   Then we would have either

$$b \cdot 3 = 13 \cdot r$$
$$r \cdot 2 = 2 \cdot 12$$

so that

$$r = 12, \ b = 52$$

or

$$b \cdot 3 = 12 \cdot r$$
$$r \cdot 2 = 2 \cdot 11$$

so that

$$r = 11, \ b = 44.$$

The next question to ask is whether there *are* any (52, 13, 12, 3, 2) designs or any (44, 12, 11, 3, 2) designs.

Cyclic Designs

Since we have as yet given no systematic methods for the construction of designs, our only method is that of trial and error.   Just as the arithmetic of integers modulo n helped us in the construction of Latin squares, here it provides us with a powerful tool for the construction of block designs.   A block design is said to be *cyclic mod n with base* **B** if

(1) The elements of V are equivalence classes mod n.
(2) **B** is a set of blocks of the design.
(3) Every block in the design may be obtained in exactly one way by adding some <u>m</u> (mod n) to each element of a block in **B**.

Example 3.1. If $V = \{0, 1, 2, 3, 4\}$, is the set of equivalence classes mod 5, and $\mathbf{B} = \{\{0, 1\}, \{2, 4\}\}$, then $\mathbf{B}$ is a base for the design

$$\{0, 1\}, \{1, 2\}, \{2, 3\}, \{3, 4\}, \{4, 0\}$$
$$\{2, 4\}, \{3, 0\}, \{4, 1\}, \{0, 2\}, \{1, 3\}.$$

Note that the first row of sets can be written

$$\{0 + 0, 0 + 1\}, \{1 + 0, 1 + 1\}, \{2 + 0, 2 + 1\}, \{3 + 0, 3 + 1\}, \{4 + 0, 4 + 1\}.$$

The second row is obtained in the same way from the set $\{2, 4\}$. Straightforward checking shows this is a BIBD with parameters $(10, 5, 4, 2, 1)$. ■

Notice that one pair of $\mathbf{B}$ consists of adjacent elements and one pair of $\mathbf{B}$ consists of two elements 2 units from each other. In particular the differences

$$1 - 0 = 1, 0 - 1 = 4, 4 - 2 = 2, 2 - 4 = 3$$

include all the nonzero integers modulo 5 exactly once. When we add 1 to each element of $\{0, 1\}$, we get another pair 1 unit apart. Continuing in this way, we get *all* pairs 1 unit apart. We also get the pair $\{4, 0\}$, which is also a pair 1 unit apart if we arrange $0, 1, 2, 3, 4$ around a circle in natural order. It is this circular visualization of distance that underlies the use of the word cyclic. In the same way, we get every pair 2 units apart by repeatedly adding 1 to the pair $\{2, 4\}$. By shifting our base blocks around the circle one place at a time, we ensure that every point appears the same number of times as every other point. The fact that each difference appears once means that as we move around the circle, each pair of numbers with difference 1 appears once, each pair with difference 2 appears once, and so on.

In general, suppose we have a base of blocks for our design. We visualize them as sets on a circle labelled by the integers modulo n. Imagine "picking up" a base block and shifting each element one unit to the right around the circle. This corresponds to adding 1 modulo n to each element of the block. Thus part 3 in the definition of a cyclic design says geometrically that we obtain all blocks by shifting the base blocks repeatedly around the circle. From this, we can guess how to arrange a bunch of sets around a circle in such a way as to be a base for a design. Namely, if a pair $\{a, b\}$ is in some block, then its block can be shifted into one of the base blocks, so for some m, $\{a + m, b + m\}$ is in the base block. We can capture this property without mentioning m by observing that whenever $\{a, b\}$ is in a block, then $a - b = a + m - (b + m)$ is a difference between two elements of a base block. If $\{a, b\}$ appears in two different blocks, then $a - b$ will appear twice as a difference between two elements in the base blocks. Our analysis suggests the following theorem.

Theorem 3.1. A set **B** of k–element sets of the integers mod n is a base for a design which is cyclic mod n if and only if each nonzero integer mod n occurs the same number of times as a difference $\underline{a} - \underline{b}$ of distinct elements chosen from the same set in **B**, and if no proper subset of **B** has this property.

Proof. Suppose **B** is a base. Then condition 3 of the definition tells us that no proper subset of **B** is a base (if a block in **B** were superfluous, then we would have more than one way to get it from a base block, namely adding $\underline{0}$ and some other $\underline{m}$ as well). Our analysis preceding the statement of the theorem shows that if the pair $\{\underline{a}, \underline{b}\}$ is in λ blocks, then $\underline{a} - \underline{b}$ is a difference between λ pairs in the base blocks.

Now suppose each integer mod n appears the same number σ of times as a difference between pairs in **B**. Then for some numbers $m_1, m_2, \ldots, m_\sigma$ the pairs $\{\underline{a} + \underline{m_i}, \underline{b} + \underline{m_i}\}$ must be distinct pairs in the blocks in **B**. Thus any pair $\{\underline{a} + \underline{m}, \underline{b} + \underline{m}\}$ will occur σ times in the blocks we generate from **B** by adding numbers. (Specifically, we add the numbers $m - m_i$ to get the pair mentioned, and these are the only numbers that give this pair.) Thus we have a family of k–element sets generated from **B** such that each pair of elements of V, the integers mod n, occurs exactly σ times among the sets. To show that this family of sets consists of all the blocks of a design, we need only show that each equivalence class mod n occurs the same number of times among the sets. However if one class $\underline{x}$ occurs r times and another class y occurs r' times, then $\underline{x}$ is in $r(k - 1)$ pairs in the blocks and y is in $r'(k - 1)$ pairs in the blocks. Since the number of pairs involving each element is the same, $r = r'$. Thus **B** is a base for a block design. ■

Example 3.2. The sets $\{\underline{0}, \underline{1}, \underline{3}\}$, $\{\underline{2}, \underline{6}, \underline{7}\}$, $\{\underline{5}, \underline{10}, \underline{12}\}$ and $\{\underline{4}, \underline{8}, \underline{11}\}$ form a base for a cyclic block design on the integers mod 13 with $\lambda = 2$. To see this, we form the differences $\underline{0} - \underline{1}, \underline{1} - \underline{0}, \underline{0} - \underline{3}, \underline{3} - \underline{0}, \underline{1} - \underline{3}, \underline{3} - \underline{1}$ for the first set in our proposed base, and the differences $\underline{2} - \underline{6}, \underline{6} - \underline{2}, \underline{2} - \underline{7}, \ldots$ and so on for the other sets in our base. Thus we get four sets of differences, one for each set in the base. The sets we get are $\{\pm \underline{1}, \pm \underline{2}, \pm \underline{3}\}$, $\{\pm \underline{4}, \pm \underline{5}, \pm \underline{1}\}$, $\{\pm \underline{5}, \pm \underline{2}, \pm \underline{7}\}$, and $\{\pm \underline{4}, \pm \underline{3}, \pm \underline{7}\}$. These sets include each equivalence class mod 13 exactly twice since $6 = -7$, $8 = -5$, etc. This design is a (52, 13, 12, 3, 2) design that *almost* meets the criteria we have for our softball league problem. If we use the blocks given in week 0 and add i (mod 13) to the elements in the base blocks to get the blocks of teams who play in week i, then each week all teams but one play once (each has one week off) and each team plays each other team exactly twice in a 13–week period. ■

Resolvable Designs

Our solution to the softball problem did not have each team play *exactly* once each week. A solution that did would be called resolvable. A (b, v, r, k, λ) design is *resolvable* if the blocks may be grouped into s sets each of which is a partition of V. (Note, we partition the blocks into sets of blocks each set of which is a partition of V into parts of size k.)

Theorem 3.2. If a design is resolvable, then $\dfrac{b}{r}$ and $\dfrac{v}{k}$ are the same integer.

Proof. Since V is a union of some number m of disjoint sets of size k, $v = km$. Thus $\dfrac{v}{k}$ is an integer. But since $bk = vr$, we get by division

$$\frac{v}{k} = \frac{b}{r}. \quad \blacksquare$$

Example 3.3. There is no perfect solution to the softball problem with 13 teams because if each team were to play in a week, the blocks for each week's play would be a partition of V and our design would be resolvable. However, $\frac{13}{3}$ is not an integer. (Note that since 13 is not a multiple of 3, we see right away that no schedule could have each team play exactly once a week with three games a night!) This is exactly what it means to say that there is no resolvable design. $\blacksquare$

∞–Cyclic Designs

We need a slightly more complicated idea to deal with a 12–team schedule. Suppose we introduce a new element ∞ to the integers mod n and define

$$\infty + m = \infty - m = \infty$$
$$m - \infty = -\infty + m = -\infty$$

for all m. We say a design is ∞–cyclic mod n if V consists of ∞ and the equivalence classes of integers mod n and ∞, and if it has a base **B** satisfying Conditions 2 and 3 of the definition of a cyclic design. Essentially, the same proof that we used for Theorem 3.1 shows that

Theorem 3.3. If **B** is a family of k–element subsets of the equivalence classes of integers mod n and ∞, then **B** is a *base* for an ∞–cyclic design if and only if ∞, $-\infty$ and each residue mod n occur the same number of times as a difference of elements of the same set in **B** and if no proper subset of **B** has this property.

Example 3.4. The sets $\{\underline{0}, \underline{1}, \underline{3}\}$, $\{\underline{4}, \underline{5}, \underline{9}\}$, $\{\underline{2}, \underline{6}, \underline{8}\}$ and $\{\infty, \underline{7}, \underline{10}\}$ form a base for a $(44, 12, 11, 3, 2)$ ∞–cyclic design mod 11. This is the kind of design

we need for our softball problem with 12 teams. Further, we can use the four
base blocks for the schedule for the first week, because each team would then go
to the field once that week. After we add 1 (mod 11) to each entry in each block,
the teams are still all represented. For each succeeding week, we again add 1 to
each entry of a block; thus we have a perfect solution to the softball problem with
12 teams. ■

The idea used to show that all teams play each week in the example may be
used to prove

Theorem 3.4 A cyclic or ∞–cyclic design is resolvable if it has a base
which is a partition of V.

Triple Systems

A design with $k = 3$ such as those we have constructed is called a *triple system*.
A design with $k = 3$ *and* $\lambda = 1$ is called a *Steiner triple system* (although these designs were first studied by Kirkman). Steiner triple systems are important in their
own right but are also useful as "building blocks" for triple systems with $\lambda = 2$.

Theorem 3.5.   If there is a triple system on v vertices, then $\lambda(v - 1)$ is even
and $\lambda v(v - 1)$ is a multiple of 6.

Proof.   We substitute $k = 3$ in the relations of Theorems 2.1 and 2.2 of this
chapter and solve for r and b, getting

$$r = \frac{\lambda(v - 1)}{2} \quad \text{and} \quad b = \frac{\lambda v(v - 1)}{6}.$$

Since r and b must be integers, the conclusion of the theorem follows. ■

Theorem 3.6.   If there is a Steiner triple system on V, then $v = 6t + 1$ or
$v = 6t + 3$ for some t.

Proof.   Either $v = 6t, 6t + 1, \ldots$, or $6t + 5$ for some t. Since 2 is a
factor of $v - 1$, the only possibilities are $v = 6t + 1, 6t + 3, 6t + 5$.
However, if $v = 6t + 5$, then $v(v - 1) = (6t + 5)(6t + 4) = 36t^2 +
54t + 20$. But 6 cannot divide this sum since it divides the first two terms
but doesn't divide 20. Thus the only possible v's have the form $6t + 1$ or
$6t + 3$. ■

Steiner and Kirkman both asked whether for each v in Theorem 3.6, there
is a Steiner triple system with this v. The answer is yes, as Kirkman showed
around 1850 (Steiner asked the question in 1857, unaware of Kirkman's work).

Hanani has shown that for any v and λ which satisfy the conditions of
Theorem 3.5, there is a triple system with this v and λ. The method is basically to

construct many families of cyclic, ∞–cyclic and related designs and then use methods like the following to build new designs from old ones.

Theorem 3.7. If S_1 is a Steiner triple system on V_1 and if S_2 is a Steiner triple system on V_2, then there is a Steiner triple system on $V_1 \times V_2 = V$.

Proof. As in Theorem 3.1, if we exhibit a family of k–element subsets of V such that each pair of distinct elements of V is in exactly one such subset, then we have exhibited a block design with $\lambda = 1$ on $V_1 \times V_2$. Consider the following subsets of $V_1 \times V_2$.

(Type 1) $\{(a_i, b)|\{a_1, a_2, a_3\}$ is a block of $S_1\}$
(Type 2) $\{(a, b_i)|\{b_1, b_2, b_3\}$ is a block of $S_2\}$
(Type 3) $\{(a_i, b_i)|\{a_1, a_2, a_3\}$ is a block of S_1 and $\{b_1, b_2, b_3\}$ is a block of $S_2\}$.

Then each pair $\{(a_1, b_1), (a_2, b_2)\}$ of distinct elements of $V_1 \times V_2$ satisfies $b_1 = b_2$, in which case $a_1 \neq a_2$, or $a_1 = a_2$, in which case $b_1 \neq b_2$, or else $a_1 \neq b_1$ and $a_2 \neq b_2$. Thus each pair lies in exactly one block of Type 1, 2, or 3. ∎

Kirkman's Schoolgirl Problem

Steiner triple systems need not be resolvable. It is far more difficult to find a *resolvable* triple system than to find just any triple system. In fact, Kirkman posed but could not solve the following problem. "If once every day for a week 15 schoolgirls are to walk in five rows of three girls each, is it possible for each girl to be in a row with each other girl exactly once?" Kirkman was asking for a resolvable (35, 15, 3, 7, 1) design. (Can you explain why?)

Kirkman's problem is especially intriguing since a triple system on 15 vertices with $\lambda = 1$ has

$$b \cdot 3 = r \cdot 15, \quad \text{so} \quad b = 5r$$

and

$$r \cdot (2) = (14) \cdot 1, \quad \text{so} \quad r = 7$$

and there are 35 blocks of size 3. Since there are five blocks walking together each day for a week, we will use exactly 35 blocks. A solution to Kirkman's problem is given in Figure 6.4. The notation is intended to suggest how the solution was constructed.

MONDAY			TUESDAY			WEDNESDAY			THURSDAY		
0	a^0	b^0	0	a^1	b^1	0	a^2	b^2	0	a^3	b^3
a^1	a^2	a^4	a^2	a^3	a^5	a^3	a^4	a^6	a^4	a^5	a^0
a^5	b^1	b^6	a^6	b^2	b^0	a^0	b^3	b^1	a^1	b^4	b^2
a^3	b^2	b^5	a^4	b^3	b^6	a^5	b^4	b^0	a^6	b^5	b^1
a^6	b^3	b^4	a^0	b^4	b^5	a^1	b^5	b^6	a^2	b^6	b^0

FRIDAY			SATURDAY			SUNDAY		
0	a^4	b^4	0	a^5	b^5	0	a^6	b^6
a^5	b^6	a^1	a^6	a^0	a^2	a^0	a^1	a^3
a^2	b^5	b^3	a^3	b^6	b^4	a^4	b^0	b^5
a^0	b^6	b^2	a^1	b^0	b^3	a^2	b^1	b^4
a^3	b^0	b^1	a^4	b^1	b^2	a^5	b^2	b^3

Figure 6.4

EXERCISES

1. Write out all the blocks of the design in Example 3.2.

2. Write out all the blocks of the design in Example 3.4.

3. Find all triple systems on four vertices.

4. For what values of λ is there a triple system on six vertices? (Hint: Exercise is in the previous section contains a useful fact.)

5. Find all resolvable triple systems on seven vertices or show there are none.

6. If there is a resolvable Steiner triple system on v–vertices, then what form must v have? (Note: this does not ask you to construct designs.)

7. An experimenter has nine suntan lotions. In order to test them, the experimenter identifies three similarly exposed patches of skin on sunbathers. Since some conditions vary from day to day, the experimenter wishes to test all nine lotions each day the experiment runs. In addition, each lotion should be tested against each other lotion on the same sunbather exactly once. Describe what kind of design is needed and construct such a design.

8. Show that any two Steiner triple systems on seven vertices are isomorphic. (This problem can be solved by gathering together bits and pieces of information you have already learned.)

9. Find a cyclic symmetric (7, 3, 1) design. Is this the design given in Section 2? Explain how this and the result of Exercise 8 fit together.

10. Is there a cyclic symmetric (11, 5, 2) design?

11. Show that any two Steiner triple systems on nine vertices must be isomorphic.

12. Are the sets {1, 3, 9} and {2, 5, 6} a base for a cyclic design mod 13?   What would the parameters of such a design be?

13. Are the sets {∞, 0, 3}, {0, 1, 4}, {0, 2, 3}, {0, 2, 7} a base for an ∞–cyclic design on ten vertices?   If so, what are its parameters?

14. Find a cyclic (13, 4, 1) design.

15. Find a cyclic (21, 5, 1) design.

16. Can the construction of Theorem 3.7 be used to create a new design with $\lambda = 1$ from two designs with $\lambda = 1$ and the same k.

17. Can the construction of Theorem 3.7 be applied when $\lambda > 1$?

18. If you are given v and k, how should you decide what values of n are worth trying for a cyclic design mod n?   How should you decide how many blocks to try in a base?

*19. Prove Theorem 3.3.

*20. If $v = 6t + 6$, show that the sets listed below form a base for a design with $\lambda = 2$ over the integers mod $6t + 5$.

$$\{\infty, 0, 3t + 2\}$$
$$\{0, i, 2t + 3 - i\} \qquad i = 1, 2, \ldots , t + 1$$
$$\{0, 2i, 3t + 3ti\} \qquad i = 1, 2, \ldots , t$$

21. (a) Consider the sets $\{\infty, a, b\}$, $\{ax, ax^2, ax^4\}$, $\{ax^5, bx, bx^6\}$, $\{ax^3, bx^2, bx^5\}$, $\{ax^6, bx^3, bx^4\}$.   Think of a and b as ax^0 and bx^0, use the rule $\infty x^i = \infty$ for all i.   Show that by adding the integers mod 7 to the *exponents* in the way we added integers mod n to elements of base blocks gives a solution to Kirkman's schoolgirl problem with these base blocks.

   *(b) There is a similar construction using seven base blocks and the integers mod 5; however, there is no element that plays a special role like ∞.   Find the base.

22. Write a computer program that uses backtracking to generate appropriate partitions of a set V and checks to see if they form a base for a cyclic resolvable block design on V.

*SECTION 4   PROJECTIVE PLANES

The Concept of a Projective Plane

In a block design with $\lambda = 1$, each pair of points lies in a unique block (because each pair of points occurs together in exactly one block), just as in geometry each pair of points determines a unique line.   In ordinary geometry, two lines intersect in either one point or no points.   In projective geometry, there are additional "points at infinity" so that two lines which would be parallel in Euclidean geometry intersect in a point at infinity.   Although these points at infinity strain the imagination for a while, they make the geometry easier.

To have a good imaginary picture of points at infinity, you might try to visualize the line at infinity which contains all these points. Try to imagine a point at the "end" of the positive x-axis, a point at the "end" of the positive y-axis, and a semicircle that starts at the "endpoint" of the x-axis, goes up through the "endpoint" of the y-axis and comes down toward the "negative end" of the x-axis without ever reaching it. This semicircle is the line at infinity. Then all lines parallel to the x-axis intersect at the right end of the x-axis, all lines parallel with the y-axis intersect at the point at its end, and all lines with slope 1 intersect at the point on the semicircle halfway between the x- and y-axis. Each possible slope determines a whole family of normally parallel lines; in this projective plane, these lines all intersect at the point we get by "projecting" lines with that slope out to the semicircle.

We define a *projective plane* π to be a set P of points and a family L of subsets of P called lines such that

(1) Each pair of points lies together in exactly one line.
(2) Each pair of lines intersects in exactly one point.
(3) There are four distinct points no three of which lie together in the same line.

Note that Postulate 2 can be turned into Postulate 1 and vice-versa by interchanging "intersect" and "lie together," and "point" and "line." We say Postulates 1 and 2 are *dual* postulates. No dual to Postulate 3 is listed because its dual is implied by postulates 1 through 3.

Theorem 4.1.   In a projective plane, there are four distinct lines no three of which intersect in the same point.

Proof.   Suppose the four points given by Postulate 3 are p_1, p_2, p_3 and p_4. Then by Postulates 1 and 3, there are four lines $L_{12}, L_{23}, L_{34}, L_{41}$ with L_{ij} containing points p_i and p_j but not the other two p's. Suppose three of these lines intersect in some point q, say

$$L_{12} \cap L_{23} \cap L_{34} = \{q\}.$$

Then $L_{12} \cap L_{23}$ must be $\{p_2\}$ by Postulate 2. Thus since

$$\{p_2\} \cap L_{34} = \{q\},$$

$p_2 = q$. Thus p_2 is in L_{34} along with p_3 and p_4, contradicting Postulate 3. ∎

You can further show that Theorem 4.1, Postulate 1 and Postulate 2 imply Postulate 3. What this means is that whenever we use only the postulates in a proof of a theorem about points and lines and then interchange the words "point" and "line," and likewise the word "intersect" and phrase "lie in," then this new statement can be proved by interchanging the same terms in the proof of the original theorem. This gives us the *duality principle* for projective planes.

> **Theorem 4.2.** A theorem which is a consequence of Postulates 1 through 3 for projective planes can be turned into another theorem by interchanging "point" for "line" and "intersect" for "lie in."

A theorem that is converted into a restatement of itself is called *self–dual*.

Projective planes on a finite set of points are block designs, as you may have suspected; the proof of this fact will be smoother if we first present certain additional information about projective planes.

Basic Facts About Projective Planes

> **Theorem 4.3.** For any two lines L_1 and L_2 in a projective plane, there is a one–to–one function from L_1 onto L_2.

> *Proof.* By Postulate 3, there must be a point q not on L_1 or L_2, or else each of these has two points $x_1, y_1 \in L_1$ and $x_2, y_2 \in L_2$ not on the other. In this case, the line joining x_1 and y_1 and the line joining x_2 and y_2 intersect in a point q not on L_1 or L_2. For each point p on L_1, there is a line L_{pq} different from L_1 and L_2 containing p and q. Define $f(p)$ to be the unique intersection of L_{pq} and L_2. Then f is a function from L_1 to L_2. If p and r are different points of L_1, then to have $f(p) = f(r)$ would mean q and $f(p)$ lie on the same unique line that q and $f(r)$ lie on. But this line intersects L_1 in only one point and contains p and r as does L_1. Thus if $f(p) = f(r)$, then $p = r$ because they are both points on the intersection of a certain line with L_1. If s is a point on L_2, then the line containing q and s must intersect L_1 in some point p; thus $f(p) = s$, so f is onto. ∎

We have just shown that for each finite projective plane, any two lines have the same number of points. (We used the correspondence principle again.) Thus there is a number n, called the *order* of the plane, such that all lines of the plane have $n + 1$ points. (The n versus $n + 1$ choice is purely technical.)

> **Theorem 4.4.** Suppose we are given a point p in a projective plane and a line L not containing p. There is a one–to–one function from the set of lines containing p onto the set of points of L.

> *Proof.* For each line L' containing p, define $f(L')$ to be the intersection of L and L'. If $f(L') = f(L'')$, then L' and L'' have two points (p and the common value of f) in common; therefore $L' = L''$. If q is on L, then p and q lie in a line L' containing p. Thus $q = f(L')$. ∎

From this theorem we can conclude that the number of points on any given line equals the number of lines containing any given point. Now we can explain the relationship between projective planes and block designs.

Projective Planes and Block Designs

Theorem 4.5. The lines of a projective plane of order n on a set V of points form an $(n^2 + n + 1, n^2 + n + 1, n + 1, n + 1, 1)$ balanced incomplete block design.

Proof. We have seen that each line has $n + 1$ points, that each point appears in $r = n + 1$ lines and that each pair of points appears together in $\lambda = 1$ lines. Thus the lines form a block design, and since $bk = vr$ and $r = k$, we get $b = v$ and since

$$r(k - 1) = \lambda(v - 1)$$

it follows that

$$(n + 1)n = v - 1,$$

or

$$v = n^2 + n + 1. \ \blacksquare$$

Theorem 4.6. The blocks of an $(n^2 + n + 1, n^2 + n + 1, n + 1, n + 1, 1)$ balanced incomplete block design with $n > 1$ on V form the lines of a projective plane on V if $n > 1$.

Proof. We know that every pair of points lies in a unique block. Now suppose two blocks B_1 and B_2 do not intersect. Let x be a point in B_1. Then for each point y in B_2, x and y are in a unique block. Also if z is in B_2, the block containing x and y must be different from the block containing x and z because B_2 is the only block containing y and z. Thus x would have to be in $n + 2$ blocks, one for each point of B_2 and B_1 as well. Since this is impossible, any two blocks must intersect. Now suppose we have two points x and y and let B_1 be the unique block containing them. There are more than $n + 1$ points, so there is another point z not in B_1. Let B_2 and B_3 be the blocks containing $\{z, x\}$ and $\{z, y\}$. Since $n \geq 2$, then $n^2 + n \geq 3n$, so $n^2 + n + 1$ is greater than the total number of points in B_1, B_2 and B_3. Thus there is a fourth point w not in any of these blocks.   That is, there are four points no three of which are in a block. ■

Example 4.1.   The block design of Examples 2.1 and 2.2 of this chapter is a (7, 7, 3, 3, 1) design and so is a projective plane of order 2.   This plane is shown schematically in Figure 6.5.   Note that six of the lines are represented as line segments; the seventh is represented by the circle. ■

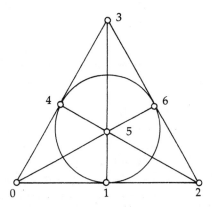

Figure 6.5

It turns out that there is at least one projective plane of order p^n for any prime number p and any positive integer n; further, no examples of projective planes with other orders are known.   One of the currently open and exciting problems of combinatorial mathematics is whether there is a projective plane of order 10.

Example 4.2.   (For those who have studied vector spaces.)   Let W be a three–dimensional vector space.   Let P be the set of one–dimensional subspaces of W and let **L** be the set of two–dimensional subspaces of V.   Then two elements of P lie in a unique member of **L**.   Further, since the dimension of the vector space sum of two members of **L** must be 3, the dimension of their intersection must be 1.   Thus two members of **L** intersect in a unique member of P.

Three linearly independent vectors and their sum generate four one–dimensional subspaces no three of which lie in the same two–dimensional space.   Thus **P** and **L** are the points and lines of a projective plane.   If you are familiar with abstract algebra, you know that any vector space is associated with a field, and that any finite field has p^n elements for some prime p and some integer n.   Further, there are $p^n + 1$ one–dimensional spaces contained in a given two–dimensional space. ■

Projective Planes and Resolvable Designs

More examples of projective planes may be obtained from sets of mutually orthogonal Latin squares. To see how this happens, we must analyze the designs associated with projective planes.

Theorem 4.7. Given a resolvable design D on V with parameters $(n^2 + n, n^2, n + 1, n, 1)$, we may adjoin one additional block with $n + 1$ additional vertices and add these additional vertices to existing blocks to obtain a design D' with parameters $(n^2 + n + 1, n^2 + n + 1, n + 1, n + 1, 1)$.

Proof. Suppose D has been resolved into families $F_1, F_2, \ldots, F_{n+1}$ of blocks each family of which is a partition of the set V.

Add the $n + 1$ symbols $f_1, f_2, \ldots, f_{n+1}$ to V to get V', adjoin f_i to each block in F_i and let D' be the design on V' whose blocks are these enlarged blocks and the block $\{f_1, f_2, \ldots, f_{n+1}\}$. Then D' consists of V' together with a family of $(n + 1)$-element sets of V' such that each point in V' lies in $n + 1$ of these sets. Further, two points of V' are either both f_i's, in which case they lie in the added block (and only that one), or are both in V, in which case they are in exactly one of the blocks created by adding an f_i to a block of D, or else one of the points is in V and the other is an f_i. But each point in V is in exactly one block of F_i. Thus these two points lie in exactly one block of D'. ∎

Theorem 4.8. Each $(n^2 + n, n^2, n + 1, n, 1)$ design D on V is resolvable.

Proof. Let B be a block of D. Let x be an element of V not in B. For each point a of B, there is a unique block of D containing a and x. Further, no two of these blocks can be the same because the only block containing two points of B is B itself. Since x is in $n + 1$ blocks, there is a unique block B_1 containing x but no points of B. For each y in B_1, the block B_1 is the unique block containing y but no points of B. Now let z be a point of V not in B or B_1. Then the unique block containing z but no points of B cannot contain any y in B_1 because B_1 is the unique block containing each of its y's and skipping B. We continue in this fashion, choosing blocks $B_1, B_2, \ldots, B_n$ until we run out of elements of V. Let F_1 be the family of blocks $B, B_1, \ldots, B_n$. Let B' be a block of D not in F_1. Then B' must intersect each block in F_1, because if B' skips every point in B_i, it is the unique block containing a certain point and missing B_i, so it would be in F_1.

We now repeat our process with B' to construct a family F_2 which is a partition of V. Repeating the process until we run out of blocks, we get a resolution $F_1, F_2, \ldots, F_n$ of D into n families of $n + 1$ blocks each. ∎

Projective Planes and Orthogonal Latin Squares

The connection between Latin squares and projective planes is a result of

> **Theorem 4.9.**   If there are $n - 1$ mutually orthogonal Latin squares of side n, then there is an $(n^2 + n, n^2, n + 1, n, 1)$ design.

> *Proof.*   Let V consist of the ordered pairs (i, j) with $1 \leq i \leq n$, $1 \leq j \leq n$. Suppose the n symbols in the Latin squares are $A_1, A_2, \ldots, A_n$ and let the Latin squares be given as matrices $M(1), M(2), \ldots, M(n - 1)$. Then let the following sets be blocks.

> (Type 1) $\{(i, j) | i$ is fixed, $j = 1, 2, \ldots, n\}$ for each i
> (Type 2) $\{(i, j) | j$ is fixed, $i = 1, 2, \ldots, n\}$ for each j
> (Type 3) $\{(i, j) | M(k)_{ij} = A_h\}$ for each k and h.

Thus we have $n^2 + n$ blocks, and since each A_h appears n times in $M(k)$, each block has size n.   Each (i, j) occurs in $n + 1$ blocks.   Suppose we are given two ordered pairs (i, j) and (r, s).   Assume they are in two different blocks B_1 and B_2.   If both B_1 and B_2 are Type 3 blocks, and if B_1 comes from $M(k)$, then

$$M(k)_{ij} = M(k)_{rs} = A_t$$

for some t, and if B_2 comes from $M(h)$, then

$$M(h)_{ij} = M(h)_{rs} = A_u$$

for some u.   Thus, the pair (A_t, A_u) appears twice among corresponding places in the Latin squares $M(k)$ and $M(h)$, and so $M(k)$ and $M(h)$ are not orthogonal, contradicting our assumption.

Now suppose B_1 is a Type 1 block.   Then $i = r$, so if B_2 is a Type 3 block, then for some k, $M(k)_{ij} = M(k)_{is}$.   Thus $M(k)$ is not a Latin square, contradicting our assumption.   If B_2 is a Type 2 block, then $j = s$ and the two pairs are not distinct, while if B_2 were a Type 1 block, then $B_2 = B_1$. All other cases for B_1 and B_2 are covered in a similar way.   Thus each pair (i, j) and (r, s) of distinct ordered pairs lies in at most one block.   However, we have n pairs per block, and $n^2 + n$ blocks, so among the blocks we will find

$$(n^2 + n) \binom{n}{2} = \frac{n^2(n + 1)(n - 1)}{2} = \frac{n^2(n^2 - 1)}{2} = \binom{n^2}{2}$$

two—element sets of pairs. Thus since none of these two—element sets of ordered pairs occurs twice in different blocks, each two—element set of ordered pairs must occur exactly once among the blocks. Thus we have a block design. ∎

Example 4.3. From Corollary 4.5, we know that if n is a prime, there are $n - 1$ orthogonal $n \times n$ Latin squares. Applying Theorems 4.9, 4.8 and 4.7, we see that, therefore, there is a projective plane of order n for each prime number n.

EXERCISES

1. What are the block design parameters of a projective plane of order 3?
2. Explain why a block design with 16 varieties and 20 blocks of size 4 must be resolvable.
3. What are the parameters of a protective plane that arises from the design of Exercise 2?
4. Given three lines in a projective plane of order n, what is the maximum number of points you can find among these lines? What about four lines?
5. Explain why in a projective plane the number of points contained in a given line equals the number of lines containing a given point.
6. Under what circumstances will a Steiner triple system be a projective plane?
7. Find a projective plane with 13 points and 4 points on a line.
8. Find a resolvable (12, 9, 4, 3, 1) design.
9. Why would a symmetric (21, 5, 1) design be a projective plane? Find one. What are the parameters of the resolvable design associated with this plane? Find such a resolvable design.
10. Can the blocks of a resolvable block design form the lines of a projective plane?
11. The "parallel postulate" states that given a point x and a line L_1 not containing x, there is a unique line L_2 containing x but not intersecting L_1. Postulate 4.3' says there are four points any two of which lie on a different line. Show that if we take the points of an $(n^2 + n, n^2, n + 1, n, 1)$ design as points and the blocks of this design as lines, then whenever $n \geq 2$, Postulates 4.1 and 4.3' and the parallel postulate are satisfied.
12. An affine plane is a collection of points together with a family of subsets of the points called lines such that axioms 4.1 and 4.3' and the parallel postulate (Exercise 11) are satisfied.
    (a) Why do two lines intersect in a single point or in the empty set?
    (b) Show that two parallel lines have the same number of points (lines are *parallel* if they do not intersect). (Hint: Use a family of parallel lines that intersect both of the two parallel lines you start with.)

    (c) Show that any point is contained in at least three lines.

    (d) Show that if two lines intersect, then they have the same number of points. (Hint: Use lines parallel to a third line through that point.)

    (e) Show that the lines of an affine plane form the blocks of $(n^2 + n, n^2, n + 1, n, 1)$ design.

    13. In the proof of Theorem 4.9, explain how to deal with the cases where B_1 is of Type 2 and B_2 is of Type 3.

    14. Show that the removal of a line and all its points from a projective plane yields an affine plane.

    15. Explain the relationship between affine planes and orthogonal Latin squares.

    16. Construct a resolvable design on four vertices using a single 2×2 Latin square.

    17. How does Exercise 21 in Section 2 relate to affine planes?

    18. Is Exercise 7 in Section 3 related to affine planes?

    19. Construct the design and then the projective plane associated with two orthogonal 3×3 Latin squares.

    20. Repeat Exercise 19 with three orthogonal 4×4 Latin squares.

SUGGESTED READING

Hall, Marshall: *Combinatorial Theory*, Blaisdell 1967.

Hall, Marshall: *Combinatorial Constructions* in *Studies in Combinatorics*, *MAA Studies in Mathematics*, Vol. 17, G. C. Rota, ed., Mathematical Association of America 1978.

John, P. W.: *Statistical Design and Analysis of Experiments*, MacMillan 1971.

Liu, C. L.: *Introduction to Combinatorial Mathematics*, McGraw-Hill 1968.

Ryser, H. J.: *Combinatorial Mathematics*, Carus Mathematical Monographs, Number 14, Mathematical Association of America 1963.

Street, P. P. and Wallis, W. D.: *Combinatorial Theory: An Introduction*, Charles Babbage Research Center, University of Winnipeg 1977.

7 *Partially Ordered Sets*

SECTION 1   PARTIAL ORDERINGS

What is a Partial Ordering?

Suppose a professor gives three examinations. Each student has a list of three grades. The professor might reasonably believe that the student with the list (g_1, g_2, g_3) is better than the student with the list (h_1, h_2, h_3) if $g_1 \geq h_1$, $g_2 \geq h_2$, $g_3 \geq h_3$ and if at least one of these inequalities is strict. Thus we have a natural "ordering" relation defined on the students. However, if for some i, $g_i > h_i$ while for some j, $g_j < h_j$, the two students with these grade lists cannot be compared. We say that the "better than" relation is an example of a *partial* ordering.

A *strict partial ordering* relation P on a set S is a relation (set of ordered pairs) such that

(1) (x, x) is not in P for any x in S.
(2) If $(x, y) \in P$, and $x \neq y$, then (y, x) is not in P.
(3) If (x, y) and $(y, z) \in P$, then $(x, z) \in P$.

The reader will recognize that Condition 3 says that P is a *transitive* relation; Condition 1 is called the *irreflexive* law and Condition 2 is called the *antisymmetry* law. Thus a strict partial ordering is an irreflexive, antisymmetric and transitive relation. In fact Condition 2 is a consequence of Conditions 1 and 3. A set S with a partial ordering P defined on it is called a *partially ordered set (poset)*.

Example 1.1.   The relation P on the set of *all subsets* of a set S given by $(X, Y) \in P$ if and only if $X \subset Y$ (but $X \neq Y$) is a partial ordering. We normally say that "$\subset$" is a partial ordering, avoiding all mention of P. Similarly, we say "$<$" is a partial ordering on the set of integers. ■

Example 1.2.   There is a relation "finer than" defined on the *partitions* of a set X. Recall that a partition of X is a collection of disjoint sets (called classes) whose union is X. We say the partition Q is finer than the partition R if every class of Q is a subset of some class of R (and $Q \neq R$). (Checking that this is a partial ordering is left to the reader as an exercise.) ■

Notice that in Example 1.1 we used the parenthetical statement "(but $X \neq Y$)," and in Example 1.2 we used the parenthetical statement "(and $Q \neq R$)." We had to include these statements to insure that the relations we were describing were irreflexive. When we think about subsets, we are used to thinking of the subset relation "$\subseteq$," i.e., when we say A is a subset of B we mean that each element of A is an element of B. Thus we allow B to be a subset of itself. In such situations we also use the idea of a *reflexive partial ordering*. This is a relation P which is reflexive (i.e., (x, x) is in P for each x in S), antisymmetric and transitive. Thus "$\subseteq$" is a reflexive partial ordering of the subsets of a set, the ordinary "$\leq$" is a reflexive partial ordering on the integers, etc. The difference between a strict and a reflexive partial ordering is normally clear in context.

Example 1.3.   The positive integers are partially ordered by the "divides" relation. We write $m|n$ if m is a factor of n. Clearly, $m|m$ is true for every positive integer m. Also if $m|n$ and $n|m$, then $n = mk$ and $m = nj$ for some positive integers j and k. Thus $n = njk$, so $jk = 1$. This is possible only if $j = k = 1$, so $m = n$. Thus if $m|n$ and $m \neq n$, then $n|m$ is false. Finally, if $m|n$ and $n|r$, then $n = mk$ and $r = nj$, so that $r = mkj$ and $m|r$. Thus the relation of "divides" is a reflexive partial order. ■

We will write $x < y$ (mod P) rather than $(x, y) \in P$ because the suggestive symbolism is helpful to our intuition. When it is clear from the context that P is the only ordering relation under consideration, we may write $x < y$ rather than $x < y$ (mod P). We write $x \leq y$ if $x < y$ or $x = y$. With this notation we can think of P as standing for both a strict partial ordering and the reflexive partial ordering formed by adding all pairs (x, x) to P. When we write $x < y$ mod P, we are considering the strict ordering; when we write $x \leq y$ mod P, we are considering the reflexive partial ordering.

Linear Orderings

The usual ordering relation of "less than or equal to" on the integers has the property that, given any two integers m and n, either $m \leq n$ or $n \leq m$. This allows us to list a set of k integers in a line as $(m_1, m_2, \ldots, m_k)$ so that $m_i < m_j$ if and only if $i < j$. By analogy, we say that a partial ordering P of a set X is *linear* if for each two elements x and y in X, either

$$x < y \text{ (mod } P) \text{ or } y < x \text{ (mod } P).$$

(Note that a linear ordering is really the same thing as a transitive tournament, a concept we studied in connection with directed graphs.) The partial ordering of subsets by inclusion, of partitions by "finer than" and of the positive integers by "divides" are all examples of nonlinear orderings. In fact, any linear ordering of a k-element set "looks like" the usual ordering on the integers between 1 and k in the following sense.

Theorem 1.1. If L is a linear ordering of the k-element set X, then there is a one-to-one function from the set $K = \{1, 2, \ldots, k\}$ onto the set X such that

$$f(i) \leq f(j) \ (\text{mod } L) \text{ if and only if } i \leq j.$$

Proof. We prove the theorem by induction on k, letting S be the set of all n such that the theorem is true for an n-element set. Clearly, 1 is in S; assume now $k - 1$ is in S.

We choose an element x_1 of X. If there are elements y in X such that $x_1 < y$ (mod L), pick one such y and let it be x_2. Continue in this way to construct a sequence; once x_{i-1} is chosen, if there is some y with $x_{i-1} < y$ (mod L), then let x_i be one such y. Now the elements of the list of x's are distinct, since $x_{i-1} < x_i$ and by successive applications of transitivity $x_{i+j} > x_i$.

Eventually we reach an x_j such that $x_j \not< y$ (mod L) for any y; since L is linear this means that $y \leq x_j$ for all $y \neq x_j$. Now let $X' = X - \{x_j\}$ and let $K' = K - \{k\}$. Since $k - 1$ is in S, there is a function f from K' onto X' such that $f(r) < f(s)$ (mod L) if and only if $r < s$. Extend the function f to K by defining $f(k) = x_j$ (see Figure 7.1). Since $x \leq x_j$ (mod L) for all x in X, $f(r) \leq f(s)$ (mod L) if and only if $r \leq s$. ∎

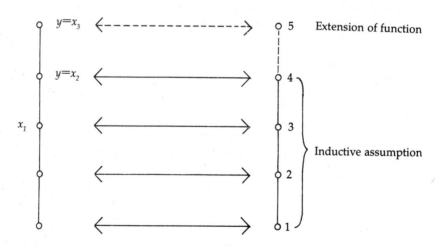

Figure 7.1

A function f (such as the one in Theorem 1.1) from the set X ordered by P to the set Y ordered by R is called an *isomorphism* if f is a one–to–one and onto function such that $f(x) \leq f(y)$ (mod R) if and only if $x \leq y$ (mod P). Thus the preceding theorem tells us that any finite, linearly–ordered set is isomorphic to a set of integers with their natural ordering.

Maximal and Minimal Elements

An element x of X is a *maximum element* relative to the partial ordering P of X if $y \leq x$ for all y in X. An element x of X is *maximal* relative to the partial ordering P if whenever $x \leq y$, in fact $x = y$. Visually, an element is maximum if it is "above everything," while an element is maximal if nothing else is "above" it. Although maximum elements are also maximal, maximal elements are not necessarily maximum. *Minimum* and *minimal* elements are similarly defined.

Example 1.4.   Let X be the set of integers from 1 to 12. Let P be the "divides" relation, i.e., $x \leq y$ (mod P) if and only if x is a factor of y. Then 7, 8, 9, 10, 11 and 12 are all maximal elements of this poset. On the other hand, 1 is the only minimal element and is also a minimum element. ∎

By using much the same method we used to find the maximum element x_j in Theorem 1.1, you can prove

Theorem 1.2.   If the finite set X is partially ordered by P, then it has at least one maximal element and at least one minimal element.

The Hasse Diagram and Covering Graph

We can draw particularly nice graphs or digraphs to represent finite posets. Given a poset X ordered by P, we use the following method, as illustrated in Figure 7.2. Let X_1 be the set of minimal elements of X relative to P. Draw dots to represent these minimal elements in a horizontal line on a piece of paper. Let $Y_1 = X - X_1$ and let P_1 be the restriction of partial ordering P to Y_1; that is P_1 contains exactly those ordered pairs (x, y) of P such that x and y are both in Y_1. Let X_2 be the set of minimal elements of Y_1 relative to P_1 and let $Y_2 = Y_1 - X_2$. Finally let P_2 be the restriction of P_1 to Y_2. Draw a horizontal line of points for the elements of Y_2 above the line for Y_1. Repeat the process, constructing sets X_i, Y_i and an ordering P_i until Y_i is empty.

To draw the digraph of P, draw an arrow from the point for x to the point for y for each (x, y) in P. This digraph is shown separately in Figure 7.3. Note that all arrows point upwards so the arrowheads can be removed without affecting our reading of the ordering from the picture. Some lines (for example, the line from 8 to 2 in Figure 7.3) are redundant in the sense that there may also be lines

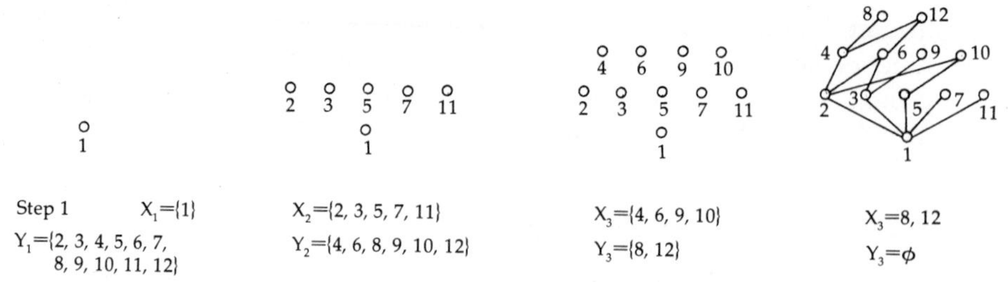

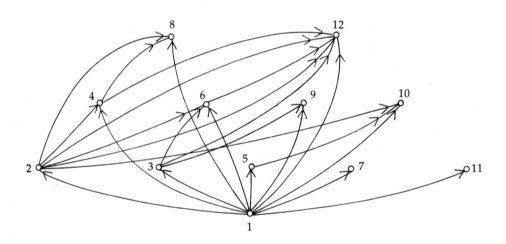

Figure 7.2

Step 1 $X_1 = \{1\}$

$Y_1 = \{2, 3, 4, 5, 6, 7,$
$\quad 8, 9, 10, 11, 12\}$

$X_2 = \{2, 3, 5, 7, 11\}$

$Y_2 = \{4, 6, 8, 9, 10, 12\}$

$X_3 = \{4, 6, 9, 10\}$

$Y_3 = \{8, 12\}$

$X_3 = 8, 12$

$Y_3 = \phi$

Figure 7.3

connecting i (8 in our example) down to j (4 in our example) and j down to k (2 in our example). These redundant lines are required by the transitive law. On the other hand, we can remove a line from i to k whenever a sequence of intersecting lines also leads down from i to k; to interpret the resulting picture, we must then apply the transitive law.   The picture we get by removing redundant lines and arrowheads in Figure 7.3 for the integers between 1 and 12 is shown at the end of Figure 7.2.   This picture gives a much clearer view of the ordering; it is called a *Hasse diagram* of the poset.   If only redundant lines are removed and the arrowheads are left in, the resulting directed graph is called a *covering graph* of the poset.

We say that the element y *covers* the element x relative to the partial ordering P if $x < y$ (mod P) and whenever $x \leq z \leq y$ (mod P) then $z = x$ or $z = y$ (i.e., there is no z such that $x < z < y$).   We can draw the Hasse (or covering) diagram of the poset by drawing the points in levels as above and then drawing a line from y down to x if y covers x.   We show the covering graph of the poset consisting of the integers {2, 4, 3, 12} ordered by divisibility in Figure 7.4.   (Note that the lines from 12 to 3 *must* skip over a level because of the way in which we decided to write down the levels.   However, for the sake of understanding the ordering, we could draw the circle for 3 anywhere below the circle for 12 and still visualize the same relation.)

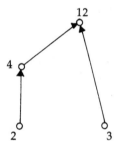

Figure 7.4

The example in Figure 7.4 is called a *restriction* or "induced subposet" of the example in Figure 7.2.   The *restriction* of the ordering P of X to Y is the set of ordered pairs in P both of whose elements are in Y.   This is the same idea of restriction used to describe the method of drawing a covering graph.   Note that a line in the covering graph of a restriction of P need not correspond to a line in the covering graph of P, as you can see by examining Figures 7.2 and 7.4.   Can you explain this?

Posets as Transitive Closures of Digraphs

In our study of digraphs, we introduced the idea of the transitive closure of a relation or diagraph.   A poset is the transitive closure of its covering graph.   In fact, if $G = (V, D)$ is a digraph whose transitive closure is the poset (X, P), then every edge in the covering graph must also be an edge in D. Thus the covering graph is the smallest digraph whose transitive closure is (X, P). This is the meaning of Theorem 1.3.

> **Theorem 1.3.**   If a digraph $G = (V, D)$ has the poset (X, P) as a transitive closure, then every edge of the covering graph of P is an edge of G.

> *Proof.*   Suppose y covers x in P, and P is the transitive closure of D. Because (x, y) is in P, we know that y is reachable from x in D. But if the path from y to x in D included another vertex z, we would have $x < z < y$ in P. Since there is no such z, (x, y) must be an edge in D. ∎

Not every digraph has a partial ordering for its transitive closure. If, however, the digraph has *no* directed cycles, then its transitive closure is a partial ordering. This can be shown by an argument using the idea of reachability. Can you see how? We will give a proof in a later section when we need to use this fact.

Trees as Posets

Recall that a *tree* is a graph with no cycles. Often, when we use a tree for some practical purpose, one of its vertices is designated as a starting place. Such a starting place is typically called a "root." This is the kind of tree we constructed for depth–first search spanning trees and breadth–first search spanning trees of a graph centered at a certain vertex. One reason these trees are so useful is that they give us a natural partial ordering of the vertices. A *rooted tree* is simply a tree with some vertex r selected and designated the root. We use (V, E, r) to stand for a rooted tree with vertex set V, edge set E, and root r. Remember that in a tree any two vertices are connected by a unique path. Let us define $x \le y$ mod P to mean that y is on the unique path connecting x to r. (Intuitively $x \le y$ means we pass y on moving from r to x.) Then P is reflexive, because x is on the unique path from r to x. P is antisymmetric, because if y is on the unique path from r to x and x is on the unique path from r to y, then following the first path from y to x and the second path from x to y would give a closed walk unless $y = x$.

Finally, the relation P is transitive, because if y is on the unique path from r to x and x is on the unique path from r to z, we obtain this second path by first taking the unique path from r to x and then the unique path from x to z. But y is on that unique path from r to x, so y is on the path from r to z. Thus if $z \le x$ and $x \le y$, we conclude that $z \le y$. Figure 7.4 shows an example of a rooted tree. The root is the vertex labelled 12. It is traditional to call x a *child* of y if y covers x. Thus 4 and 3 are the children of 12 and 2 is the child of 4. Vertex 2 and vertex 3 are said to have no children; they are called *terminal* vertices or *leaves*. Whenever we regard a rooted tree as a partially ordered set, the Hasse diagram of the partial ordering is isomorphic to the original tree. The Hasse diagram makes it easier to visualize properties of the tree that are related to the natural ordering. Two more rooted trees are shown in Figure 7.5. The second tree (marked (b)) in Figure 7.5 is an example of a *binary* tree.

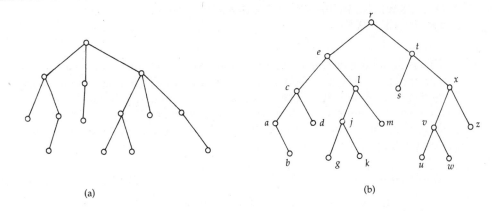

(a)   (b)

Figure 7.5

A binary tree is a special kind of rooted tree in which each vertex has 0, 1 or 2 children. Further, each child is designated as either a right child or a left child of a given vertex and no child is simultaneously left and right. In Figure 7.5, the vertex marked *b* is a right child of the vertex marked *a*; the vertex marked *a* is a left child of vertex *c* and vertex *s* is a left child of vertex *t*. In computer science, binary trees have proved especially useful as tools for keeping track of data. For example, to determine if a certain letter is in the tree of Figure 7.5(b), we start at the root, comparing the letter with *r*. Of course, if the letter is *r*, we are done. If the letter comes before *r*, it will be in the rooted tree that begins with the left child of *r*; otherwise, it will be in the rooted tree that begins with the right child of *r*. How can we describe the process of searching for a letter in the tree? We want a way of describing concisely the idea of looking for our letter at the root, then looking at each child and so on. By giving our procedure a name, we can give the "and so on" a precise interpretation. We shall call our procedure Tree-search(Tree,letter), thus indicating that it is a procedure which searches a given tree for a letter. The procedure is as follows.

PROCEDURE TREESEARCH(TREE,LETTER):
IF "LETTER" IS ROOT OF "TREE," RETURN "YES."
IF "LETTER" COMES BEFORE ROOT OF "TREE," THEN
    IF ROOT OF "TREE" HAS NO LEFT CHILD, RETURN "NO."

OTHERWISE, REPEAT TREESEARCH (LEFT CHILD'S
TREE,LETTER).
IF "LETTER" COMES AFTER ROOT OF "TREE," THEN
IF ROOT OF "TREE" HAS NO RIGHT CHILD, RETURN "NO."
OTHERWISE, REPEAT TREESEARCH (RIGHT CHILD'S
TREE, LETTER).

This is another example of a "recursive" algorithm, one that "calls upon" itself.
The tree examined is called a "binary search tree;" it is an excellent way of in-
dexing a computer file. This topic is taken up in detail in courses on data structures and the analysis of algorithms. In particular, there are a variety of ways to
store a tree in a data structure; the actual structure used depends on the application and the programming language.

EXERCISES

1. Let X be a set of individuals and let P be the relation "is older than" defined on
X. Show that P is a partial ordering. If no two individuals in X have the same age, what
kind of ordering relation is P?

2. Let X be a set of individuals and let $x \le y$ stand for "the weight of person x is
less than or equal to the weight of person y." Show that the relation described by $\le$ is a
partial ordering. Is the relation reflexive or strict?

3. Draw the Hasse diagram for the $\subseteq$ ordering on the set of all subsets of $\{1, 2\}$. Do
the same for all subsets of $\{1, 2, 3\}$.

4. Draw the Hasse diagram for the "is a factor of" partial ordering on $\{1, 2, 4, 6, 9,$
$12, 18, 36\}$.

5. In a simplified notation for partitions, $ab|cde$ stands for the partition of $\{a, b, c,$
$d, e\}$ into the two sets $\{a, b\}$ and $\{c, d, e\}$, while $d|bc|ae$ stands for the partition into the sets
$\{d\}$, $\{b, c\}$, and $\{a, e\}$. Using this kind of notation, write down all partitions of $\{1, 2, 3\}$ and
draw the Hasse graph of the partial ordering "is finer than" on these partitions.

6. Using the notation described in Exercise 5, draw a labelled Hasse graph
showing all partitions of $\{1, 2, 3, 4\}$ ordered by the "finer than" relation. (Hint: There are
15 such partitions.)

7. Show that the subset relation described in Example 1.1 is a partial ordering.

8. Show that the "finer than" relation described in Example 1.2 is a partial ordering.

9. An ordering P of X is called a *weak linear ordering* (or a weak ordering for short)
if whenever $x < y$ (mod P), then for any other z in X, either $x < z$ (mod P) or $z < y$ (mod
P). Show that if P is a weak linear ordering of the k–element set X, then there is a function
(not necessarily one–to–one) from X to a set $J = \{1, 2, \ldots, j\}$ with $j \le k$ such that

$$f(x) < f(y) \quad \text{if and only if} \quad x < y.$$

Explain what this means about the Hasse diagram of a weak linear order.

10. Prove Theorem 1.2.

11. Draw a Hasse diagram of the restriction of the "subset of" relation on the subsets of $\{1, 2, 3, 4\}$ to subsets of size 1 and 3.

12. Show that Condition 2 in the definition of a strict partial ordering is implied by Conditions 1 and 3.

13. If X is a poset ordered by the reflexive partial ordering P, we say a pair (x, y) of elements of X is a *cover* if y covers x. Define a relation on the set C of covers of (X, P) by $(x, y) < (z, w)$ if and only if $(y, z) \in P$. Show that the relation given by $<$ is a partial ordering of C. Is this relation strict? Do we still get a partial ordering if P is strict?

14. Show how the relationship between an equivalence relation and its equivalence class partition can be used to prove that the partially ordered set of all partitions of Y with the "finer than" relation is isomorphic to the partially ordered set of all equivalence relations (considered as sets of ordered pairs) on Y ordered by the "subset" relation.

15. Draw a Hasse diagram for the rooted–tree partial ordering of the vertices of Figure 7.6 obtained with vertex a as root and the ordering obtained with vertex b as a root.

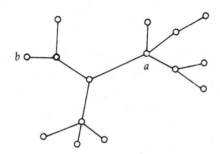

Figure 7.6

16. Make a list of all the roots checked in the order specified by the algorithm Treesearch in searching for vertex c, vertex j, vertex x, and vertex w in the tree of Figure 7.5(b).

17. (a) Extend the tree in Figure 7.5(b) to one with 26 nodes by adding the remaining letters of the alphabet in order. Follow the procedure Treesearch until you reach a point where you would say "no" because a vertex did not have the child you were looking for. Create a child and put this letter there.

    (b) Now do the same thing by adding the letters of the alphabet in reverse order.

    (c) Should we have expected to get the same result in parts (a) and (b)?

18. Why is the transitive closure of a directed graph without directed cycles a partially ordered set?

SECTION 2   LINEAR EXTENSIONS AND CHAINS

The Idea of a Linear Extension

From the pictures of partially ordered sets, it would seem clear that we could line up the points in a straight vertical line in such a way that if $x < y \pmod{P}$, then x

would be lower than y. We simply list all the points on the bottom level in a covering graph in any order we choose, then line up the points on the next level above those from the bottom level, and so on.

In this way we construct a picture of a linear ordering L of the set X which is compatible with P in the sense that whenever $a \leq b$ (mod P), then $a \leq b$ (mod L) as well. Of course unless P were linear to start with, there would be elements x and y in X with $x \leq y$ (mod L), but $x \nleq y$ (mod P). Note that saying that $a \leq b$ (mod L) whenever $a \leq b$ (mod P) is the same as saying that (understood as sets of ordered pairs) $P \subseteq L$. In this case, we say that L is a *linear extension* of P. In computer science, linear extensions are sometimes called "topological sortings" of P. We have seen geometrically that each partial ordering has a linear extension. In fact, a partial ordering P has so many linear extensions that their intersection is P itself. (This will be a corollary of the next theorem.)

Theorem 2.1. Suppose P is a strict partial ordering of a finite set X and let x and y be elements of X such that $x \nleq y$ and $y \nleq x$. Then P has a linear extension in which $x < y$.

Proof. We show first that given a partial ordering P of X and two elements x_1 and y_1 not compared by P (i.e., neither (x_1, y_1) nor (y_1, x_1) is in P), there is a *partial* ordering P_1 containing P and (x_1, y_1). Geometrically speaking, the idea is that to put y_1 above x_1, we must also put each element over y_1 above each element under x_1. Symbolically, we let

$$P_1 = P \cup \{(z, w) \mid z \leq x_1 \quad \text{and} \quad y_1 \leq w \ (\text{mod } P)\}.$$

(In words, we get P_1 from P by adding all pairs (z, w) such that $z \leq x_1$ and $y_1 \leq w$, i.e., the pairs formed by a point under x_1 and a point over y_1.) Since no pair of the form (x, x) is in P_1, it is irreflexive. Now to show P_1 is transitive, suppose (x, y) and (y, t) are in P_1. If they are both in P, then (x, t) is in P and thus P_1. If (x, y) is not in P, then $x \leq x_1$ and $y_1 \leq y$. Thus $y \nleq x_1$ because $y \leq x_1$ would imply $y_1 \leq x_1$. Since y is not less than or equal to x_1, (y, t) was not in the set to the right of the union in the definition of P_1. Therefore $(y, t) \in P$. Thus $y_1 \leq y \leq t$ (mod P), and so $y_1 \leq t$ (mod P). Thus (x, t) *is* in the second set in the definition of P_1. Therefore (x, t) is in P_1, which is what the transitive law requires. In a similar way, it is possible to show that if (y, t) is not in P, then (x, t) is in P_1. For this reason, P_1 is transitive. To see that P_1 is antisymmetric, note that if (x, y) and (y, x) were in P_1, then by transitivity (x, x) would be in P_1—which we already know is impossible. Thus P_1 is a partial ordering.

Now let x_1 and y_1 be the elements x and y referred to in the statement of the theorem. If P_1 is not linear, we can find a pair of elements x_2 and y_2 such that $x_2 \nleq y_2$ and $y_2 \nleq x_2$. Using these elements, we can construct a

P_2 where they are compared.   Since there are only a finite number of ordered pairs of elements of X, we can continue to repeat the process of constructing orderings P_i until P_i has no pair of elements x_{i+1} and y_{i+1} such that $x_{i+1} \nleq y_{i+1} \pmod P$ and $y_{i+1} \nleq x_{i+1} \pmod{P_i}$.   Thus P_i will be linear.   This completes the proof.   ■

Corollary 2.2.   A partial ordering of a finite set X is the intersection of the collection of all its linear extensions.

Proof.   The proof is an elementary application of Theorem 2.1 and is left as an exercise.   ■

Dimension of a Poset

The size of the smallest set of linear orderings whose intersection is P is called the *dimension* of P.

Example 2.1.   The partitions of $\{1, 2, 3\}$ are $\{\{1\}, \{2\}, \{3\}\}$, $\{\{1, 2\}, \{3\}\}$, $\{\{1, 3\}, \{2\}\}$, $\{\{1\}, \{2, 3\}\}$ and $\{\{1, 2, 3\}\}$.   We use the shorthand notation 1/2/3, 12/3, 13/2, 1/23, 123 to stand for these five partitions.   Three linear orderings L_1, L_2 and L_3 which are linear extensions of the refinement ordering are given by

and

$$L_1: 1/2/3 < 12/3 < 13/2 < 23/1 < 123$$
$$L_2: 1/2/3 < 23/1 < 13/2 < 12/3 < 123$$
$$L_3: 1/2/3 < 12/3 < 23/1 < 13/2 < 123.$$

Note that $L_1 \cap L_2$ is the refinement ordering on the partitions of $\{1, 2, 3\}$.   However, $L_1 \cap L_3$ is not the refinement ordering because the ordered pair (12/3, 13/2) is in $L_1 \cap L_3$, meaning that $12/3 < 13/2$ in $L_1 \cap L_3$; however, 12/3 is not a refinement of 13/2.   Since nevertheless there are two orderings (L_1 and L_2) whose intersection is the refinement ordering, and the refinement ordering is not linear, the dimension of the refinement ordering on the partitions of $\{1, 2, 3\}$ is 2.   ■

Example 2.2.   A professor has six students A, B, C, D, E and F.   On test number 1, the six students are rated

$$L_1: A < B < D < C < E < F.$$

On test 2, the six students are rated

$$L_2: C < A < F < B < D < E,$$

and on test 3, they are rated

$$L_3: B < C < E < A < F < D.$$

Assume the professor wishes to order X above Y if X is above Y in all three tests. Then the professor's ordering is the intersection of L_1, L_2 and L_3. Let us draw the Hasse diagram for the resulting ordering of the six students and find the dimension of the ordering. Since A is lower than any other student in the first ordering, A cannot be above any other student in the intersection. (To see why, note that if (X, A) is in the intersection, then (X, A) is in each linear ordering, but no (X, A) is in L_1, so no (X, A) is in the intersection.) Thus A is a minimal element of the intersection, and by the same argument, B and C must be minimal elements too. The same kind of argument shows that D, E and F are maximal elements of the intersection ordering. A is below D and F in all three orderings, but is below E in L_1 and L_2 and above E in L_3. Thus A is below D and F only. Similarly, B is below D and E only and C is below E and F only. Since two minimal elements are incomparable—that is neither is above the other—and two maximal elements are incomparable, we have found all the pairs in the intersection ordering. Thus we get the Hasse diagram of Figure 7.7.

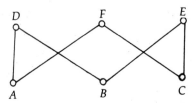

Figure 7.7

Now we know the dimension of our ordering is no more than 3 since the ordering is an intersection of three linear orderings. The dimension cannot be 1, for then the ordering would be linear. Since the dimension can be only 2 or 3, let us see whether the ordering shown in Figure 7.7 could be an intersection of two linear orderings. Suppose this is the case. In at least one of the orderings, A must be above E; in at least one of the orderings, B must be above F; and also in at least one of the orderings, C must be above D (otherwise (A, E), (B, F) or (C, D) would be in the intersection). Now A is above E in one linear ordering, therefore, it is above C in this ordering. Thus C cannot be above D in this ordering, because then C would be above A (because D is over A). Similarly, B cannot be above F

in this ordering.   Also B must be above F in one ordering and the same argument as above shows that C cannot be above D in this ordering.   Thus the three pairs (E, A), (F, B) and (D, C) must occur in three different linear orderings.   Therefore any set of linear orderings whose intersection is the ordered set shown in Figure 7.7 must have at least three linear orderings, so the dimension of our poset is three.   In terms of the interpretation of test results, this means that the professor must have at least three tests to come to the opinion about the six students shown in Figure 7.7. ■

Topological Sorting Algorithms

There are so many applications of the idea of a linear extension that the study of "topological sorting" algorithms has become an important topic in computer science.   Our first geometric construction of a linear extension gives the following easy algorithm for finding linear extensions.

Find a minimal element.
Put it on top of anything already in the linear ordering.
Remove it from the poset.
Repeat the process until the poset is empty.

This procedure is not the best way to do things, for the process of finding a minimal element could lead us along all the edges of the poset.   Then the number of steps in the algorithm execution would be approximately equal to the number n of vertices times the number of edges, which product would likely be a significant fraction of n^3.

To follow the procedure outlined in the proof of Theorem 2.1, we have to systematically check all pairs of vertices for incomparability, and whenever we add a pair of vertices, check only the edges incident with them.   In this procedure, no more than $\binom{n}{2}$ pairs of vertices have to be checked; further, edges are checked only when they impinge on vertices being compared.   Thus the number of steps in this algorithm would be $n^2 + e$, more or less.   This is an improvement over $n \cdot e$.   There are even better algorithms with number of steps proportional to $n + e$.   These algorithms can be modified to guarantee an extension with $x < y$, as in Theorem 2.1.   One such algorithm is outlined in the exercises.

Chain Decompositions of Posets

In organizing a major project with many individual tasks to be completed, we are naturally faced with a partial ordering.   If one task must be carried out before some other task, then it must come before the second task in the partial ordering.

If there is no rush to complete the project, then we can simply schedule the individual tasks in a linear order that extends the partial order. Then we carry out the tasks according to the schedule and thus finish the project. If there is time pressure, however, we might like to schedule the tasks in such a way that if one task neither precedes nor follows a second task, the two tasks could be carried out at the same time by different people. (There are other ways to speed up schedules; however, let us concentrate on this technique alone.) If there are k tasks, no two of which are comparable, then k different people could perform them simultaneously. Of course, if one task comes after another, we could assign both to the same person. For each person involved in the project, we could create a list of tasks such that each task on the list follows (in the partial ordering) its predecessors in the list. For example, for the first person we could choose a "minimal" task, i.e., a task that follows no others, then a task that follows it, and another that follows it and so on until there are no further tasks. Now we can forget about these already assigned tasks, take a second person and create a list for this person in the same way from the remaining tasks. A natural question to ask is "How many people will have to be involved?" We saw that if there are k tasks, no two of which are related, then we need at least k people. In fact, if the biggest set of unrelated tasks has k tasks, there will always be a way to give k people lists of jobs to be carried out according to their linear order on the lists so that all jobs get done.

In a partially ordered set, a subset whose elements are linearly ordered (relative to each other) by the partial ordering is called a *chain*. (This terminology reflects the fact that a diagram of just the part of the ordered set represented by the points in question would look like a chain whose links are the lines connecting adjacent points.) A set of points in a partially ordered set, no two of which are compared by the ordering, is called an *antichain*.

Our analysis of job assignments above showed that if X is partially ordered by P and is a union of k sets X_i such that each X_i is a chain (i.e., each X_i is linearly ordered by P), then k is at least the size of a maximum–sized antichain of (X, P). A theorem originally discovered by R. P. Dilworth says that in fact such a chain decomposition using k chains will always exist. The proof we give is due to M. Perles.

Theorem 2.3. If X is partially ordered by P and if the largest antichain(s) have k elements, then there is a set of k chains whose union is X.

Proof. If X has one element, then X is linearly ordered by P, so X is a union of one chain, namely itself! We will use induction; let S be the set of all integers n such that if X has n elements and has a maximum–sized antichain with k elements, then X is a union of k chains. We see that 1 is in S and now we assume that all positive integers less than j are in S. Suppose now X has j elements.

We consider two cases. Suppose first that the only maximum–

sized antichain is the set of all maximal elements or the set of all minimal elements or that these two sets are the only maximum–sized antichains. Let x be a maximal element, y a minimal element less than or equal to x (mod P) and X' the set $X - (\{x\} \cup \{y\})$ obtained by removing x and y from X. Let P' be the ordering obtained from P by removing ordered pairs in which x or y occur. Any antichain of (X', P') is an antichain of (X, P) and cannot contain all the maximal elements or all the minimal elements of X. Thus maximum–sized antichain of (X', P') has $k - 1$ elements, so X' is a union of $k - 1$ chains because it has fewer than j elements. Since chains for P' are chains for P, X is the union of these $k - 1$ chains together with the chain $\{x, y\}$.

Now we assume that X contains an antichain A with k elements some of which are not maximal and some of which are not minimal. Let U (for upper) consist of those elements of X greater than or equal (mod P) to some element of A and let L consist of those elements less than or equal (mod P) to some element of A. Then U and L are partially ordered by the restriction $P|_U$ of P to U (which consists of pairs of P with elements only of U) and $P|_L$, the restriction of P to L. Since the sizes of U and L are integers in S, $(U, P|_U)$ and $(L, P|_L)$ are each unions of k chains because A is a k-element antichain in each. Each of these chains C_i of U has an element of A at its bottom, and this element of A is also at the top of exactly one chain D_i of L. Since each element of D_i is below that element of A, which in turn is below each element of C_i, $C_i \cup D_i$ is a chain. Thus we have k chains whose union is X.

Thus by the principle of mathematical induction, S contains all positive integers and so the theorem is true. ∎

The size of a maximum–sized antichain is called the *width* of (X, P). The *height* of *an element* of X is the size of the largest chain of elements strictly below (and not equal to) it. The *height* of (X, P) is the largest height of any element of X and thus is 1 less than the size of the largest chain of P.

Example 2.3. The width of the partially ordered set of all subsets of the three–element set $\{1, 2, 3\}$ is 3; the height is also 3. A maximum–sized antichain is $\{\{1\}, \{2\}, \{3\}\}$ and a maximum–sized chain is $\{\emptyset, \{1\}, \{1, 2\}, \{1, 2, 3\}\}$. Three chains whose union is X (the set of all subsets of our three–element set) are

$$\{\emptyset, \{1\}, \{1, 2\}\}$$
$$\{\{2\}, \{2, 3\}, \{1, 2, 3\}\}$$
$$\{\{3\}, \{1, 3\}\} \quad \blacksquare$$

Example 2.4. Show that among a group of $rs + 1$ people, there is either a list of $r + 1$ people such that each person on the list is a descendent of the next person on the list or a set of $s + 1$ people no two of which are descendents of each other.

Consider the relation "is a descendent of." If this relation is a union of m, but not fewer, chains, then a maximum–sized antichain has m elements. If the longest chain(s) has length n, then the set has no more than mn elements. Thus, we cannot have both n less than or equal to r and m less than or equal to s. A chain gives a list of descendents and an antichain is a set of people no two of which are descendents of each other. Note that we could have interchanged the roles of r and s and proved the corresponding result. ∎

*Chain Decomposition and Network Flows

As yet we have no systematic way of determining a decomposition of a partially ordered set into chains. Since Dilworth's theorem turns out to be most useful as a theoretical tool, the lack of a method is not a serious problem but an interesting one. There are methods of finding chain decompositions based on network flows. Given a decomposition of a partially ordered set into n chains, we could add a source and sink to the poset and then use the n chains to send n units to flow from the source to the sink. However, since we want to find a minimal decomposition of the poset into chains, we will have some sort of minimal–flow problem rather than the usual problem of finding a maximal flow. There is an entire theory of finding minimum flows satisfying certain conditions (see the books by Even and by Ford and Fulkerson in the suggested reading); here we shall outline a small part of the theory that provides a method of finding chain decompositions.

We construct a network from our poset (X, P) as follows. We replace each vertex x of X by two vertices x_1 and x_2 together with the edge (x_2, x_1), so that $x_2 \le x_1$. Then for any y less than or equal to x mod P, we make $y_1 \le x_2$ mod P'. We add a source s connected to all the minimal vertices of our new set X' of vertices and a sink t connected to all the maximal vertices. We now seek a flow that has value at least 1 along each of the special edges (x_2, x_1); this is the kind of flow we would get by sending one unit of flow along each chain in a chain decomposition. By finding a flow F of minimum value subject to the condition that the flow is at least 1 along each special edge (x_2, x_1), we can then find a chain decomposition. The process is as follows.

(1) Before deciding on any capacities, construct an f of value F from the source to the sink that gives each edge (special or otherwise) at least one unit of flow. (There are straightforward ways to do this; inspection often suffices.)

(2) For each edge e that is not an (x_2, x_1), let $c(e) = f(e)$. For each (x_2, x_1), let $c(x_2, x_1) = f(x_2, x_1) - 1$.

(3) Find a maximum flow g of value G relative to these capacities.

(4) Now let $c'(e) = f(e) - g(e)$ for each edge.

(5) Replace each edge such that $c'(e) > 1$ with $c'(e)$ edges in parallel, each with capacity 1.

(6) Construct a maximum flow through this network.

As in the proof of Menger's theorem, the flow in this network will give F minus G disjoint paths from the source to the sink.   These paths will correspond to the chains of a chain decomposition.

EXERCISES

1. Write down a linear extension of the eight–element partially ordered set consisting of all subsets of the set $\{1, 2, 3\}$.

2. Write down three linear extensions of the partially ordered set of Exercise 1 whose intersection is the partial ordering.

3. Can the partial ordering of all subsets of a three–element set be an intersection of two linear extensions?

4. What are the height and width of the set of all subsets of a four–element set?

5. Write down a set of chains whose union is the set of all subsets of $\{1, 2, 3, 4\}$.   Is your set as small as possible?

6. Draw the six–element partially ordered set consisting of a, b, c, d, e and f with $a < b$, $c < b$, $a < d < e$, $c < f < e$.   Find its width.

7. Write down three linear extensions of the poset of Exercise 6 whose intersection is the partial ordering.

8. What is the dimension of the poset of Exercises 6 and 7?

9. Show that the conclusion of Dilworth's theorem may be modified to state that the k chains are disjoint.   (Note: it is not necessary to reprove the entire theorem.)

10. What are the height and width of the poset of all numbers between 1 and 12 ordered by "divisibility?"

11. What is the dimension of the partially ordered set of Exercise 10?

12. Apply the construction process of Theorem 2.1 to the poset in Figure 7.7 using B and F as the incomparable pair.   Draw a Hasse diagram for the resulting partial ordering.

13. Show that adding an element b to (X, P) and putting (b, x) in P for all x in X does not change the dimension of P.

14. Show how the following problem (due to Dantzig and Fulkerson) is related to Dilworth's theorem.   A bus company has n different routes (numbered 1 through n).   A schedule set in advance gives the starting time s_i and finishing time f_i for each route.   Let t_{ij} be the time required to move a bus from the destination of route i to the origin of route j. This partially orders the routes: we put (i, j) in P if $f_i + t_{ij} < s_j$.   What is the minimum number of buses needed?

15. Sperner's theorem says that the width of the partially ordered set of all subsets of an n–element set is the largest value of $\binom{n}{k}$.   Prove this theorem.

*16. Given the flow in step 6 of the network flow method, explain why the following technique gives a minimum chain decomposition.   Start at the source with an edge of flow 1.   From its opposite end, take another edge with flow 1 and repeat this until you reach the sink.   Remove these edges and their flows from the network and you will have a *flow* of value one less on the resulting network.   (Why will it be a flow?)   Repeat the process until there is no more flow.

*17. Apply the flow method to get a chain decomposition with the poset consisting of 2, 4, 8, 5, 10, 15, and 20 ordered by "is a factor of".

*18. Show that if flows f and g are constructed as described for a partially ordered set, then the value of $f - g$ is the size of a maximum antichain.

*19. Suppose we use the network of step 4 in our description of the use of network flows for chain decompositions. Will flow–augmenting paths for this network correspond to chains of a chain decomposition?

20. Show that the following algorithm traverses each edge of a poset at most once in each direction and gives a numbering of the poset that corresponds to a topological sorting (linear extension). Analyze the approximate number of steps used.

(0) Set $N = 0$.
(1) Find a minimal element x.
(2) Do a breadth–first search from x to find all elements over x and, in the process, keep track of the longest path from x to each element.
(3) Number the elements from $N + 1$ and up according to their maximum distance from x first and their breadth–first numbering second.
(4) Delete all edges and vertices traversed from the poset and repeat the process.

*21. Write a computer program that computes the chain decomposition of a partially ordered set using the network flow method.

22. The following (recursive) algorithm prints out a special linear extension of a rooted binary tree. Apply it to the tree of Figure 7.5(b). How does this suggest an application of trees as tools for sorting a list of numbers (or words) in increasing order?

Procedure Inorder (root, tree)

(1) If root has a left child carry out Procedure Inorder (left child, left child's tree) before going to step 2.
(2) Print root.
(3) If root has a right child carry out Procedure Inorder (right child, right child's tree).

SECTION 3   LATTICES

What is a Lattice?

In the partially ordered set of all subsets of a set X with the subset ordering, there is a great deal of mathematical structure that is not typical of ordered sets in general. The union of two subsets X and Y is a set greater than or equal to both X and Y in the subset ordering. (The subsets of a 3–element set are shown in Figure 7.9(b).) In the integers from 1 to 12, ordered by "m is less than or equal to n mod P if and only if m is a factor of n," 3 and 4 are both less than 12, so the element 12 is greater than or equal to both of them. However, 3 and 5 have no element greater than or equal to both of them. Thus for 3 and 5 there is no element that plays the role of a union. In the poset in Figure 7.8, the element c is "right above" both a and b as is the element d. On the other hand, *any* set that contains two sets X and Y in the subset ordering also contains the set $X \cup Y$, and so in Figure 7.9(b) $X \cup Y$ is the only set "right above" X and Y. In this sense, both c and

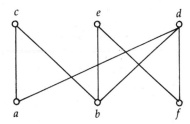

Figure 7.8

d fail to play the role played by a union of two sets, because neither is the *only* element right above c and d.

The set $X \cup Y$ has two fundamental properties. First it is above both X and Y in the subset ordering; we say it is an "upper bound" for X and Y. Second, given any other set above both X and Y, this other set must be above $X \cup Y$; we say $X \cup Y$ is the "least upper bound" for X and Y. In any partially ordered set, we say that u is an *upper bound* for x and y if u is greater than or equal to x and y. We say this upper bound u is the *least upper bound* of x and y if for any v with $x \leq v$ (mod P) and $y \leq v$ (mod P), the element u is less than or equal to v (mod P). (Note that x and y could not have more than one least upper bound.) We use the suggestive notation

$$u = x \vee y$$

and also call u the *"join"* of x and y when u is the least upper bound of x and y; we read $x \vee y$ as "x join y". Similarly, we say w is the *greatest lower bound* of x and y and we write

$$w = x \wedge y$$

if $w \leq x$ (mod P) and $w \leq y$ (mod P), and whenever $v \leq x$ (mod P) and $v \leq y$ (mod P), then $v \leq w$ (mod P) as well. We also call $x \wedge y$ the *"meet"* of x and y; we read $x \wedge y$ as "x meet y". Figure 7.8 shows that in general a pair of elements in a partially ordered set should not be expected to have a meet or a join.

We say that a partially ordered set is a *lattice* if every pair of elements has a least upper bound (join) and a greater lower bound (meet).

Example 3.1. The subsets of a set X, ordered by set inclusion, form a lattice. ■

Example 3.2. If you have had linear algebra you will be able to see that the subspaces of a vector space V form a lattice in which the meet of two subspaces is their intersection and the join of two subspaces is their subspace sum. To show this is so, we must show that the intersection is a greatest lower bound and the sum is a least upper bound. We shall consider the subspace sum here and save the intersection as an exercise. The sum $S + T$ of two subspaces is by definition the set $\{s + t \,|\, s \in S, t \in T\}$. Now $S = \{s + 0 \,|\, s \in S\}$ and $T = \{0 + t \,|\, t \in T\}$, so that $S \subseteq S + T$ and $T \subseteq S + T$. Thus $S + T$ is an upper bound for S and T. Now suppose $S \subseteq U$ and $T \subseteq U$ for a subspace U of V. Then if $s \in S$ and $t \in T$, $s + t \in U$ because U is a subspace. Thus every element of $S + T$ is in U and therefore $S + T \subseteq U$. Therefore $S + T$ is the least upper bound of S and T. ∎

Example 3.3. The posets in Figures 7.8, 7.7, 7.2, 7.4 and 7.5 are not lattices. All the posets shown in Figure 7.9 are lattices. ∎

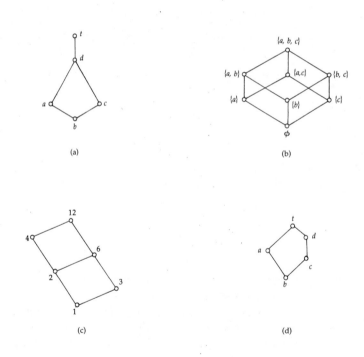

Figure 7.9

The Partition Lattice

Still another example of a lattice is the partially ordered set of all partitions of a set Y. The greatest lower bound of two partitions $\mathbf{P} = \{P_1, P_2, \ldots, P_k\}$ and $\mathbf{Q} = \{Q_1, Q_2, \ldots, Q_m\}$ is easy to describe. Since each element x of Y is in one of the P_i's and one of the Q_j's, each x is in an intersection $P_i \cap Q_j$. Either a set $P_i \cap Q_j$ is empty or else it has no element in common with any of the other sets $P_r \cap Q_s$. Thus the nonempty sets $P_r \cap Q_s$ form the classes of a partition of Y. In any partition below both $\mathbf{P}$ and $\mathbf{Q}$ in the refinement ordering, each block is a subset of one of the P_i and subset of one of the Q_j. Thus each block is a subset of a $P_i \cap Q_j$. Therefore, the partition whose blocks are the nonempty intersections of P_i's and Q_j's is the greatest lower bound for $\mathbf{P}$ and $\mathbf{Q}$. It turns out that in finite lattices we can use the meets to describe the joins. From this it will follow that the partitions of Y form a lattice—we won't have to describe the join to prove it exists. To see the structure we need in order to use meets to describe joins, we introduce the concept of a semilattice.

A partially ordered set is called a *meet–semilattice* when any pair of elements has a greatest lower bound. Correspondingly, a poset is called a *join–semilattice* when each pair of elements has a least upper bound. Each poset in Figure 7.5 is a join–semilattice but not a lattice. The result of the last paragraph may be restated as follows:

Theorem 3.1. The partitions of a set form a meet–semilattice.

Now if any two elements of X have a greatest lower bound mod P and if F is a finite subset of X, then F has a greatest lower bound. In other words, there is an element x such that $x \le y$ for all $y \in F$, and if $z \le y$ for all $y \in F$, then $z \le x$. The proof for general sets F is quite analogous to the proof for the case when $F = \{a, b, c\}$. In this case, we let $x = (a \wedge b) \wedge c$. Then $x \le a \wedge b$ and $x \le c$. But since $a \wedge b$ is a lower bound for a and b, $x \le a$ and $x \le b$. Thus x is less than or equal to all three elements of F. Now suppose z is less than or equal to a, b and c. Then $z \le a \wedge b$ (because $a \wedge b$ is the greatest lower bound for a and b) and $z \le c$. Therefore, $z \le (a \wedge b) \wedge c$, since the right–hand side of the inequality is the greatest lower bound of $a \wedge b$ and c. However, we have just shown that x satisfies the definition of a greatest lower bound of F.

Lemma 3.2. If F is a finite set in a meet–semilattice, then F has a greatest lower bound.

Proof. Let S be the set of all nonnegative integers n such that if F has n elements, then F has a greatest lower bound. Then 2 is in S by definition. Assume k is in S and let F have $k + 1$ elements $x_1, x_2, \ldots, x_{k+1}$. Let $F' = \{x_1, x_2, \ldots, x_k\}$ and let x be the greatest lower bound of F'. Then

$x \wedge x_{k+1}$ is less than or equal to x_{k+1} and x, and since x is less than or equal to all elements of F', then $x \wedge x_{k+1}$ is less than or equal to all elements of F. But if z is less than or equal to all elements of F, it is less than or equal to all elements of F', and so is less than or equal to x (why?) and x_{k+1}. Thus it is less than or equal to $x \wedge x_{k+1}$. Therefore, $x \wedge x_{k+1}$ satisfies the two conditions for a greatest lower bound to F. ■

Thus if (X, P) is a finite meet–semilattice, such as the partitions of a finite set, any nonempty subset of X has a meet. In particular, if x and y are in X and if some elements of X are above both x and y, then the set of all elements above both x and y is a nonempty set F. The meet of F is above both x and y, so it is an upper bound to x and y. Any element above x and y is in F, so it is greater than or equal to the meet of F. In other words, the meet of F is the least upper bound of x and y. This observation yields our next theorem.

Theorem 3.3. If a finite meet–semilattice (X, P) has an element t such that $t \geq x$ for all $x \in X$, then it is a lattice.

Proof. Since the "top element" t is above each x and y, the set F of elements above x and y is nonempty. Thus, as above, the meet of F is a least upper bound for x and y, so every pair of elements has a least upper bound. ■

Example 3.4. If X is the set of partitions of a set Y and if P is the refinement ordering, then (X, P) is a lattice.

This is a direct consequence of Lemma 3.2 and Theorem 3.3 because the partition of Y with one class—Y itself—is greater than or equal to all partitions relative to the refinement ordering. ■

The disadvantage of this method of showing that the poset of partitions of Y is a lattice is that given two partitions **P** and **Q**, we have no concrete description of the least upper bound, or join of **P** and **Q**. Let us try to find such a concrete description. Because **P** and **Q** must be finer than **P** $\vee$ **Q**, every class of **P** must be a subset of some class of **P** $\vee$ **Q** and every class of **Q** must be a subset of some class of **P** $\vee$ **Q**. Now suppose a class C of **P** and a class K of **Q** have an element x in common. Then both C and K must be subsets of the class of **P** $\vee$ **Q** containing x. Thus it would appear that the classes of **P** $\vee$ **Q** are formed by taking unions of overlapping classes of **P** and **Q**. This will be the case if we interpret "union of overlapping classes" properly. For such a proper interpretation, we must realize that while C and K may have x in common, C may have an element y in common with still another class K' of **Q**, and K' may have another element z in common with a class C' of **P** and so on. Thus we form unions until we run out of overlaps at which point we have a class of **P** $\vee$ **Q**.

The Bond Lattice of a Graph

Example 3.5. If $G = (V, E)$ is a (multi)graph, there are some special partitions of V that give rise to an interesting lattice that will be useful later. A partition **P** of V is called a *bond* of G if for each class C of **P**, all pairs of vertices of C are connected by a walk whose vertices are all in C. Thus for the graph in Figure 7.10, the partition $\{\{1, 4\}, \{2, 3\}\} = 14/23$ is not a bond, because 1 and 4 are not connected by a walk in the set $\{1, 4\}$. On the other hand, 12/34 is a bond because 1 and 2 are connected by the edge $\{1, 2\}$ and 3 and 4 are connected by the edge $\{3, 4\}$. The partial ordering on bonds is the usual partial ordering on partitions. The bond lattice of G is shown next to G in Figure 7.10. Notice that the meet of 124/3 and 134/2 is *not* the partition 14/2/3 (as in the partition lattice) because 14/2/3 is *not* a bond.

The bond partitions of a graph *do* form a lattice when ordered by the usual ordering on partitions. Certainly we can check that any two bonds have a greatest lower bound and a least upper bound in the picture in Figure 7.10.

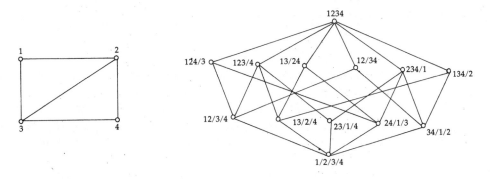

Figure 7.10

The meet of two bonds may be described rather cleverly as follows. Suppose **P** and **Q** are bonds and let F be the set of edges of G both of whose vertices lie in the same class of $\mathbf{P} \wedge \mathbf{Q}$ in the partition lattice. Let **R** be the connected component partition of (V, F). (Note that some vertices grouped together by $\mathbf{P} \wedge \mathbf{Q}$ might not be joined by a walk whose edges are in F.) Though it is not a natural guess to make, it is straightforward to show that **R** is the meet of **P** and **Q** in the bond lattice.

The join of each pair **P** and **Q** in the lattice shown in Figure 7.10 is their join in the partition lattice. It is not too difficult to use the "overlapping classes" description of the join of partitions to show that the join of two bonds in the partition lattice is their join in the bond lattice as well. ■

The Algebraic Description of Lattices

The notation $x \lor y$ for the join of x and y and $x \land y$ for the meet of x and y is suggestive of the algebraic operations of $x + y$ or $x \cdot y$. The set union and intersection operations have many of the arithmetic properties of "plus" and "times." To a lesser extent, the join and meet operations share some of these properties. The most useful properties are listed in the theorem below.

Theorem 3.4.   The operations $\lor$ and $\land$ in a lattice L satisfy the rules

(1) $(x \lor y) \lor z = x \lor (y \lor z)$ (associative law)
$\quad\;\; (x \land y) \land z = x \land (y \land z)$
(2) $x \lor x = x \land x = x$ (idempotent law)
(3) $x \lor y = y \lor x$ (commutative law)
$\quad\;\; x \land y = y \land x$
(4) $x \lor (x \land y) = x$ (absorptive law)
$\quad\;\; x \land (x \lor y) = x$

Proof.   Rules 2 and 3 follow immediately from the definition of least upper bound and greatest lower bound. The first part of rule 4 states that x is the least upper bound of x and $x \land y$. However, regardless of what y is, $x \land y \le x$. Thus x is greater than or equal to both x and $x \land y$. If z is greater than or equal to both x and $x \land y$, then z is greater than or equal to x. Then by definition, x is the least upper bound of x and $x \land y$; this proves the first absorptive law. The proof of the second is similar.
To prove the first associative law, note that if $(x \lor y) \lor z = w$, then $w \ge z$ and $w \ge x \lor y$. Then by the definition of $\lor$, $w \ge x$ and $w \ge y$. Then since $w \ge y$ and $w \ge z$, it follows that $w \ge y \lor z$, and since $w \ge x$ as well, it follows that $w \ge x \lor (y \lor z)$. Thus we get

$$(x \lor y) \lor z \ge x \lor (y \lor z).$$

A similar argument shows that

$$(x \lor y) \lor z \le x \lor (y \lor z)$$

and by the antisymmetry of $\leq$, we get the desired equality.   The second associative law is proved similarly.   ■

It is tempting to expect the "distributive" law $x \wedge (y \vee z) = (x \wedge y) \vee (x \wedge z)$ to hold since it holds for intersections and unions of sets. (Check that it holds for sets if you don't remember why!)   However, it does not hold in general for lattices, and so you must take care in making computations using standard arithmetic rules.

Note by the way that $(x \wedge y) \wedge z$ is exactly the greatest lower bound of x, y and z, as we already showed must be the case.   Thus the associative law says that either natural interpretation of the symbol $x \wedge y \wedge z$ is the greatest lower bound of x, y and z.   A valuable observation is that rules 1–4 give us an alternate description of a lattice.

Theorem 3.5.   Any mathematical system L with two operations $\wedge$ and $\vee$ defined on all pairs of elements in L and satisfying rules 1–4 of Theorem 3.4 is a lattice with the partial ordering P defined by $x \leq y \bmod P$ if and only if $x = x \wedge y$.

Proof.   It is necessary to verify that the relation P is a partial ordering. This is a straightforward application of rules 1–3.   Then it is necessary to verify that $x \wedge y$ is the greatest lower bound of x and y.   This is also a straightforward application of rules 1–3.   To show that $x \vee y$ is the least upper bound of x and y, we first use rule 4 to show that $x \leq y$ if and only if $y = x \vee y$; then we apply rules 1–3 again.   ■

EXERCISES

1. Let X consist of the integers from 1 to 10 and let P be the usual less than or equal to relation.   Show that two elements of X have a least upper bound, namely the larger of the two numbers.   Show that (X, P) is a lattice.

2. Let X consist of the positive factors of 12 (1 and 12 included) and let P be the relation "is a factor of" or "divides."   Does every pair of elements have a least upper bound?

3. Let X consist of all the positive integers between 1 and 12 (inclusive) and let P be the relation "is a factor of."   Does every pair of elements have a greatest lower bound?

4. Every positive integer can be factored in one and only one way into a product of powers of prime integers.   Use this fact to show that every pair of positive integers has a least upper bound relative to the partial ordering "is a factor of."   (Hint: Consider all primes that occur in the unique factorization of two integers.   Multiply together all prime powers that occur in one but not both factorizations and multiply this by the higher power of each prime appearing in both of the factorizations.)

5. (Continuation of Exercise 4).   Show that each pair of elements in the partially ordered set of Exercise 4 has a greatest lower bound.

6. Each equivalence relation on a set Y may be regarded as a set of ordered pairs.

For two equivalence relations E and R on Y, let us say that $E \leq R$ if E is a subset of R. Show that the equivalence relations on Y form a meet–semilattice.

7. (Continuation of Exercise 6). Show that the equivalence relations on Y form a lattice.

8. The lattice of Exercise 7 is isomorphic to another lattice studied in the text. Discover which one it is and show why the two lattices are isomorphic.

9. Find all partitions of $\{1, 2, 3, 4\}$ that are not bonds in Example 3.5 and show that they are in fact not bonds. (One such partition was described in Example 3.5). How do you know you have found them all?

10. Let G be the graph on the vertices $\{1, 2, 3, 4\}$ with edges $\{1, 2\}, \{2, 3\}, \{3, 4\}$ and $\{1, 4\}$. Find the bonds of G and draw the bond lattice of G.

11. Draw the bond lattice of the graph on the vertices $\{1, 2, 3, 4\}$ with edges $\{1, 2\}$, $\{2, 3\}, \{1, 3\}$ and $\{1, 4\}$.

12. What is the bond lattice of the complete graph on four vertices, the graph with all possible edges connecting four vertices? (Hint: this is a lattice whose name you already know.)

13. Recall that a tree is a connected graph with no cycles. Give as complete a description as you can of the bond lattice of a tree on four vertices.

14. What can you say about the bond lattice of a complete graph? Of a tree?

15. (a) Show that the join of two bonds may be computed as described in Example 3.5.

   (b) Show that the meet of two bonds may be computed as described in Example 3.5.

16. Prove the second associative law in Theorem 3.4.

17. Prove the second absorptive law in Theorem 3.4.

18. We say b is the *bottom* of a poset (X, P) if $b \leq x$ for all $x \in X$ and t is the *top* if $t \geq x$ for all $x \in X$. Prove that if a lattice has a bottom and a top, then

$$x \vee b = t \wedge x = x \qquad \text{for all } x \text{ in the lattice } (X, P)$$
$$t \vee b = t \quad \text{and} \quad b \wedge x = b \qquad \text{for all } x \text{ in the lattice } (X, P).$$

19. Show that the lattice of subsets of a set satisfies the distributive laws
   (a) $X \cap (Y \cup Z) = (X \cap Y) \cup (X \cap Z)$
   (b) $X \cup (Y \cap Z) = (X \cup Y) \cap (X \cup Z)$.

20. Show that the lattice of partitions of $\{1, 2, 3\}$ satisfies neither of the distributive laws
   (a) $X \wedge (Y \vee Z) = (X \wedge Y) \vee (X \wedge Z)$
   (b) $X \vee (Y \wedge Z) = (X \vee Y) \wedge (X \vee Z)$.

21. Show that a finite lattice has a bottom and a top (see Exercise 15).

22. Show that the set of all partial orderings of X, ordered by inclusion, is a poset. Is it a lattice? A semilattice?

23. (a) Show that the relation P defined in Theorem 3.5 is a partial ordering.
   (b) Show that $x \wedge y$ is the greatest lower bond of x and y in the partial ordering of Theorem 3.5.
   (c) Show that $x \vee y$ is the least upper bond of x and y in the partial ordering of Theorem 3.5.

*24. Review the concept of a matroid.   An element x of a matroid is said to depend on the independent set I if $I \cup \{x\}$ is *not* independent.   The *flat* $F(I)$ determined by I consists of I together with all points that depend on I.   Show that the flats $F(I)$ of a matroid form a lattice.   (This kind of lattice is called a geometric lattice and is very important in theoretical combinatorics.   The bond lattice of a graph turns out to be isomorphic to a geometric lattice, for example.)

SECTION 4   BOOLEAN ALGEBRAS

The Idea of a Complement

We began our study of lattices with the lattice of subsets of a set.   We soon saw that the lattice of subsets of a set "behaves" much more nicely than other lattices.   We have, for example, the distributive laws

$$A \cap (B \cup C) = (A \cap B) \cup (A \cap C)$$

and

$$A \cup (B \cap C) = (A \cup B) \cap (A \cup C).$$

We saw these laws don't hold in all lattices, but they do in some.   For example, in the lattice of integral divisors of 12 shown in Figure 7.11, we have that

$$p \wedge (q \vee r) = (p \wedge q) \vee (p \wedge r) \quad \text{and} \quad p \vee (q \wedge r) = (p \vee q) \wedge (p \vee r). \quad (4.1)$$

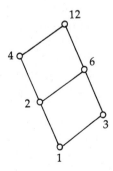

Figure 7.11

A lattice satisfying the two rules given in Equations (4.1) is called a *distributive lattice*. There is, however, another property that makes subset lattices, such as the one shown in Figure 7.12, stand out from other distributive lattices such as the one in Figure 7.11. Each set A in Figure 7.12 can be paired with one other subset A^{-1} in such a way that

$$A \cup A^{-1} = \{a, b, c\} \quad \text{and} \quad A \cap A^{-1} = \emptyset.$$

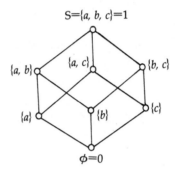

Figure 7.12

On the other hand, in Figure 7.11 there is no element 3^{-1} such that $3 \vee 3^{-1} = 12$.

The set A^{-1} is called the complement of A in the set $S = \{a, b, c\}$. For example, $\{a\}^{-1} = \{b, c\}$ and $\{a, b\}^{-1} = \{c\}$. Thus A^{-1} is just the set we have been denoting by $S - A$ all along. We have known all along that

$$(S - A) \cup A = S \quad \text{and} \quad (S - A) \cap A = \emptyset.$$

The complementation property occurs in other areas as well, for example in the negation of statements in elementary logic or programming languages and in inverter circuits in circuit design. It turns out that both these important examples (which we shall discuss later) also give rise to distributive lattices. In order to give a uniform description of the use of complementation in distributive lattices, we introduce some new terms. In this context, it has become quite standard to use 0 to stand for the bottom element of a lattice and 1 to stand for the top element. We adopt this standard for this section.

In an (arbitrary) lattice L with bottom 0 and top 1, an element c is said to be a *complement* to a if

$$a \vee c = 1$$

and

$$a \wedge c = 0.$$

A lattice L is said to be *complemented* if every element of L has a complement.

Boolean Algebras

A *Boolean algebra* is a complemented distributive lattice.

Example 4.1.   Discuss which lattices in Figures 7.11, 7.12 and 7.13 are distributive, which are complemented and which are Boolean algebras.

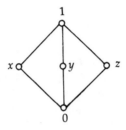

Figure 7.13

We have already observed that the lattice in Figure 7.11 is distributive and that the lattice in Figure 7.12 is distributive because it is a subset lattice.   In Figure 7.13, however, $(x \vee y) \wedge z = 1 \wedge z = z$ and $x \wedge z = 0$ and $y \wedge z = 0$, so $(x \wedge z) \vee (y \wedge z) = 0$.   Thus the lattice is not distributive.   We have observed that 3 has no complement in Figure 7.11; in fact neither do 2, 4 or 6.   We have observed that Figure 7.12 represents the subsets of a set S and that for each A, the set $S - A$ is its complement.   Thus Figure 7.12 is a complemented lattice.   In fact, Figure 7.13 is also a picture of a complemented lattice because 0 has 1 for a complement, 1 has 0 for a complement, x has two complements y and z, y has two complements x and z, and z has two complements x and y.   Among the three lattices, only the lattice in Figure 7.12 is both distributive and complemented.   Thus the lattice in Figure 7.12 is the only Boolean algebra among the three lattices.   ■

Boolean Algebras of Propositions

Given a collection of arrangements, we can make a variety of statements about them.   For example, our arrangements could be graphs on five vertices.   Some statements we make are statements which, for certain of the arrangements, will be false. For example, the statements, "This graph is connected and has four edges" will be true for the trees in our collection of graphs on five vertices and will be false for all other graphs in the collection. Other statements are ambiguous—for example, "This graph is pretty," is ambiguous. A statement that is true for a certain set of arrangements in our collection and false for all other arrangements is called a *proposition*. A proposition can be true for all arrangements, it can be false for all arrangements or true for some and false for all others. The important point is that for each given arrangement, the proposition is either a true statement or a false statement.

The idea of the complement to a statement is a simple one. The complementary statement to "This graph is connected" is "It is not the case that this graph is connected" or, more briefly, "This graph is not connected". Given a certain set $p_1, p_2, \ldots, p_n$ of propositions, we can form compound propositions such as "p_1 and p_2" or "(p_1 or p_2) and it is not the case that p_3". To avoid the clumsy use of English, it is traditional to say "not p_i" and use $\sim p_i$ (rather than p_i^{-1}) to stand for "It is not the case that p_i". We use $p_i \wedge p_j$ to stand for "p_i and p_j" and $p_i \vee p_j$ to stand for "p_i or p_j or both p_i and p_j". In English we ordinarily read $p_i \vee p_j$ as "p_i or p_j." This symbolism for "and" and "or" suggests that there is a lattice involved, and in fact there is.

To see that compound propositions form a lattice, we must check the rules of Theorems 3.4 and 3.5. These rules involve the notion of equality, though, and we haven't said what it means for two statements to be "equal," i.e., equivalent. For example, in a discussion of graphs on five vertices, the two statements

$$G \text{ is connected and } G \text{ has four edges}$$
$$G \text{ has four edges and } G \text{ is connected}$$

both mean the same thing, namely, that G is a tree on five vertices. We will say that two propositions about a certain collection of arrangements are equivalent if they are true for exactly the same arrangements. We will write $p = q$ to mean p is equivalent to q. Clearly, we would agree that $p \vee q$ is equivalent to $q \vee p$. This is the lattice law $p \vee q = q \vee p$. How should we go about checking a more complicated lattice law like $(p \vee q) \wedge p = p$?

Truth Tables

The method of truth table analysis allows us to quickly search for equivalences among statements.

In Table 7.1 we show possible assignments of F (false) and T (true) to each of the statements p and q and the corresponding truth values normally applied to the compound statements $p \vee q$, $p \wedge q$ and $\sim p$.

p	q	p	$\wedge$	q	p	$\vee$	q	$\sim$	p
T	T	T	T	T	T	T	T	F	T
T	F	T	F	F	T	T	F	F	T
F	T	F	F	T	F	T	T	T	F
F	F	F	F	F	F	F	F	T	F
		1	2	1	1	2	1	2	1

Table 7.1

A truth table has a column for every symbol in an expression (except for parentheses if there are any). First, in the column under each of the original propositions, we copy the truth value of the original proposition from the left–hand columns. These columns are marked at the bottom with 1's meaning this was our first step. Next, we look for each operator ($\wedge$, $\vee$ or $\sim$) whose truth value we can determine, and fill it in. These columns are marked with 2's.

Such a step–by–step process can be used to analyze more complex propositions. For example, in Table 7.2, we analyze the two statements $(p \wedge q) \wedge p$ and $(p \wedge q) \vee q$ that appear in the absorptive laws for lattices.

p	q	$(p$	$+$ $\vee$	$q)$	$\wedge$	$+$ p	$(p$	$+$ $\wedge$	$q)$	$\vee$	$+$ q
T	T	T	T	T	T	T	T	T	T	T	T
T	F	T	T	F	T	T	T	F	F	F	F
F	T	F	T	T	F	F	F	F	T	T	T
F	F	F	F	F	F	F	F	F	F	F	F
		1	2	1	3	1	1	2	1	3	1

Table 7.2

In steps 1 and 2, we proceed as above; in step 3, we use the truth values in the two columns marked with a plus sign to combine the truth values of $(p \vee q)$ with those of p and to combine the truth values $p \wedge q$ with those of q. Now we note that the column marked 3, which has the truth values for the proposition $(p \vee q) \wedge p$, has exactly the same truth values as does the column for p. Thus $(p \vee q) \wedge p$ is equivalent to p. Column 3 under $(p \wedge q) \vee q$ has the same entries as the column for q. Thus $(p \wedge q) \vee q$ is equivalent to q. We have proved that the lattice laws $(p \vee q) \wedge p = p$ and $(p \wedge q) \vee q = q$ apply to propositions.

In Table 7.3 we give the truth table analysis of $(p \vee q) \vee r$ and $p \vee (q \vee r)$. This shows that the associative lattice identity holds for propositions.

				+		+		+	+		+	
p	q	r	(p	∨	q)	∨	r	p	∨	(q	∨	r)
T	T	T	T	T	T	T	T	T	T	T	T	T
T	T	F	T	T	T	T	F	T	T	T	T	F
T	F	T	T	T	F	T	T	T	T	F	T	T
T	F	F	T	T	F	T	F	T	T	F	F	F
F	T	T	F	T	T	T	T	F	T	T	T	T
F	T	F	F	T	T	T	F	F	T	T	T	F
F	F	T	F	F	F	T	T	F	T	F	T	T
F	F	F	F	F	F	F	F	F	F	F	F	F
			1	2	1	3	1	1	3	1	2	1

Table 7.3

Again for convenience, we mark with a plus sign the two columns used in stage 3.

In order to study complements, we need statements to act as 0 and 1 in our lattice. Given our collection of arrangements, we can use for 0 the statement "This arrangement is not in our collection". This statement is false for all arrangements; it could be called "logically false." For 1 we can use the statement "This arrangement is in our collection". Note that $\sim 1 = 0$ and $\sim 0 = 1$, so that 1 and 0 are complements of each other. The statement $p \wedge \sim p$ is never true, so it is equivalent to 0, i.e., $p \wedge \sim p = 0$. The statement $p \vee \sim p$ is always true, so $p \vee \sim p = 1$. Using truth table analyses we can check the other lattice laws and the distributive laws, so we get the theorem:

> **Theorem 4.1.** The propositions that may be made about a collection of arrangements (with "=" interpreted as "is equivalent to") form a Boolean algebra.

Combinatorial Gate Networks

Boolean algebras provide a convenient framework for studying electrical networks built from "two–state devices." A spot on a magnetic tape can be either magnetized or not, a switch can be either on or off, a voltage in a wire can be either low or high and so on. These are all examples of two–state devices.

Several standard devices can be built to electronically combine voltages $x_1, x_2, \ldots, x_n$ on n different input wires. One is an "inverter," shown first in Figure 7.14. An inverter turns a high voltage into a low one and turns a low voltage into a high one. An "and gate," shown next in Figure 7.14, takes two voltages as input and produces a high voltage if both are high, and otherwise a low voltage. An "or gate," shown third in Figure 7.14, takes two voltages as input, and produces a low voltage if both are low and otherwise produces a high voltage.

Table 7.4 shows the voltages produced by each gate using the convention 0 for low voltage and 1 for high voltage.

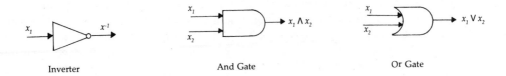

Figure 7.14 Standard symbols for logic gates

Inputs		Devices		
x_1	x_2	x_1-Inverter	And Gate	Or Gate
1	1	0	1	1
1	0	0	0	1
0	1	1	0	1
0	0	1	0	0

Table 7.4

Because the 0's and 1's in this table mirror exactly the F's and T's in the table for $\sim$, $\wedge$, and $\vee$, we may use $x_1 \vee x_2$ to stand for the result of feeding voltages x_1 and x_2 into an "or" gate, $x_1 \wedge x_2$ to stand for the result of feeding voltages x_1 and x_2 into an "and" gate, and x_1^{-1} (read as x_1 inverted) for the result of feeding x_1 into an "inverter" gate. Then any expression such as $(x_1 \vee x_2) \wedge (x_3 \vee x_1^{-1})$ corresponds to a circuit which can be built with an input wire for each x_i and one overall output wire. In Figure 7.15 we show the circuit represented by the expression $(x_1 \vee x_2) \wedge (x_3 \vee x_1^{-1})$. Such a circuit is called a *combinatorial network*.

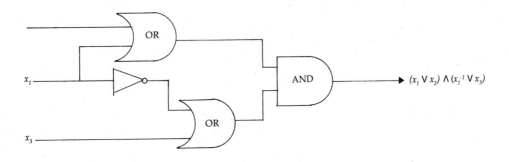

Figure 7.15

We can use the symbol 0 to stand for a device that accepts any number of input wires and produces low voltage regardless of the input. (A wire with a piece cut out or a circuit for $x_1^{-1} \wedge x_1$ is such a device.) We can use the symbol 1 to stand for a device that takes any number of inputs and produces high voltage regardless of the input. (The circuit for $x_1^{-1} \vee x_1$ is such a device.) The voltage table (7.4) is in a one–to–one correspondence with the truth tables we used to show that propositions form a lattice. Using voltage tables in the same way as we used truth tables and considering two circuits equal if, based on the same inputs, they produce the same outputs, we get Theorem 4.2.

> **Theorem 4.2.** The circuits built from inverters, "and" gates and "or" gates with inputs $x_1, x_2, \ldots, x_n$ form a Boolean algebra. (Note: Implicit in this theorem is that we are allowed to make the usual series and parallel circuit connections as in Figure 7.15.)

Boolean Polynomials

We give no proof for Theorem 4.2 since it uses the same ideas as Theorem 4.1. The meaningful expressions that can be built up from $x_1, x_2, \ldots, x_n$ using parentheses, $\vee$, $\wedge$, and inversion are called *Boolean polynomials* in the symbols $x_1, x_2, \ldots, x_n$. When we substitute the values 1 or 0 for the x_i's and use the rules $1 \vee 1 = 1$, $1 \wedge 1 = 1$, $0 \vee 0 = 0$, $0 \wedge 0 = 0$, $0 \vee 1 = 1 \vee 0 = 1$, $1 \wedge 0 = 0 \wedge 1 = 0$, $0^{-1} = 1$ and $1^{-1} = 0$ of Boolean arithmetic, we are using our polynomials as *Boolean functions* of n variables. Thus Boolean polynomials represent circuits and Boolean functions represent the relationship between the input and output of such circuits.

How can we determine whether two circuits accomplish exactly the same thing—i.e., are equivalent? Determining such equivalences is one reason we develop algebraic manipulation rules for Boolean algebras. Our eventual goal is to have a standard or *canonical form* of writing polynomials; two polynomials are then equivalent if and only if they have the same canonical form.

In the same way that the circuits we may build up from certain wires using Boolean operators form a Boolean algebra, we see that the set of Boolean polynomials in certain variables form an isomorphic Boolean algebra, an abstract version of the algebra we can build from circuits. Although it is clear what it means to put an inverter in front of a circuit, it is perhaps not so clear what it means to form the complement $(p(x_1, x_2, \ldots, x_n))^{-1}$ of a Boolean polynomial. It turns out that whenever we have an expression such as $(x \vee y)^{-1}$ or $(x \wedge y)^{-1}$, it can be re-expressed in terms of x^{-1} and y^{-1}. *DeMorgan's laws* tell us that in any Boolean algebra

$$(x \vee y)^{-1} = x^{-1} \wedge y^{-1}$$

and

$$(x \wedge y)^{-1} = x^{-1} \vee y^{-1}.$$

Theorem 4.3.   DeMorgan's laws are valid in every Boolean algebra.

Proof.   We must show that $x \vee y$ has $x^{-1} \wedge y^{-1}$ as its complement.   Thus we form the join

$$\begin{aligned}
(x \vee y) \vee (x^{-1} \wedge y^{-1}) &= ((x \vee y) \vee x^{-1}) \wedge (x \vee y) \vee y^{-1})) \\
&= (x \vee x^{-1} \vee y) \wedge (x \vee y \vee y^{-1}) \\
&= 1 \wedge 1 = 1.
\end{aligned}$$

Now we form the meet

$$\begin{aligned}
(x \vee y) \wedge (x^{-1} \wedge y^{-1}) &= (x \wedge x^{-1} \wedge y^{-1}) \vee (y \wedge x^{-1} \wedge y^{-1}) \\
&= 0 \vee 0 \\
&= 0.
\end{aligned}$$

DeMorgan's other law is proved similarly.   ∎

By induction, we may extend DeMorgan's laws to read

$$(x_1 \vee x_2 \vee \cdots \vee x_n)^{-1} = x_1^{-1} \wedge x_2^{-1} \wedge \cdots \wedge x_n^{-1}$$

and

$$(x_1 \wedge x_2 \wedge \cdots \wedge x_n)^{-1} = x_1^{-1} \vee x_2^{-1} \vee \cdots \vee x_n^{-1}. \tag{4.2}$$

By applying DeMorgan's laws (4.2) repeatedly, we may take any Boolean polynomial and express it as a polynomial in the symbols $x_1, x_1^{-1}, x_2, x_2^{-1}$, $\ldots, x_n, x_n^{-1}$ such that the only operations used to build up the polynomial are join and meet.   This is analogous to applying the binomial and multinomial theorems to ordinary polynomials built up from multiplication, sum, and nonnegative integer exponentiation to obtain polynomials built up from the symbols $x_1^{i_1}, x_2^{i_2}, \ldots, x_n^{i_n}$ by addition and multiplication.   For ordinary polynomials, we have a canonical form: namely, every polynomial in the variables x_1, x_2, $\ldots, x_n$ that may be built up using sum, product (and exponentiation) may be written as a sum of monomials of the form $x_1^{i_1} \cdot x_2^{i_2} \cdot \ldots \cdot x_n^{i_n}$ where, of course, x_i^0 is interpreted as 1.

Disjunctive Normal Form

Our canonical form for Boolean polynomials, which goes by the name *disjunctive normal form*, is given in Theorem 4.5. A preliminary result is Theorem 4.4 which just formalizes the fact that we can distribute meet over join in the same way as we distribute multiplication over sum.

Theorem 4.4. Every Boolean polynomial may be written as a join of monomials of the form

$$x_1^{i_1} \wedge x_2^{i_2} \wedge \cdots \wedge x_n^{i_n} = \bigwedge_{i=1}^{n} x_j^{i_j}$$

where $i_j = +1, -1$, or 0, $x_i^1 = x_i$, x_i^{-1} is the complement of x_i and $x_i^0 = 1$.

Proof. By applying DeMorgan's laws, we may rewrite our polynomial as a polynomial in the terms $x_1, x_2, \ldots, x_n$ and $x_1^{-1}, x_2^{-1}, \ldots, x_n^{-1}$ built up from meet and join alone. Just as ordinary multiplication distributes over ordinary addition, meet distributes over join in a Boolean algebra. Thus, just as we may express an ordinary polynomial made from sum and product operations as a sum of products, we may write a Boolean polynomial made from join and meet operations as a join of meets. Since $x_i^0 = 1$, any meet of just some terms of the form x_i and x_j^{-1} may be written in the form $x_1^{i_1} \wedge x_2^{i_2} \wedge \cdots \wedge x_n^{i_n} = \bigwedge_{i=1}^{n} x_j^{i_j}$. ∎

Now in each monomial of the form

$$x_1^{i_1} \wedge x_2^{i_2} \wedge \cdots \wedge x_n^{i_n}$$

in which x_j^0 appears, we may replace x_j^0 by $x_j \vee x_j^{-1}$. We then distribute meet over join again as in

$$x_1 \wedge x_2 \wedge x_3^0 = x_1 \wedge x_2 \wedge (x_3 \vee x_3^{-1}) = (x_1 \wedge x_2 \wedge x_3) \vee (x_1 \wedge x_2 \wedge x_3^{-1}).$$

In this way, we may rid our expression of all terms with an exponent of 0 to get Theorem 4.5.

Theorem 4.5. (Disjunctive Normal Form). Every Boolean polynomial may be written as a join of monomials of the form

$$x_1^{i_1} \wedge x_2^{i_2} \wedge \cdots \wedge x_n^{i_n} = \bigwedge_{j=1}^{n} x_j^{i_j}$$

where each i_j is either -1 or 1.

Proof. Replace each $x_j{}^0$ by $x_j \vee x_j{}^{-1}$ and distribute as shown above. ∎

It may be surprising that each Boolean algebra is isomorphic to a Boolean algebra of sets. From results used to prove this fact, we shall conclude that each of the joins of monomials is different from all the rest, i.e., that the disjunctive normal form is unique. Then to tell if two polynomials (or two circuits, or two logical expressions) are equivalent, we put the two polynomials in disjunctive normal form and determine if they are joins of the same monomials.

All Boolean Algebras are Subset Lattices

An *atom* in a lattice is an element which covers the bottom element 0. We shall show that in a Boolean algebra, each element can be represented uniquely as a join of atoms. From this, it will follow that a Boolean algebra whose atoms are a set A is isomorphic to the lattice of subsets of A. Our next three theorems are special cases of Garret Birkhoff's unique representation theorems for distributive lattices in general, and Stone's representation theorem for Boolean Algebras in general.

> **Theorem 4.6.** Every nonzero element of a finite Boolean algebra B is a join of atoms.
>
> *Proof.* Let x be nonzero and let $a_1, a_2, \ldots, a_k$ be the atoms below x (there must be at least one or else x would be 0). Let $y = a_1 \vee a_2 \vee \cdots \vee a_k$. Now since $y \vee y^{-1} = 1$ and $x \geq y$, $x \vee y^{-1} = 1$. If there are any atoms below $x \wedge y^{-1}$, they are also below x, but no atom below x is below y^{-1} (since $a_i \wedge y^{-1} = 0$ by the definition of y^{-1}). But if $x \wedge y^{-1}$ has no atoms below or equal to it, then $x \wedge y^{-1} = 0$. Now we apply the distributive law to

$$x = x \wedge 1 = x \wedge (y \vee y^{-1})$$

to get

$$x = (x \wedge y) \vee (x \wedge y^{-1}) = (x \wedge y) \vee 0 = y.$$

Thus x must be the join of all the atoms below it. ∎

We have just shown that each nonzero x is the join of the set of all atoms below it. If we show that x is not a join of a proper subset of these atoms, then we have shown that x is expressed uniquely as a join of atoms.

> **Theorem 4.7.** Each nonzero element can be represented in only one way as a join of atoms, namely as the join of all atoms below it.

Proof. Let $x = a_1 \vee a_2 \vee \cdots \vee a_k$ where the a_i's are all the atoms below x. Suppose $x = a_1 \vee \cdots \vee a_{k-1}$ as well. Then

$$
\begin{aligned}
a_k = a_k \wedge x &= a_k \wedge (a_1 \vee a_2 \vee \cdots \vee a_{k-1}) \\
&= (a_k \wedge a_1) \vee (a_k \wedge a_2) \vee \cdots \vee (a_k \wedge a_{k-1}) \\
&= 0 \vee 0 \vee \cdots \vee 0 \\
&= 0
\end{aligned}
$$

because the meet of two distinct atoms is 0. This contradiction means that x is not the join of any $k - 1$ of the atoms below it and thus not the join of any proper subset of the atoms below it. ■

Theorem 4.8. The map $f(x) = $ "the set of all atoms below x" is an isomorphism from a Boolean algebra with atoms A onto the lattice of subsets of A.

Proof. Theorems 4.5 and 4.6 show the map is one–to–one and onto. Both f and f^{-1} are order–preserving, so f is an isomorphism between the lattices, understood as partially ordered sets; thus f is an isomorphism of lattices and, therefore, of Boolean algebras. ■

Corollary 4.9. If a Boolean algebra has n atoms, it has 2^n elements.

Corollary 4.10. Every Boolean polynomial has a unique representation in disjunctive normal form.

Proof. In the lattice of Boolean polynomials, the monomials $x_1^{i_1} \wedge x_2^{i_2} \wedge \cdots \wedge x_n^{i_n}$, where $i_n = \pm 1$, are the atoms. To see this, note that

$$
(x_1^{i_1} \wedge x_2^{i_2} \wedge \cdots \wedge x_n^{i_n}) \wedge (x_1^{j_1} \wedge x_2^{j_2} \wedge \cdots \wedge x_n^{j_n}) = 0
$$

unless each i_k and j_k are equal; if $i_k \neq j_k$, then $x_k \wedge x_k^{-1} = 0$ is in this meet. By the distributive law, the monomial $M = x_1^{i_1} \wedge x_2^{i_2} \wedge \cdots \wedge x_n^{i_i}$ will meet with a join of monomials to give 0 unless M is a monomial in that join. Then by Theorem 4.5, $M \wedge p(x_1, x_2, \ldots, x_n) = 0$ unless $M \leq p(x_1, x_2, \ldots, x_n)$. Thus there is no polynomial $q(x_1, x_2, \ldots, x_n)$ with $M > q(x_1, x_2, \ldots, x_n) > 0$. Therefore, each monomial is an atom. Since, by Theorem 4.4 everything else is a join of monomials, the monomials are the only atoms. Then by Theorem 4.7, this representation is unique. ■

Corollary 4.11. The number of nonequivalent Boolean polynomials in n variables is 2^{2^n}.

Proof. We have 2^n atoms, the monomials. Apply Corollary 4.8. ∎

Corollary 4.12. There are 2^{2^n} nonequivalent combinatorial networks (gating networks) with n inputs.

An important topic not covered in this section is the efficient representation of Boolean polynomial, i.e., ways of representing a polynomial with as few connectives as possible. This is especially important in electrical network design. The books in the suggested reading by Birkhoff and Bartee, by Fisher, and by Prather all contain discussions of this problem, as do books on the design of digital circuits.

EXERCISES

1. Analyze which of the lattices in Figure 7.16 are distributive.

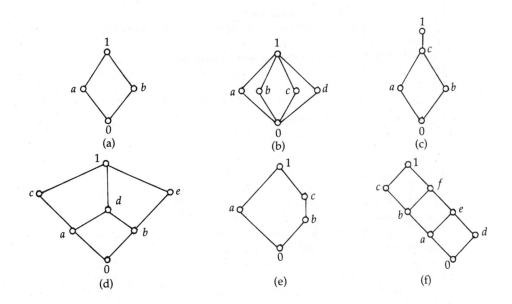

Figure 7.16

2. Analyze which of the lattices in Figure 7.16 are complemented.
3. Which lattices in Figure 7.16 are Boolean algebras and why?
4. A lattice is "uniquely complemented" if, for each element a, there is a unique

element c such that

$$a \vee c = 1$$

$$a \wedge c = 0.$$

Which lattices in Figure 7.16 are uniquely complemented?

5. Draw the Boolean algebra of all subsets of a four–element set.

6. By Corollary 4.12, there are 2^{2^2} combinatorial networks that can be built up with two input wires. Some of these are trivial consisting of x_1 and x_2 going in and only x_1 coming out for example. The Boolean algebra has 16 elements. Draw the Boolean algebra and sketch two typical combinatorial networks corresponding to each of two elements of height 1, height 2 and height 3 in the lattice respectively.

7. Draw and label the Boolean algebra of all Boolean polynomials in two variables x and y.

8. Construct a truth table to show that the associative law for $\wedge$ holds for propositions.

9. Construct a truth table to show that the distributive law $p \wedge (q \vee r) = (p \wedge q) \vee (p \wedge r)$ holds for propositions.

10. Construct a truth table to show that the distributive law $p \vee (q \wedge r) = (p \vee q) \wedge (p \vee r)$ holds for propositions.

11. Explain why the lattice of nonnegative factors of 30 is a Boolean algebra, while the lattice of nonnegative factors of 36 is not. (Recall the ordering in both cases is by ''is a factor of.'')

12. Construct circuits that correspond to the following Boolean polynomials.
   (a) $(x_1 \vee x_2) \wedge x_3$
   (b) $x_1^{-1} \vee x_2$
   (c) $(x_1 \vee x_2)^{-1}$
   (d) $(x_1 \wedge x_2^{-1}) \vee (x_1^{-1} \wedge x_2)$
   (e) $(x_1 \vee x_2) \wedge x_3^{-1}$
   (f) $(x_1 \wedge x_2 \wedge x_3) \vee (x_1 \wedge x_2^{-1} \wedge x_3^{-1}) \vee (x_1^{-1} \wedge x_2 \wedge x_3^{-1}) \vee (x_1^{-1} \wedge x_2^{-1} \wedge x_3)$.

13. It is conventional to use $x_i \cdot x_j$ to stand for $x_i \wedge x_j$. Rewrite each of the Boolean polynomials in Exercise 12 using this convention.

14. The truth set of a statement about a collection of arrangements is the set of arrangements that makes the statement true. In Figure 7.17, we show Venn diagrams of truth sets 1 and 2 corresponding to statements x_1 and x_2 and of truth sets 1, 2 and 3 corresponding to the statements x_1, x_2 and x_3. Copy the diagrams and shade in the truth sets of the statements corresponding to the polynomials in Exercise 12.

15. An element p of a lattice is called prime or *join–irreducible* if whenever $p = x \vee y$ either $p = x$ or $p = y$.
   (a) Show that in a finite distributive lattice, every element is a join of join–irreducibles.
   (b) Is the word ''distributive'' necessary in part (a)?
   (c) An element p_i in a subset $\{P_1, P_2, \ldots, P_n\}$ of a lattice is redundant if $p_i \leq p_j$ for some $j \neq i$. Show that each element of a finite distributive lattice may be represented uniquely as a join of a set of join–irreducible elements

which has no redundant elements. (This is the general result of Birkhoff referred to in the text.)

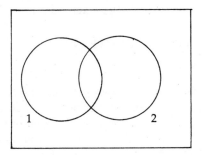 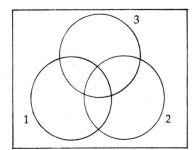

Figure 7.17

16. Prove the DeMorgan law $(x \wedge y)^{-1} = x^{-1} \vee y^{-1}$.

17. Write out the DeMorgan laws for two variables x and y in the notation of propositions and verify them by means of truth tables.

18. Addition in a Boolean algebra can be defined by

$$a \oplus b = (a \wedge b^{-1}) \vee (a^{-1} \wedge b).$$

(a) Show that addition is commutative and associative and that $a \oplus 0 = a$ and $a \oplus a = 0$.

(b) Show that meet distributes over addition.

(c) Does join distribute over addition?

19. We implictly assumed (but did not actually use) the fact that each element of a Boolean algebra has exactly one complement. Prove that if an element of a distributive lattice has a complement, it has a unique complement.

20. (For those who have studied linear algebra with finite fields.) Show that a Boolean algebra is a vector space over the integers mod 2 with the addition defined in Exercise 18 and with $1 \cdot a = a$, $0 \cdot a = 0$.

21. The truth tables for the two propositions $p \rightarrow q$ and $p \leftrightarrow q$ (read "if p then q" and "p if and only if q") are

p	q	$p \rightarrow q$		p	q	$p \leftrightarrow q$
T	T	T		T	T	T
T	F	F		T	F	F
F	T	T		F	T	F
F	F	T		F	F	T

Write down propositions involving p, q, $\sim$, $\wedge$, and $\vee$ that have each of these truth tables. Explain (in words) why the propositions you just wrote down involving the three basic connectives "should be" equivalent to $p \rightarrow q$ and $p \leftrightarrow q$.

22. Write down schematic diagrams of electric circuits which are the analogs of the propositions "if x_1 then x_2" and "x_1 if and only if x_2" (see Exercise 21).

SECTION 5   MÖBIUS FUNCTIONS

A Review of Inclusion and Exclusion

One of the important enumeration principles we developed in Chapter 3 was the principle of inclusion and exclusion. A similar basic principle called Möbius inversion has been thoroughly developed in number theory. In solutions of various combinatorial problems there have been a number of similar computations. In the 1960's, G.–C. Rota developed an all–inclusive formulation of a general method known as Möbius inversion that unified all these results and has had far–reaching applications to partially ordered sets. We shall present the elements of this generalization and a few of its applications here.

In computing the number of functions from a set J onto a set K, we studied, for each subset I of K the number $N_{\geq}(I)$ of functions from J to K that "skipped" I or some bigger subset of K. We also studied the number $N_{=}(I)$ of functions from J to K that skipped exactly the set I (in other words, functions that skip all of I but don't skip anything else). The number $N_{=}(\emptyset)$ is the number of functions that skip the empty set and nothing else; thus it is the number of functions from J onto K. We found $N_{=}(\emptyset)$ by applying the principle of inclusion and exclusion. We could also find $N_{=}(\emptyset)$ by solving systems of linear equations. Note that a function that skips I or some bigger set must skip exactly a certain set S with $S \supseteq I$. This lets us write the symbolic equation

$$N_{\geq}(I) = \sum_{S=I}^{J} N_{=}(S) \tag{5.1}$$

for each subset I of J. This yields a system of 2^j equations in 2^j unknowns; the unknowns are the numbers $N_{=}(S)$. We can see without too much effort that the number $N_{\geq}(I)$ is just 2^{j-i}. The principle of Möbius inversion for sets will tell us that whenever we have a system of equations of the form (5.1), regardless of the actual values of $N_{\geq}(I)$ and $N_{=}(S)$, the system of equations can be solved for $N_{=}(I)$ in terms of $N_{\geq}(I)$ by writing

$$N_{=}(I) = \sum_{S=I}^{J} (-1)^{|S|-|I|} N_{\geq}(S). \tag{5.2}$$

This is the same formula that we got for the principle of inclusion and exclusion.

The difference is that there is no mention of properties here; instead the equations in (5.1) *define* how $N_=$ and $N_\geq$ are related.

Notice that Equation (5.1) has a counterpart for any finite partially ordered set (X, P) with top element t; namely, if N is a numerical function defined on the set X, we can write

$$N_\geq(x) = \sum_{y=x}^{t} N(y) = \sum_{y: y \geq x} N(y). \qquad (5.3)$$

Here $y \geq x$ means $y \geq x$ mod P. (From here on we will use N rather than $N_=$ to remind us that it is Equation (5.3) that defines the relation between N and $N_\geq$, rather than some set of properties.) The second sum in Equation (5.3) is read "the sum of $N(y)$ over all y such that y is greater than or equal to x," and the first sum is read as "the sum of $N(y)$ over all y from x to t." Since the notation of the second sum is more explicit and does not force us to assume that (X, P) has a top element t, we shall adopt it. Equation (5.3) then actually determines the values $N_\geq(x)$ of a numerical function $N_\geq$ defined on X. Just as we saw in the case of onto functions, $N_\geq(x)$ will often be a number we can compute while $N(x)$ is difficult to compute directly. Thus a result analogous to Equation (5.2) would be desirable. It is just this kind of result which was found by Rota.

The Zeta Matrix

We show how to convert the equations shown in (5.3) into a matrix equation. By solving this matrix equation, we will get the desired analog to (5.2). In our discussion of graph theory, we presented the adjacency matrix of a graph. Since a partial ordering P of a set X is a directed graph on X, it has an adjacency matrix; in the context of partially ordered sets, it is traditional to call this the Zeta matrix of P. Specifically, given a listing $x_1, x_2, \ldots, x_n$ of X, we let

$$Z_{ij} = \begin{cases} 1 & \text{if } x_i \leq x_j \\ 0 & \text{otherwise.} \end{cases}$$

This is the matrix version of the function "Zeta" given by

$$\zeta(x, y) = \begin{cases} 1 & \text{if } x \leq y \\ 0 & \text{otherwise.} \end{cases}$$

Example 5.1. Compute the Zeta matrix for the lattice of partitions of a three–element set.

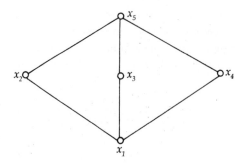

Figure 7.18

If we number the elements of the lattice of partitions of a three–element set as shown in Figure 7.18, we get for that poset the Zeta matrix

$$
\begin{array}{c@{\quad}c}
 & \begin{array}{ccccc} 1 & 2 & 3 & 4 & 5 \end{array} \\
\begin{array}{c} 1 \\ 2 \\ 3 \\ 4 \\ 5 \end{array} &
\begin{bmatrix}
1 & 1 & 1 & 1 & 1 \\
0 & 1 & 0 & 0 & 1 \\
0 & 0 & 1 & 0 & 1 \\
0 & 0 & 0 & 1 & 1 \\
0 & 0 & 0 & 0 & 1
\end{bmatrix}. \blacksquare
\end{array}
$$

Given a column vector C of length n, the product ZC is another column vector D whose i–th entry is given by

$$D_i = \sum_{j=1}^{n} Z_{ij} C_j. \tag{5.4}$$

Since Z_{ij} is 1 exactly when $x_i \le x_j$, the sum in equation (5.4) is the sum of the values C_j that correspond to all the x_j's greater than or equal to x_i. In symbols,

$$D_i = \sum_{x_j : x_j \ge x_i} C_j. \tag{5.5}$$

This leads us to a restatement of the equations in (5.3) as follows.

Theorem 5.1.   Let C be the column vector whose entries are $N(x_1)$, $N(x_2)$, . . . , $N(x_n)$.   Let D be the column vector whose entries are $N_\ge(x_1)$, $N_\ge(x_2)$, . . . , $N_\ge(x_n)$.   Then

$$D = ZC.$$

Proof.   For C_j we substitute $N(x_j)$ in formula (5.5).   For D_i we substitute $N_{\geq}(x_i)$.   Then equation (5.5) reads

$$N_{\geq}(x_i) = \sum_{x_j : x_j \geq x_i} N(x_j). \tag{5.6}$$

Equation (5.6) is the same as Equation (5.3) with x_i playing the role of x and x_j playing the role of y.   Thus the equation $D = ZC$ is just a restatement of the relation between N and $N_{\geq}$. ■

The Möbius Matrix

Note that the matrix Z we computed for the lattice in Figure 5.1 is upper triangular and has 1's along the main diagonal.   For any poset, there is a way to number the elements of the poset so that the Zeta matrix will be upper triangular.   From this, we can show that Z is invertible.

Theorem 5.2.   For any poset (X, P), the Zeta matrix Z has an inverse M. Further, there is a way to list X as $x_1, x_2, \ldots, x_n$ so that Z and M are both upper triangular.

Proof.   List X in the order of a linear extension of P.   Use this listing to construct Z.   Then if $i > j$, we know that $x_i \nleq x_j$ so $Z_{ij} = 0$.   Thus Z is upper triangular.   Further, $Z_{ii} = 1$ because $x_i \leq x_i$ for each i.   Since Z is upper triangular and has 1's on the main diagonal, it has an inverse M.   The inverse of an upper triangular matrix is upper triangular, so M is upper triangular. ■

The matrix M is called the *Möbius matrix* of the partially ordered set.

Example 5.2.   What is the Möbius matrix for the lattice of partitions of a three–element set?   By row reduction, we may compute that

$$M = \begin{bmatrix} 1 & -1 & -1 & -1 & 2 \\ 0 & 1 & 0 & 0 & -1 \\ 0 & 0 & 1 & 0 & -1 \\ 0 & 0 & 0 & 1 & -1 \\ 0 & 0 & 0 & 0 & 1 \end{bmatrix}.$$

(You may wish to check that $ZM = I$.) ■

The fact that Z has an inverse allows us to solve our system of equations.

Theorem 5.3.   Let C be the column vector whose entries are $N(x_1)$, $N(x_2)$, $\ldots$, $N(x_n)$.   Let D be the column vector whose entries are

$N_\ge(x_1)$, $N_\ge(x_2)$, . . . , $N_\ge(x_n)$. Then

$$C = MD.$$

Proof. By Theorem 5.1, $D = ZC$. Thus

$$
\begin{aligned}
MD &= M(ZC) \\
&= (MZ)C \\
&= IC \\
&= C. \ \blacksquare
\end{aligned}
$$

The Möbius Function

The *Möbius function* μ (the Greek mu) of a partially ordered set (X, P) is a function of two variables given by

$$\mu(x_i, x_j) = M_{ij}.$$

Thus μ is defined from M in the same way that ζ is defined from Z. The analog of the inclusion–exclusion theorem is Theorem 5.4, the *principle of Möbius inversion*.

> **Theorem 5.4.** Suppose (X, P) is a partially ordered set and let the functions N and $N_\ge$ be related by
>
> $$N_\ge(x) = \sum_{\substack{y:y \ge x \\ \mathrm{mod}\ P}} N(y).$$

Then

$$N(x) = \sum_{\substack{y:y \ge x \\ \mathrm{mod}\ P}} \mu(x, y)N_\ge(y).$$

Proof: The fact that $N(x) = \sum_{y:y \in X} \mu(x, y)N_\ge(y)$ is a translation of Theorem 5.3 from matrix notation. The remainder of the proof consists in showing that $\mu(x, y)$ is 0 unless $y \ge x$. Recall that Z_{ij} is 0 unless $x_i \le x_j$; in fact, M_{ij} is 0 unless $x_i \leq x_j$. To see why this is so, note that

$$MZ = I$$

is equivalent to saying

$$\sum_{k=1}^{n} M_{ik} Z_{kj} = \begin{cases} 1 & \text{if } i = j \\ 0 & \text{otherwise.} \end{cases} \tag{5.7}$$

Assume there is a pair (i, k) such that $M_{ik} \neq 0$ but $x_i \nleq x_k$. Since the inverse of an upper triangular matrix is upper triangular, M_{ik} must be 0 if $i > k$, so we may assume $i \leq k$. If it is the case that $x_i < x_k$, then $i \neq k$. Thus, if $M_{ik} \neq 0$ but $x_i \nleq x_k$, we may assume $i < k$. Among all pairs (i, k) with these properties, pick a pair with $k - i$ as small as possible. In other words, pick i and k so that if $i \leq j < k$ and $M_{ij} \neq 0$, then $x_i \leq x_j$. Now by Equation (5.7) and the fact that M is upper triangular

$$0 = \sum_{j=i}^{k} M_{ij} Z_{jk} = \sum_{j=i}^{k-1} M_{ij} Z_{jk} + M_{ik} Z_{kk}$$

$$= \sum_{j=i}^{k-1} M_{ij} Z_{jk} + M_{ik}$$

so that

$$\sum_{j=1}^{k-1} M_{ij} Z_{jk} = -M_{ik} \neq 0.$$

Thus for some j, both M_{ij} and Z_{jk} must be nonzero, so that $x_i \leq x_j$ since $M_{ij} \neq 0$ and $j - i < k - i$; in addition $x_j \leq x_k$ since $Z_{jk} \neq 0$. Then by transitivity $x_i < x_k$, a contradiction. Thus for all i and j, $M_{ij} \neq 0$ implies $x_i \leq x_j$. For this reason, we can rewrite the conclusion of Theorem 5.3 as

$$N(x_i) = \sum_{j=1}^{n} M_{ij} N_{\geq}(x_j) = \sum_{x_j : x_i \leq x_j} M_{ij} N_{\geq}(x_j). \tag{5.8}$$

Now we define $\mu(x_i, x_j)$ to be M_{ij}; this allows us to rewrite Equation (5.8) as

$$N(x) = \sum_{\substack{y:x\leq y \\ \bmod P}} \mu(x, y)N_{\geq}(y). \quad \blacksquare$$

Theorem 5.5.   Suppose (X, P) is a finite partially ordered set and μ is the Möbius function of (X, P).   Whenever two functions N and $N_{\leq}$ are related by the equations

$$N_{\leq}(x) = \sum_{\substack{y:y\leq x \\ \bmod P}} N(y)$$

for all x in X, then

$$N(x) = \sum_{\substack{y:y\leq x \\ \bmod P}} \mu(y, x)N_{\leq}(y).$$

Proof.   The proof is analogous to that of Theorem 5.4. $\blacksquare$

Equations that Describe the Möbius Function

It would appear that the function μ depends on the order in which we list the elements of X, since M depends on that order.   The next theorem shows this is not the case.

Theorem 5.6. μ is the unique function such that

(a) $\mu(x, x) = 1$ for all x

(b) $\sum\limits_{z:x\leq z\leq y} \mu(x, z) = 0$ for all x and y with $x < y$

or

(b') $\sum\limits_{z:x\leq z\leq y} \mu(z, y) = 0$ for all x and y with $x < y$

(c) $\mu(x, y) = 0$ if $x \nleq y$.

Proof. We show first that μ satisfies these three conditions. We have already proved that $M_{ij} = 0$ unless $x_i \le x_j$, so $\mu(x_i, x_j) = 0$ unless $x_i \le x_j$, which proves Statement (c). Since $MZ = I$,

$$I_{ii} = \sum_{k=1}^{n} M_{ik}Z_{ki}$$

$$= \sum_{x_k : x_i \le x_k \le x_i} M_{ik}Z_{ki} = M_{ii}Z_{ii} = M_{ii} = \mu(x_i, x_i).$$

Since $I_{ii} = 1$, this proves Statement (a). Also,

$$0 = I_{ij} = \sum_{k=1}^{n} M_{ik}Z_{kj} = \sum_{x_k : x_i \le x_k \le x_i} M_{ik}Z_{kj}$$

$$= \sum_{x_k : x_i \le x_k \le x_i} \mu(x_i, x_k) \cdot 1 = \sum_{z : x_i \le z \le x_j} \mu(x_i, z).$$

Substituting x and y for x_i and x_j gives Statement (b).

By using any listing of X in a linear extension of P, we may prove by induction on $j - i$ that the value of $\mu(x_i, x_j)$ is determined by equations (5.1), (5.2) and (5.3). The idea behind this inductive proof is that the equation

$$\sum_{z : x \le z \le y} \mu(x, z) = 0$$

may be rewritten as

$$\mu(x, y) + \sum_{z : x \le z < y} \mu(x, z) = 0,$$

so

$$\mu(x, y) = - \sum_{z : x \le z < y} \mu(x, z).$$

The construction of the proof is left as an exercise. ∎

Example 5.3. Compute the Möbius functions for the posets shown in Figure 7.19.

For the first poset we note that $\mu(b, b) = \mu(t, t) = 1$ by Theorem 5.6, Part

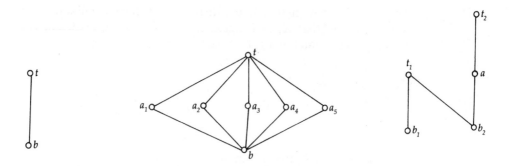

Figure 7.19

(a). From Part (b), note that $\displaystyle\sum_{z:b\le z\le t} \mu(b, z) = 0$, so $\mu(b, b) + \mu(b, t) = 0$ yields

$\mu(b, t) = -1$.

For the second poset, again $\mu(x, x) = 1$ for $x = b$, t or a_i. Also, $\mu(b, b) + \mu(b, a_i) = 0$ by Part (b) of Theorem 5.6, so $\mu(b, a_i) = -1$. Again applying Part (b) of Theorem 5.6, we get

$$\mu(a_i, a_i) + \mu(a_i, t) = 0$$

so $\mu(a_i, t) = -1$. Still another application of Part (b) gives us

$$\mu(b, b) + \mu(b, a_1) + \mu(b, a_2) + \mu(b, a_3) + \mu(b, a_4) + \mu(b, a_5) + \mu(b, t) = 0$$

or

$$1 - 1 - 1 - 1 - 1 - 1 + \mu(b, t) = 0,$$

so

$$\mu(b, t) = 4.$$

Since $a_i \not\le a_j$ unless $i = j$, $\mu(a_i, a_j) = 0$ by Part (c) of Theorem 5.6 whenever $i \ne j$.

For the third poset, $\mu(x, x) = 1$ once again for each x, and $\mu(b_1, t) = \mu(b_2, a) = \mu(a, t_2) = -1$ as above. Another application of Theorem 5.6, Part

(b) gives

$$\mu(b_2, b_2) + \mu(b_2, a) + \mu(b_2, t_2) = 0$$
$$1 - 1 + \mu(b_2, t_2) = 0,$$

so

$$\mu(b_2, t_2) = 0.$$

Note that this is our first example of an x and y with $x \le y$ but $\mu(x, y) = 0$. Thus as in Theorem 5.6, $\mu(x, y) \ne 0$ means that $x \le y$, but $x \le y$ *need not mean that* $\mu(x, y) \ne 0$. ∎

For small posets, the above method is usually simpler than matrix inversion for computing μ.

The Number of Connected Graphs

Example 5.4. Show how to use the μ function of the partition lattice to compute the number of connected graphs on a vertex set V.

Each graph is a union of its connected components; thus, we can associate the connected component partition $P(G)$ with each graph G. For each partition P of V, let $N(P)$ stand for the number of graphs G such that $P(G) = P$. The number of connected graphs is $N(\{V\})$. The equation below in which the P on the left hand side runs over all partitions P of V

$$\sum_P N(P) = 2^{\binom{v}{2}}$$

holds because the right–hand side is the total number of graphs on a vertex set V with v elements while the left–hand side sums the number of graphs having a certain connected component partition over all possible connected component partitions. This gives us one equation in quite a number of unknowns; one of the unknowns, $N(\{V\})$, in which $\{V\}$ is the partition with just one part, is the number of connected graphs. If we could get the same number of equations, perhaps we could solve for all the unknowns, including the one we are interested in. The Möbius inversion theorems suggest that we look for equations involving summing over all partitions less than or equal to a given one or all partitions greater than or equal to a given one. If we add up $N(P)$ for all partitions P contained in the partition Q, we should get the total number of graphs whose connected component partitions are contained in Q; that is, the number of graphs each of whose connected components is a subset of some class of Q. Thus, this sum is the number of graphs all of whose edges connect two points in one of the classes C_i of Q. This, by the multiplication principle, is the number of graphs all of whose edges are in C_1 multiplied by the number of graphs all of whose edges are in C_2 multiplied by . . . , etc. The number of graphs all of whose edges connect vertices in C_i is just the number of graphs whose vertex set is C_i. Thus if C_i has size c_i, this number is

$$2^{\binom{c_i}{2}}.$$

Putting this all together gives

$$\sum_{\substack{\text{Partitions } P: \\ P \leq Q = \{C_1, C_2, \ldots, C_k\}}} N(P) = \prod_{i=1}^{k} 2^{\binom{c_i}{2}}$$

By Theorem 5.5,

$$N(Q) = \sum_{\substack{\text{Partitions } P: \\ P \leq Q \\ P = \{B_1, B_2, \ldots, B_j\}}} \mu(P, Q) \prod_{i=1}^{j} 2^{\binom{b_i}{2}}$$

where b_i stands for the size of the class B_i of P. The number we want is $N(\{V\})$. Also, all partitions P of V are finer than $\{V\}$, so

$$N(V) = \sum_{\substack{\text{Partitions } P \text{ of } V \\ P = \{B_1, B_2, \ldots, B_j\}}} \mu(P, \{V\}) \prod_{i=1}^{j} 2^{\binom{b_i}{2}}. \quad \blacksquare$$

This completes the example and suggests that we should try to compute $\mu(P, \{V\})$ for the partition lattice. There are many methods that have been developed for computation of Möbius functions, and we have only space to "scratch the surface" here.

A General Method of Computing Möbius Functions

Theorem 5.7. Let L be a lattice with an element b below all elements of L and an element t above all elements of L. Then for any element a of L

$$\mu(b, t) = - \sum_{\substack{x: x \wedge a = b \\ \text{and } x \neq b}} \mu(x, t).$$

Proof.   We prove that

$$\sum_{x:x\wedge a=b} \mu(x, t) = 0.$$

The statement of the theorem follows because $x = b$ is one of the x's such that $x \wedge a = b$. For each $y \leq a$ in L, let $N(y) = \sum_{x:x\wedge a=y} \mu(x, t)$. Then since for each x, $x \wedge a$ is a specific $y \leq a$, we may write

$$\sum_{y\leq a} N(y) = \sum_{y\leq a} \sum_{x:x\wedge a=y} \mu(x, t)$$
$$= \sum_{x\in L} \mu(x, t) = 0$$

by Theorem 5.6. Then by applying Theorem 5.5 to the poset of all elements between 0 and a, we get

$$N(y) = \sum_{y\leq a} \mu'(y, a) \cdot 0 = 0,$$

where μ' is the μ function for the poset of all elements in the interval between 0 and a. The substitution of b for y gives the desired equation. ∎

The Möbius Function of the Partition Lattice

Theorem 5.8.   Let μ_n stand for the Möbius function of the lattice of partitions of an n-element set. Then

$$\mu_n(b, t) = (-1)^{n-1} \cdot (n - 1)!$$

Proof: Let $S = \{n \mid \mu_n(b, t) = (-1)^{n-1}(n - 1)!\}$. Then 1 is in S because then the lattice has just one element. Suppose $n - 1$ is in S. If the set being partitioned is $N = \{1, 2, \ldots, n\}$, let P be the partition with two classes given by $P = \{\{1, 2, \ldots, n - 1\}, \{n\}\}$. Theorem 5.7 tells us that

$$\mu_n(b, t) = - \sum_{\substack{\text{Partitions } Q\neq b \\ Q\wedge P=b}} \mu_n(Q, t). \qquad (5.9)$$

Now if $Q \wedge P = b$, all intersections of classes of P and Q have size 1. However, for a set C to intersect $\{1, 2, \ldots , n - 1\}$ in a set of size 1, C must be $\{i\}$ or $\{i, n\}$ for some $i \neq n$. Thus any class C of Q of size more than 1 must be $\{i, n\}$, and since n can't be in two classes, Q can have exactly one nontrivial class $\{i, n\}$.

Theorem 5.6 tells us that $\mu(Q, t)$ depends only on the ordered set consisting of partitions of N containing Q. However, the poset of partitions of N containing Q is isomorphic to the poset of partitions of $N' = \{1, 2, \ldots , n - 1\}$. (To see this, for each R containing Q, let R' be the partition obtained from R by deleting n from the class of R containing i. This mapping is one–to–one and onto the partitions of N', and $R \leq S$ if and only if $R' \leq S'$; therefore the mapping is an isomorphism.) Theorem 5.6 may be used to show that isomorphic posets have the same Möbius function values, so

$$\mu_n(Q, t) = \mu_{n-1}(b, t). \tag{5.10}$$

Now there are $n - 1$ choices for the element i of Q, so there are $n - 1$ choices for Q. Thus substituting Equation (5.10) into Equation (5.9) gives us

$$\begin{aligned}
\mu_n(b, t) &= -(n - 1)\mu_{n-1}(b, t) \\
&= -(n - 1)(-1)^{n-2}(n - 2)! \\
&= (-1)^{n-1}(n - 1)!
\end{aligned}$$

since $n - 1$ is in S. Thus n is in S and so the theorem is proved by the principle of mathematical induction. ∎

In the proof of the theorem, we used an isomorphism between the interval from Q to the top t of the partition lattice of a set with the partition lattice of a smaller set. In fact, any interval from *any* partition P to the top t of a partition lattice is isomorphic to a partition lattice on a smaller set. In particular, if P has k classes, then any partition above P has unions of these classes as its classes. In particular, these unions partition the *classes* of P. Two classes of P are partitioned together by R if their union lies entirely in a class of R. This partitioning of classes of P gives an isomorphism between the partitions of $K = \{1, 2, \ldots , k\}$ and the partitions of N containing our partition P with k classes. We have proved the following theorem.

Theorem 5.9. If P has k classes, then

$$\mu(P, t) = (-1)^{k-1}(k - 1)!$$

Proof.   We have seen that the interval from P to t is isomorphic to the partitions of K, and thus $\mu(P, t)$ must be $\mu_k(b, t)$.   ∎

EXERCISES

1. Write down four equations involving the Möbius function $\mu(\emptyset, A)$ for subsets of a two–element set using Condition 2 of Theorem 5.6.   Use these four equations to compute $\mu(\emptyset, A)$ for each subset A.

2. Use the method of Exercise 1 to compute $\mu(\emptyset, A)$ for the subsets A of a three–element set.

3. What is $\mu(x, y)$ for each x and y in $\{1, 2, 3, 4\}$ with the usual ordering?

4. Compute the Zeta matrix and its inverse for the poset of subsets of $\{1, 2, 3\}$ ordered by inclusion.   (Note: These are 8×8 matrices.)

5. Complete the proof of Theorem 5.6.

6. Write out a proof of Theorem 5.6 in which Condition 2 is replaced by Condition 2'.

7. Find $\mu(x, y)$ for all x and y in the partially ordered sets of Figure 7.20.

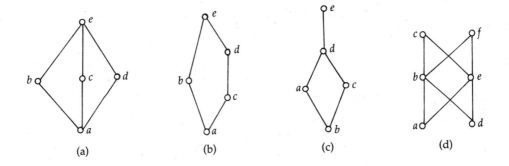

Figure 7.20

8. Write down a Zeta matrix for each poset in Exercise 7.   Check your Mobius function from Exercise 7 by matrix multiplication.

9. Use an isomorphism argument similar to the one we used in discussing the Möbius function of the partition lattice to explain why for subsets S and T of N with $S \leq T$, $\mu(S, T) = \mu(\emptyset, T - S)$.

10. (Exercise 9 continued)   Prove that $\mu(S, T) = (-1)^{|T|-|S|}$ whenever $S \subseteq T$.

11. To do this exercise, review the *bond lattice* of a graph.   Given a coloring of a graph with x colors, the *bond* of the coloring is the equivalence class partition of the relation "x is related to y if x and y are connected by a path all of whose vertices have the same color."

(a) Show that the bond of a coloring is a bond in the sense of the bond lattice.

(b) What is the bond of a proper coloring (i.e., a coloring in which adjacent vertices get different colors)?

(c) Given a bond B whose partition has k classes, how many colorings have a bond which is B or which contains B?

(d) Find a formula for the number of proper colorings of the graph using x colors. (Your formula should make use of the Möbius function of the bond lattice and will involve summing over the bond lattice.)

(e) Show the number you have computed is a polynomial of the form $\Sigma a_i x^i$. What is the coefficient a_i? (This polynomial is called the *chromatic polynomial* of the graph.)

12. State and prove an analog of Theorem 5.7 that involves joins.

13. Determine the Möbius function of a linearly ordered set. What does the Möbius inversion theorem say in this case?

14. Determine the Möbius function of the positive factors of 12, ordered by "is a factor of."

*15. Show that for the bond lattice of a connected graph $(-1)^{h(B)}\mu(b, B) > 0$ where b is the bottom element of the lattice and $h(B)$ is the height of the bond B in the bond lattice. What does this say about the coefficients of the chromatic polynomial (Exercise 11)?

16. What can you say about $\mu(b, t)$ in a poset with a bottleneck, i.e., a pair of points $x \neq b$, $y \neq t$ such that y covers x and every other element of the poset is either above y or below x?

17. Determine the Möbius function of the positive factors of an integer n, ordered by "is a factor of".

SECTION 6   PRODUCTS OF ORDERINGS

The Idea of a Product

We began our study of ordered sets with the idea of developing a single ordering based on several other orderings. In particular, a professor with several rankings of students would rank person A above person B if and only if A ranks above B on each test. This same kind of construction can be used to put together several partially ordered sets into a new one. Given partially ordered sets (X_1, P_1), (X_2, P_2), . . . , (X_n, P_n), we define their *product*

$$\prod_{i=1}^{n} (X_i, P_i) = \left(\prod_{i=1}^{n} X_i, \prod_{i=1}^{n} P_i \right)$$

to be the set X of all lists $(y_1, y_2, \ldots, y_n)$ with y_i in X_i ordered by the rule

$$(y_1, y_2, \ldots, y_n) \leq (z_1, z_2, \ldots, z_n)$$

if and only if $y_i \leq z_i \pmod{P_i}$ for all i. We call this ordering the *product* ordering.

(To explain this terminology note that the set X of lists is the Cartesian product of the sets X_i.)

The Euclidean plane gives an example that shows how to visualize products of orderings. If we think of the ordinary x– and y–axes as ordered sets with the usual ordering for numbers, then the set of pairs (x, y) is an ordered set in which a point A is less than or equal to a point B if and only if A is below and to the left of B in the plane.

When we have just two partially ordered sets, we normally use the notation $(X_1, P_1) \times (X_2, P_2)$ to stand for $\prod_{i=1}^{2} (X_i, P_i)$.

Example 6.1. If $P = \{M, N\}$ is a partition of the set S into two classes M and N of size m and n, then the ordered set of all partitions contained in P (finer than P) is isomorphic to $L_m \times L_n$, the product of the lattice of partitions of M and the lattice of partitions of N.

To see this, note that if Q is finer than P, then every block of Q is a subset of M or N. Thus if $f(Q)$ is the ordered pair consisting of a partition of M and a partition of N given by

$$f(Q) = \left(\left\{ C \middle| \begin{array}{l} C \quad \text{a class of} \\ Q \text{ and } C \subseteq M \end{array} \right\}, \left\{ D \middle| \begin{array}{l} D \quad \text{a class of} \\ Q \text{ and } D \subseteq N \end{array} \right\} \right),$$

then f is a function from the poset of partitions finer than P to $L_M \times L_N$. We shall show f is an isomorphism. First we show that $Q \leq R$ if and only if $f(Q) \leq f(R)$. If Q is finer than R (which is finer than P), then every class of Q is a subset of a class of R, which in turn is a subset of M or N. Thus $f(Q) \leq f(R)$ in the product ordering. Further, if $f(Q) \leq f(R)$ and if Q and R are both finer than P, then every class of Q and every class of R is a subset of either M or N. But each class of Q in M must be a subset of some class of R in M if $f(Q)$ is to be less than or equal to $f(R)$ in the first coordinate. Similarly, since the only other classes of Q must be subsets of N, they must be contained in classes of R if $f(Q)$ is to be less than or equal to $f(R)$ in the second coordinate. We conclude that if $f(Q) \leq f(R)$, then $Q \leq R$. In particular, if $f(Q) = f(R)$, then $Q \leq R$ and $R \leq Q$, so $Q = R$. Therefore the function f is one–to–one. Finally, f is onto, so we have a one–to–one onto function f such that $f(x) \leq f(y)$ if and only if $x \leq y$. Thus f is an isomorphism. ∎

Recall that we have found $\mu(P, t)$ for any partition P, but still don't know $\mu(b, P)$ for a partition P. If $P = \{M, N\}$, then the interval of all partitions between b and P is isomorphic to $L_M \times L_N$. Thus $\mu(b, P)$ is the same as the value of $\mu(b, t)$ in $L_M \times L_N$. One of the advantages gained from an understanding of products of posets is that these products help us understand Möbius functions.

Products of Posets and Möbius Functions

Theorem 6.1.   If $(X, P) = (Y, Q) \times (Z, R)$, and if μ_P, μ_Q and μ_R stand for the corresponding Möbius functions, then

$$\mu_P((y_1, z_1), (y_2, z_2)) = \mu_Q(y_1, y_2)\mu_R(z_1, z_2).$$

Proof.   Note that

$$\sum_{\substack{(t_1, t_2):\\ y_1 \le t_1 \le y_2 \\ z_1 \le t_2 \le z_2}} \mu_Q(y_1, t_1)\mu_R(z_1, t_2) = \sum_{\substack{t_1:\\ y_1 \le t_1 \le y_2}} \mu_Q(y_1, t_1) \sum_{\substack{t_2:\\ z_1 \le t_2 \le z_2}} \mu_R(z_1, t_2)$$

$$= 0 \cdot 0 = 0, \text{ if } y_1 < y_2 \text{ and } z_1 < z_2.$$

In fact, this sum is 0 if $y_1 = y_2$ and $z_1 < z_2$ or if $y_1 < y_2$ and $z_1 = z_2$ because one of the factors of the product will be 0. Thus the function μ_P defined by

$$\mu_P((y_1, z_1), (y_2, z_2)) = \mu_Q(y_1, y_2)\mu_R(z_1, z_2)$$

satisfies Condition 2 of Theorem 5.6 applied to $P = Q \times R$. But Conditions 1 and 3 are immediate consequences of the formula used to define μ_P. Thus by Theorem 5.6, μ_P must be the Möbius function of (X, P). ∎

Corollary 6.2.   Let μ be the Möbius function of the partition lattice. Then for a partition P with k blocks of size $i_1, i_2, \ldots, i_k$,

$$\mu(b, P) = \prod_{j=1}^{k} (-1)^{i_j - 1}(i_j - 1)!$$

Proof.   By induction on the result of Theorem 6.1, the Möbius function of a product of k posets will be the product of their individual Möbius functions. Thus $\mu(b, P)$ is a product of the Möbius functions of k partition lattices. ∎

In the exercises you will see explicitly what you may have guessed already: that the lattice of subsets of a set is the product of two–element lattices. Up to isomorphism, there is only one two–element lattice, the set $\{0, 1\}$ with $0 < 1$, which is denoted by **2**. In **2**, $\mu(0, 1) = -1$. Thus for any set K of size k,

$$\mu(\emptyset, K) = \prod_{i=1}^{k} \mu_i(0, 1) = (-1)^k.$$ This can be applied to the Möbius inversion theorem in Section 5 to obtain the inclusion–exclusion theorem of Chapter 2.

Products of Posets and Dimension

The notion of products is closely related to the dimension of a partial ordering (recall that the dimension of P is the minimum number of linear orderings whose intersection is P). Suppose P is the intersection of the orderings $L_1, L_2, \ldots, L_n$ of X. Then there is a natural way to embed (X, P) in the product

$$\prod_{i=1}^{n} (X, L_i),$$

namely, send the element x to the n-tuple $(x, x, \ldots, x) = f(x)$. Then $y \leq x$ (mod P) if and only if $(y, x) \in L_i$ for all i so $x \geq y$ (mod P) if and only if $f(x) \geq f(y) \left(\text{mod} \prod_{i=1}^{n} L_i \right)$. A *subposet* (X, P) of a poset (Y, Q) is a subset X of Y ordered by a relation P consisting of all pairs in Q that relate only points of X. Thus it consists of X and the restriction Q to X. Thus if a poset is an intersection of n linear orderings, it is isomorphic to a subposet of the product of n linearly–ordered sets by the map f described above. (It is only necessary to check that the map f described above is one–to–one to show that it is an isomorphism from (X, P) to the image of f.) We summarize these remarks in a theorem.

> **Theorem 6.3.** If (X, P) has dimension n, then it is isomorphic to a subposet of a product of n linearly ordered sets.
>
> *Proof.* Given above. ∎

In this theorem, all the linear orderings are orderings of X. However, it turns out that if we can embed (X, P) in a product of n linear orderings of n sets, then it happens that (X, P) has dimension no more than n. Thus the dimension of (X, P) is the smallest n such that (X, P) can be embedded in a product of n linear orderings. Pictorially, this means a finite poset has dimension n if and only if we can embed it into n–dimensional Euclidean space with the "componentwise" partial ordering.

To simplify the proof of the converse of Theorem 6.3, we review the idea of the *transitive closure* $T(R)$ of a relation R. Recall that $T(R)$ is the intersection of all transitive relations containing R. We proved that (x, y) is in $T(R)$ if and only if y is reachable from x in the digraph of R.

> **Theorem 6.4.** If a reflexive relation R has no directed cycles, then $T(R)$ is a reflexive partial ordering.
>
> *Proof.* $T(R)$ is transitive, and $T(R)$ is reflexive because R is. If (x, y) and (y, x) were in $T(R)$ and $x \neq y$, then R would have a directed cycle because x

would be reachable from y and y would be reachable from x. Thus, $T(R)$ must be antisymmetric as well. ∎

Theorem 6.5. If there is an isomorphism f from a poset (X, P) onto a sub-poset of

$$\prod_{i=1}^{n} (X_i, P_i),$$

with each P_i linear, then (X, P) has dimension n or less.

Proof. All partial orderings referred to in this proof are assumed to be re-flexive partial orderings. We must prove that P is an intersection of n linear orderings of X. We prove that there are n partial orderings Q_i of X, each containing P, such that if $x \not\leq y \pmod{P}$, then $y \leq x \pmod{Q_i}$ for some i. Then we choose arbitrary linear extensions L_i of the partial orderings Q_i. Their intersection contains exactly those ordered pairs in P, because any pair (v, w) of incomparable elements has $v \leq w \pmod{Q_j}$ and $w \leq v \pmod{Q_i}$ for some i and j. Thus, we will know P has dimension no more than n.

Note that since the range of f is a subset of a product,

$$f(x) = (f_1(x), f_2(x), \ldots, f_n(x)),$$

is an n–tuple of elements with $f_i(x)$ from X_i. Now if $x \leq y$, $f_i(x) \leq f_i(y)$ for all i, and if $f_i(x) \leq f_i(y)$ for all i, then $x \leq y$ since f is an isomorphism. Consequently, if $x \not\leq y$, then $f_i(y) < f_i(x)$ for some i. Now let

$$Q_i = T(P \cup \{(u, v) \mid f_i(u) < f_i(v)\}).$$

By our remarks immediately above, if $u \not\leq v$, then $(v, u) \in Q_i$ for some i. However, the relation

$$P \cup \{(u, v) \mid f_j(u) < f_j(v)\}$$

has no directed cycles in its digraph. Assume to the contrary that $x_1, (x_1, x_2), \ldots, (x_n, x_1), x_1$ were such a directed cycle. In what follows, adopt the convention that (x_n, x_{n+1}) means (x_n, x_1). Then for each i, either $x_i \leq x_{i+1} \pmod{P}$ or $f_j(x_i) \leq f_j(x_{i+1}) \pmod{P_j}$.

However since f is order–preserving, if $x_i \leq x_{i+1} \pmod{P}$, then $f_j(x_i) \leq f_j(x_{i+1}) \pmod{P_j}$. (It could be that $x_i \leq x_{i+1}$ and $f_j(x_i) = f_j(x_{i+1})$, though.) Thus either $f_j(x_i) = f_j(x_{i+1})$ for all i, or the elements $f_j(x)$ form a cycle in the linear ordering P_j. But then since the linear ordering P_j has no cycles, all the ordered pairs (x_i, x_{i+1}) in the assumed cycle must be in P, and the graph of P has no cycles. Therefore by Theorem 6.4, Q_i is a partial ordering. By the remarks preceding the construction of Q_i, the fact that Q_i is a partial ordering proves the theorem. ∎

Width and Dimension of Partially Ordered Sets

This result lets us prove an interesting theoretical constraint on the dimension of a partially ordered set. The *width* of a poset is the largest size of any antichain of the poset.

Theorem 6.6.   If (X, P) is a finite partially ordered set, then the dimension of P is less than or equal to its width.

Proof.   Let P have width w. Assume first that X has an element $b \leq x$ (mod P) for all $x \in X$. By Dilworth's theorem (Theorem 2.3 in this chapter), X is a union of w chains C_i, each linearly ordered by P. Let $X_i = C_i \cup \{b\}$ and choose L_i to be the linear ordering of X_i which orders X_i in the same way as does P.

Now define a function f_i by

$$f_i(x) = \text{the highest element of } X_i \text{ below } x \text{ (mod } P).$$

Thus if $x \leq y, f_i(x) \leq f_i(y)$. Define

$$f(x) = (f_1(x), f_2(x), \ldots , f_n(x)).$$

Then f is an order–preserving map of (X, P) onto a subset of $\prod_{i=1}^{n} X_i$. Suppose now that $f_k(x) = f_k(y)$ for all k. Let C_i be the chain containing x and C_j be the chain containing y. Then using $k = i$, we see that x is the highest element of X_i below y (mod P) and using $k = j$, y is the highest element of X_j below x (mod P). Thus $x \leq y$ and $y \leq x$, giving $y = x$. Therefore, f is one–to–one. To show that f is an isomorphism from (X, P) to a subposet of the product, we must show that when $f_i(x) \leq f_i(y)$ for all i, then $x \leq y$.

However, if C_i is the chain containing x, then $f_i(x) = x$, and thus $x \leq f_i(y) \leq y$ (mod P), so $x \leq y$ (mod P). Thus f is an isomorphism from (X, P) onto a subposet of $\prod_{i=1}^{n} (X_i, L_i)$.

Now if (X, P) has no bottom element $b \leq x$ (mod P) for all $x \in X$, adding such an element will not change the dimension of (X, P). (The proof of this is elementary.) Thus it suffices to show the theorem is true for posets with bottom elements. Then by Theorem 6.5, we have shown that the dimension of (X, P) is less than or equal to w. ■

Theorem 6.6 is just one illustration of how Dilworth's theorem may be used as a theoretical tool in the study of partially ordered sets.

EXERCISES

1. Show explicitly that the lattice of subsets of $\{1, 2, 3\}$ is the product of the lattices of subsets of $\{1\}$, $\{2\}$ and $\{3\}$.

2. Recall the function f_A given by

$$f_A(i) = \begin{cases} 0 & \text{if } i \notin A \\ 1 & \text{if } i \in A \end{cases}$$

we used in Chapter 1 to compute the number of subsets of a set. Show that the function f given by

$$f(A) = (f_A(1), f_A(2), \ldots, f_A(n))$$

is an isomorphism from the lattice of subsets of a set $N = \{1, 2, \ldots, n\}$ to a product of the poset $\{0, 1\} = \mathbf{2}$ with itself n times.

3. Show that the poset of all positive factors of 12 ordered by "is a factor of" is a product of two smaller posets.

4. If the number $n = pqr$, where p, q and r are primes, show that the lattice of positive factors of n, ordered by "is a factor of," is a product of three two–element posets.

5. How did we use the assumption that $b \leq x$ for all x in Theorem 6.6?

6. Let G be a graph with two conected components. Prove that the bond lattice of G is the product of the bond lattices of its two connected components.

7. Generalize Exercise 6 appropriately.

8. Draw a Hasse graph of the product of a chain of length 2 and a chain of length 3.

9. Using "is a factor of," give another description of a product of k chains of length $i_1, i_2, \ldots,$ and i_k.

10. Show that if the top element of L_1 is t_1 and the bottom element of L_2 is b_2, then the element $x = (t_1, b_2)$ of $L_1 \times L_2$ satisfies the distributive laws

$$x \wedge (y \vee z) = (x \wedge y) \vee (x \wedge z)$$

and

$$y \wedge (x \vee z) = (y \vee x) \wedge (y \vee z).$$

11. Discuss the relationship between products of positive integers and lattices of factors of positive integers.

12. If the poset P has height h and the poset Q has height k, what is the height of $P \times Q$?

13. What is the Möbius function of the lattice of factors of the positive integer n ordered by "is a factor of"?

14. Find an example of a poset whose dimension is 2 and width is 3.

15. One of the examples of partial orderings in the text has its width and dimension equal to half its number of elements. Which one is it?

*16. Find all six–element posets whose dimension is equal to 3.

17. Show that the poset of 1– and $(n - 1)$–element subsets of an n–element poset has its dimension equal to its width.

*18. Show that if M is a matroid which is the disjoint union of two matroids M_1 and M_2 (i.e., its independent sets are exactly those sets which are disjoint unions of an independent set of M_1 and an independent set of M_2), then the lattice of flats of M is the product of the lattice of flats of M_1 and the lattice of flats of M_2. (See Exercise 22, Section 3.)

Suggested Reading

Bender, E. A. and J. R. Goldman: "On the Applications of Möbius Inversion in Combinatorial Analysis", *American Mathematical Monthly* 1975, p. 789.

Birkhoff, Garret: *Lattice Theory*, 3rd Ed., American Mathematical Society 1967.

Birkhoff, G. and T. Bartee: *Modern Applied Algebra*, McGraw-Hill 1970.

Crawley, P. and R. Dilworth: *Algebraic Theory of Lattices*, Prentice Hall 1973.

Even, S.: *Graph Theory Algorithms*, Computer Science Press 1979.

Fisher, J. L.: *Application Oriented Algebra*, IEP, Crowell 1977.

Ford, L. and D. Fulkerson: *Flows in Networks*, Princeton University Press 1962.

Prather, R.: *Discrete Mathematical Structures for Computer Science*, Houghton Mifflin 1976.

APPENDIX:
Fundamentals of Matrix Algebra

SECTION 1   MATRICES AND VECTORS

Row and Column Vectors

A list of numbers in a horizontal row is called a row vector. Examples of row vectors are

$$[1, 2, 3, 4], [-1, 0, 1], [x, y], [40, 20, 30, 60, 30], [2 + a, 3 - b, c, 4].$$

Notice that symbols like x and y which may stand for numbers are also allowed to appear in vectors. A list of numbers in a vertical column is called a column vector. Examples of column vectors are

$$\begin{bmatrix} 1 \\ 2 \\ 3 \end{bmatrix} \begin{bmatrix} x + 3 \\ y \\ 4 - z \end{bmatrix} \begin{bmatrix} -1 \\ 0 \\ 1 \\ -1 \end{bmatrix} \begin{bmatrix} 1 \\ 0 \\ 1 \\ 0 \\ 1 \end{bmatrix} \begin{bmatrix} .35 \\ .40 \\ .30 \\ .35 \\ .45 \end{bmatrix}.$$

We often use row and column vectors to keep track of related quantities.

Example 1.1. The local distributor for a Florida fruit cooperative sells grapefruit, lemons, limes, temple oranges, navel oranges and juice oranges to local grocery stores. These products are sold to stores in standard size boxes, and each week a representative of the distributor visits a store to take its order. In a given week, our grocery store might order four boxes of grapefruit, one box each of lemons and limes, two boxes of temple oranges, four boxes of navel oranges and four boxes of juice oranges. The representative might keep track of this order by means of an expression of the form

$$\text{Store 1: } [4, 1, 1, 2, 4, 4],$$

writing down the numbers in the same order we listed the fruits. Perhaps there are two other stores in the area with orders

$$\text{Store 2: } [3, 1, 0, 3, 3, 2]$$
$$\text{Store 3: } [5, 2, 1, 4, 6, 4].$$

The the representative could keep track of the total number of boxes of each kind of fruit to be sent to our area on the delivery truck by the expression

$$\text{Total: } [12, 4, 2, 9, 13, 10]. \blacksquare$$

Addition of Vectors

The example suggests a natural operation of addition on vectors, namely, we add vectors by adding corresponding components. In symbols, if R and S are vectors

$$R = [R_1, R_2, \ldots , R_n]$$
$$S = [S_1, S_2, \ldots , S_n],$$

then $R + S$ is the vector whose j-th entry, which we denote by $(R + S)_j$, is given by

$$(R + S)_j = R_j + S_j.$$

It is customary to use R_j to denote the j-th entry of the row vector denoted by the symbol R. We use C_i to denote the i-th entry of the column vector C and similarly define the sum of two column vectors C and D with the same number of entries by

$$(C + D)_i = C_i + D_i.$$

Note that we only define sums of row or column vectors when they have the same length, and do not define the sum of a row vector and a column vector.

This kind of vector addition will satisfy all the usual rules of arithmetic such as $R + S = S + R$ and $(R + S) + T = R + (S + T)$ because each component or entry of a sum of two vectors R and S is just the ordinary sum of the corresponding components of R and S.

Example 1.2. The owner of Store 1 has decided at the last minute, because of favorable prices, to have a big sale on citrus fruit. The store owner calls the representative and says "triple my order." Symbolically we write

$$3[4,\ 1,\ 1,\ 2,\ 4,\ 4] = [12,\ 3,\ 3,\ 6,\ 12,\ 12]$$

to represent the process of tripling an order. ■

Multiplying a Vector by a Number

This suggests another arithmetic operation on vectors, namely, numerical multiplication. Specifically, given a vector V and a real number r, we write

$$(rV)_k = rV_k$$

to say that the k–th entry of rV is r times the k–th entry of V. Thus

$$4 \cdot \begin{bmatrix} 3 \\ 2 \\ 1 \end{bmatrix} = \begin{bmatrix} 12 \\ 8 \\ 4 \end{bmatrix}$$

and

$$\tfrac{3}{2}[6,\ 2] = [9,\ 3].$$

Again, because it is defined component by component, this operation will satisfy rules of arithmetic we might expect it to satisfy, such as $r(V + W) = rV + rW$ and $rs(V) = (rs)V$.

Matrix and Dot Products of Vectors

Another arithmetic operation on vectors arises naturally as an important bookkeeping device.

    Example 1.3. Our representative has a price list each week. If a box of grapefruit costs four dollars, a box of lemons six dollars, a box of limes eight dollars, a box of temple oranges five dollars, a box of naval oranges six dollars and a box of juice oranges four dollars, the price list could be written.

$$\begin{bmatrix} 4 \\ 6 \\ 8 \\ 5 \\ 6 \\ 4 \end{bmatrix}.$$

What is the cost to Store 1 of its order?

Of course the cost of an order of one item is the quantity ordered times the price per unit; if we have a vector of quantities in a row and another vector of corresponding prices in a column, we write symbolically

$$Cost = Quantity \text{ times } Price = [4,\ 1,\ 2,\ 3,\ 4,\ 4] \begin{bmatrix} 4 \\ 6 \\ 8 \\ 5 \\ 6 \\ 4 \end{bmatrix}$$

$$= 4 \cdot 4 + 1 \cdot 6 + 1 \cdot 8 + 2 \cdot 5 + 4 \cdot 6 + 4 \cdot 4$$
$$= 16 + 6 + 8 + 10 + 24 + 16 = 80.$$

Thus Store 1's original order would cost $80. ■

The kind of product described in Example 1.3 arises in many situations and is called the *matrix product* of two vectors. The *matrix product* of R and C is defined in this way only when R is a row vector and C is a column vector. This product is *not* defined if R and C do not have the same length. In symbols we write

$$[R_1, R_2, \ldots, R_n] \begin{bmatrix} C_1 \\ C_2 \\ \cdot \\ \cdot \\ \cdot \\ C_n \end{bmatrix} = R_1 C_1 + R_2 C_2 + \cdots + R_n C_n$$

or in a more compact notation, we may write

$$RC = \sum_{k=1}^{n} R_k C_k.$$

Thus,

$$[4, 3, 2, 2] \begin{bmatrix} \frac{1}{2} \\ \frac{1}{3} \\ \frac{1}{2} \\ \frac{1}{3} \end{bmatrix} = 4 \cdot \tfrac{1}{2} + 3 \cdot \tfrac{1}{3} + 2 \cdot \tfrac{1}{2} + 2 \cdot \tfrac{1}{3} = 4\tfrac{2}{3}.$$

Notice that we write just RC for the matrix product of a row and column vector rather than $R \times C$ or $R \cdot C$ or introducing some special symbol for the multiplication. One reason for this is to help us distinguish between the matrix product defined on a row vector and a column vector and another parallel product defined analogously but defined either on two row vectors or two column vectors. This product, called the dot product, is given by

$$[a_1, a_2, \ldots, a_n] \cdot [b_1, b_2, \ldots, b_n] = a_1 b_1 + a_2 b_2 + \cdots + a_n b_n$$

or

$$\begin{bmatrix} a_1 \\ a_2 \\ \cdot \\ \cdot \\ \cdot \\ a_n \end{bmatrix} \cdot \begin{bmatrix} b_1 \\ b_2 \\ \cdot \\ \cdot \\ \cdot \\ b_n \end{bmatrix} = a_1 b_1 + a_2 b_2 + \cdots + a_n b_n.$$

Conceptually, there is no difference between a dot product and a matrix product. However, they come up naturally in different kinds of situations—the matrix product is a special case of a more general operation called matrix mutiplication, so we make a technical distinction by using a different notation.

Transposes

In fact, by using the concept of the *transpose* of a vector, we can express matrix products in terms of dot products. Namely, the transpose of a row vector R is the column vector with the same entries in the same order. We use R^t to stand for the transpose of R. Thus

$$[1, 2, 3]^t = \begin{bmatrix} 1 \\ 2 \\ 3 \end{bmatrix}.$$

Similarly,

$$\begin{bmatrix} 1 \\ 2 \\ 3 \end{bmatrix}^t = [1, 2, 3].$$

The transpose of a column vector is just the row vector with the same entries in the same order. Thus we get, for example,

$$[a_1, a_2, a_3] \begin{bmatrix} b_1 \\ b_2 \\ b_3 \end{bmatrix} = a_1 b_1 + a_2 b_2 + a_3 b_3$$

and

$$[a_1, a_2, a_3]^t \cdot \begin{bmatrix} b_1 \\ b_2 \\ b_3 \end{bmatrix} = \begin{bmatrix} a_1 \\ a_2 \\ a_3 \end{bmatrix} \cdot \begin{bmatrix} b_1 \\ b_2 \\ b_3 \end{bmatrix} = a_1 b_1 + a_2 b_2 + a_3 b_3$$

and

$$[a_1, a_2, a_3] \cdot \begin{bmatrix} b_1 \\ b_2 \\ b_3 \end{bmatrix}^t = [a_1, a_2, a_3] \cdot [b_1, b_2, b_3] = a_1 b_1 + a_2 b_2 + a_3 b_3.$$

In general, we get for the product of a row vector R and a column vector C of the same length,

$$RC = R^t \cdot C = R \cdot C^t.$$

Matrices

A *matrix* is a rectangular array of numbers. Examples of matrices are

$$[1, 2], \quad \begin{bmatrix} 1 \\ 3 \end{bmatrix}, \quad \begin{bmatrix} 1 & 2 \\ 3 & 4 \end{bmatrix}, \quad \begin{bmatrix} 3 & 2 \\ 1 & 3 \\ 1 & 2 \end{bmatrix}, \quad \begin{bmatrix} 1 & 2 & 4 \\ 1 & 3 & 9 \end{bmatrix}$$

Example 1.4.   We make a matrix from the three row vectors in Example 1.1 by writing one on top of the other without commas as

$$\begin{bmatrix} 4 & 1 & 1 & 2 & 4 & 4 \\ 3 & 1 & 0 & 3 & 3 & 2 \\ 5 & 2 & 1 & 4 & 6 & 4 \end{bmatrix}.$$

Thus, for each week, we have an order matrix which gives us the quantities of each kind of fruit ordered by the stores in the area. We say the matrix has three rows

$$R1 = \begin{bmatrix} 4 & 1 & 1 & 2 & 4 & 4 \end{bmatrix}$$
$$R2 = \begin{bmatrix} 3 & 1 & 0 & 3 & 3 & 2 \end{bmatrix}$$

and

$$R3 = \begin{bmatrix} 5 & 2 & 1 & 4 & 6 & 4 \end{bmatrix}.$$

and six columns

$$C1 = \begin{bmatrix} 4 \\ 3 \\ 5 \end{bmatrix} \quad C2 = \begin{bmatrix} 1 \\ 1 \\ 2 \end{bmatrix} \quad C3 = \begin{bmatrix} 1 \\ 0 \\ 1 \end{bmatrix} \quad C4 = \begin{bmatrix} 2 \\ 3 \\ 4 \end{bmatrix} \quad C5 = \begin{bmatrix} 4 \\ 3 \\ 6 \end{bmatrix} \quad \text{and} \quad C6 = \begin{bmatrix} 4 \\ 2 \\ 4 \end{bmatrix}. \quad \blacksquare$$

A Matrix Times a Column Vector

Recall that we had a column vector C of the six prices of the six different kinds of fruit. Intuitively, if we were to multiply our quantity matrix by this price matrix, we should get a cost matrix. Is there such a product? If we have a matrix M with rows $R1, R2, \ldots , Rm$ all of length n and a column vector of C of length m, then we define the *product* of M with C to be the column vector given by

$$MC = \begin{bmatrix} R1\,C \\ R2\,C \\ \cdot \\ \cdot \\ \cdot \\ Rm\,C \end{bmatrix}.$$

Example 1.5.   The product of the quantity matrix of Example 1.4 and the price matrix of Example 1.3 is

$$\begin{bmatrix} 4 & 1 & 1 & 2 & 4 & 4 \\ 3 & 1 & 0 & 3 & 3 & 2 \\ 5 & 2 & 1 & 4 & 6 & 4 \end{bmatrix} \begin{bmatrix} 4 \\ 6 \\ 8 \\ 5 \\ 6 \\ 4 \end{bmatrix} = \begin{bmatrix} 16 + 6 + 8 + 10 + 24 + 16 \\ 12 + 6 + 0 + 15 + 18 + 8 \\ 20 + 12 + 8 + 20 + 36 + 16 \end{bmatrix} = \begin{bmatrix} 80 \\ 59 \\ 112 \end{bmatrix}.$$

Thus when we multiply the matrix of quantities ordered by the stores times the column vector of prices, we get the column vector of costs to the individual stores. ■

Just as we can define a product of a matrix on the left times a column vector on the right when the number of entries in each row in the matrix is equal to the number of entries in the column, we can also define a product of a row vector on the left times a matrix on the right.   In fact, it is possible to define a very useful product operation, the product of *two matrices,* which includes all of our other matrix products as special cases.

Example 1.6.   Suppose we are given the four–column vectors $C1$, $C2$, $C3$ and $C4$ which represent the citrus fruit prices in four successive weeks.   Assume store i orders fruit according to the standing order given by row Ri in Example 1.4 each week and write down the matrix whose entry in row i and column j is the total cost of fruit to store i in week j.

We are given

$$C1 = \begin{bmatrix} 4 \\ 6 \\ 8 \\ 5 \\ 6 \\ 4 \end{bmatrix} \quad C2 = \begin{bmatrix} 7 \\ 6 \\ 8 \\ 5 \\ 5 \\ 4 \end{bmatrix} \quad C3 = \begin{bmatrix} 4 \\ 5 \\ 8 \\ 6 \\ 5 \\ 4 \end{bmatrix} \quad C4 = \begin{bmatrix} 5 \\ 5 \\ 8 \\ 5 \\ 5 \\ 3 \end{bmatrix}.$$

Now we already know from Example 1.1 that to get the cost to store i in week j we form the matrix product of row vector Ri with column vector Cj.   For example, for store 1 in week 2 we form the product $R1C2$, and for the cost to store 3 in week 4 we form $R3C4$.   Thus the matrix asked for in the problem is:

$$\begin{bmatrix} R1C1 & R1C2 & R1C3 & R1C4 \\ R2C1 & R2C2 & R2C3 & R2C4 \\ R3C1 & R3C2 & R3C3 & R3C4 \end{bmatrix} = \begin{bmatrix} 80 & 79 & 77 & 75 \\ 59 & 59 & 60 & 56 \\ 112 & 110 & 108 & 105 \end{bmatrix}. ■$$

The Product of Two Matrices

In general, to *define* a product of a matrix M with rows $R1, R2, \ldots, Rm$ by a matrix N with columns $C1, C2, \ldots, Cn$, we must have the length of each row of M equal to the length of each column of N. Then we define MN to be the matrix whose entry in row i and column j is $RiCj$. We *define* A_{ij} to stand for the entry of the matrix A in row i and column j. Thus we write

$$(MN)_{ij} = RiCj.$$

In an expanded form, we may write schematically

$$
\begin{bmatrix} \text{------} R1 \text{------} \\ \text{------} R2 \text{------} \\ \\ \text{------} Rm \text{------} \end{bmatrix}
\begin{bmatrix} | & | & & | \\ C1 & C2 & \cdots & Cn \\ | & | & & | \end{bmatrix}
=
\begin{bmatrix} R1C1 & R1C2 & \cdots & R1Cn \\ R2C1 & R2C2 & \cdots & R2Cn \\ & \vdots & & \\ RmC1 & \cdots & \cdots & RmCn \end{bmatrix}
$$

Notice that by using M_{ij} to stand for the entry in row i and column j of M, we have

$$R1 = [M_{11}, M_{12}, \ldots, M_{1k}]$$

and

$$Ri = [M_{i1}, M_{i2}, \ldots, M_{ik}].$$

Since we have assumed that the columns of N have the same length as the rows of M, the columns of N each have k entries. Therefore, we get for the columns of N

$$
C1 = \begin{bmatrix} N_{11} \\ N_{21} \\ \vdots \\ \\ N_{k1} \end{bmatrix}
\quad \text{and} \quad
Cj = \begin{bmatrix} N_{1j} \\ N_{2j} \\ \vdots \\ \\ N_{kj} \end{bmatrix}
$$

Thus, in this detailed and useful notation, the formula

$$(MN)_{ij} = RiCj$$

becomes

$$(MN)_{ij} = M_{i1}N_{1j} + M_{i2}N_{2j} + \cdots + M_{ik}N_{kj}$$
$$= \sum_{h=1}^{k} M_{ih}N_{hj}.$$

Example 1.7. Compute the matrix products

$$\begin{bmatrix} 2 & 1 \\ 3 & 0 \end{bmatrix}\begin{bmatrix} 1 & 1 \\ 1 & 0 \end{bmatrix}, \quad \begin{bmatrix} 1 & 2 \end{bmatrix}\begin{bmatrix} 3 \\ 4 \end{bmatrix} \quad \text{and} \quad \begin{bmatrix} 1 \\ 2 \end{bmatrix}\begin{bmatrix} 3 & 4 \end{bmatrix}.$$

We have

$$\begin{bmatrix} 2 & 1 \\ 3 & 0 \end{bmatrix}\begin{bmatrix} 1 & 1 \\ 1 & 0 \end{bmatrix} = \begin{bmatrix} 2 \cdot 1 + 1 \cdot 1 & 2 \cdot 1 + 1 \cdot 0 \\ 3 \cdot 1 + 0 \cdot 1 & 3 \cdot 1 + 0 \cdot 0 \end{bmatrix} = \begin{bmatrix} 3 & 2 \\ 3 & 3 \end{bmatrix}$$

$$\begin{bmatrix} 1 & 2 \end{bmatrix}\begin{bmatrix} 3 \\ 4 \end{bmatrix} = \begin{bmatrix} 1 \cdot 3 + 2 \cdot 4 \end{bmatrix} = \begin{bmatrix} 3 + 8 \end{bmatrix} = \begin{bmatrix} 11 \end{bmatrix}$$

and

$$\begin{bmatrix} 1 \\ 2 \end{bmatrix}\begin{bmatrix} 3 & 4 \end{bmatrix} = \begin{bmatrix} 1 \cdot 3 & 1 \cdot 4 \\ 2 \cdot 3 & 2 \cdot 4 \end{bmatrix} = \begin{bmatrix} 3 & 4 \\ 6 & 8 \end{bmatrix}.$$

EXERCISES

For Problems 1–20, let

$$R = [1, 2, 3, 1]$$

$$S = [1, 3, 3, 1] \quad C = \begin{bmatrix} 1 \\ 2 \\ 3 \\ 4 \end{bmatrix} \quad D = \begin{bmatrix} 2 \\ 0 \\ 2 \\ 1 \end{bmatrix} \quad E = \begin{bmatrix} 1 \\ 3 \\ 3 \\ 1 \end{bmatrix}$$

$$T = [1, 0, 2, 0]$$
$$0 = [0, 0, 0, 0].$$

1. Find the sum $R + S$ and the sum $D + E$.
2. Find the differences $R - S$ and $D - E$ by constructing an appropriate definition of subtraction for row and column vectors.
3. Verify that $(R + S) + T = R + (S + T)$.
4. Verify that $R + 0 = R$.
5. What is RC? RD?
6. What is OC? OD?
7. Write down S^t. Write down D^t.

For Problems 8–20, let I, M and N be the matrices

$$I = \begin{bmatrix} 1 & 0 & 0 & 0 \\ 0 & 1 & 0 & 0 \\ 0 & 0 & 1 & 0 \\ 0 & 0 & 0 & 1 \end{bmatrix} \quad M = \begin{bmatrix} 1 & 3 & -1 & 0 \\ -1 & 2 & 0 & 1 \\ 1 & -1 & 0 & 5 \\ 5 & 0 & 2 & 1 \end{bmatrix} \quad N = \begin{bmatrix} 1 & 2 & 3 & 0 \\ 0 & 1 & 2 & 3 \\ 0 & 0 & 1 & 2 \\ 0 & 0 & 0 & 1 \end{bmatrix}.$$

8. Find the product MC.
9. Find the product MD.
10. Find the products RM and SM.
11. Verify that $(R + S)M = RM + SM$.
12. Find the product OM.
13. Find the product $MM = M^2$.
14. The *transpose* A^t of a matrix A is the matrix whose rows have the same entries as the columns of A all in the same order. Find M^t and the product MM^t.
15. Find the products ES and SE.
16. Compute $M \begin{bmatrix} x_1 \\ x_2 \\ x_3 \\ x_4 \end{bmatrix}$ and $[x_1, x_2, x_3, x_4]M$.
17. Compute NM.
18. Compute $(NM)C$ and $N(MC)$. What law does this suggest?
19. Compute $N^2 = NN$.
20. Compute MI and NI. What law does this suggest? What would you expect IM and IN to be?

SECTION 2   REDUCTION OF MATRICES AND
SYSTEMS OF EQUATIONS

Writing a System of Equations as a Matrix Equation

Matrices are intimately related to systems of linear equations.

Example 2.1.   A dairy produces three kinds of cottage cheese—creamed cottage cheese, dry curd cottage cheese and cottage cheese with pineapple chunks. The ingredients for a pound of pineapple–chunk cheese are 10 ounces of cottage cheese, two ounces of cream, and four ounces of pineapple. The ingredients in a pound of creamed cottage cheese are 12 ounces of cottage cheese and four ounces of cream. The ingredients for a pound of dry curd cottage cheese are simply 16 ounces of cottage cheese. If we are to produce x_1 pounds of dry curd cheese, x_2 pounds of creamed cheese, and x_3 pounds of pineapple cottage cheese, how many ounces of each ingredient will we need?

Note that in the x_3 pounds of pineapple cheese, there are $10x_3$ ounces of cheese, in the x_2 pounds of creamed cheese, there are $12x_2$ ounces of cheese, while there are $16x_1$ ounces of cheese required for the dry curd cheese. Thus we need

$$16x_1 + 12x_2 + 10x_3 = \begin{bmatrix} 16 & 12 & 10 \end{bmatrix} \begin{bmatrix} x_1 \\ x_2 \\ x_3 \end{bmatrix}$$

ounces of cottage cheese. Similarly, we need

$$0x_1 + 4x_2 + 2x_3 = \begin{bmatrix} 0 & 4 & 2 \end{bmatrix} \begin{bmatrix} x_1 \\ x_2 \\ x_3 \end{bmatrix}$$

ounces of cream and

$$0x_1 + 0x_2 + 4x_3 = \begin{bmatrix} 0 & 0 & 4 \end{bmatrix} \begin{bmatrix} x_1 \\ x_2 \\ x_3 \end{bmatrix}$$

ounces of pineapple. Thus the first entry of the matrix product

$$\begin{bmatrix} 16 & 12 & 10 \\ 0 & 4 & 2 \\ 0 & 0 & 4 \end{bmatrix} \begin{bmatrix} x_1 \\ x_2 \\ x_3 \end{bmatrix}$$

is the number of ounces of cottage cheese we will use, the second entry is the number of ounces of cream, and the third entry is the number of ounces of pineapple. Suppose on a given day we have 2000 ounces of cottage cheese, 400 ounces of cream, and 400 ounces of pineapple.   Is there a way to use up all of our ingredients in making our three products?   Using the matrix product above, we can say

$$\begin{bmatrix} 16 & 12 & 10 \\ 0 & 4 & 2 \\ 0 & 0 & 4 \end{bmatrix} \begin{bmatrix} x_1 \\ x_2 \\ x_3 \end{bmatrix} = \begin{bmatrix} 2000 \\ 400 \\ 400 \end{bmatrix}$$

or

$$\begin{bmatrix} 16x_1 + 12x_2 + 10x_3 \\ 0 + 4x_2 + 2x_3 \\ 0 + 0 + 4x_3 \end{bmatrix} = \begin{bmatrix} 2000 \\ 400 \\ 400 \end{bmatrix}$$

giving us the three equations

$$16x_1 + 12x_2 + 10x_3 = 2000 \tag{1}$$

$$4x_2 + 2x_3 = 400 \tag{2}$$

$$4x_3 = 400. \tag{3}$$

Solving Equations by Gaussian Elimination

There are two easy ways to solve these equations.   One is to solve the last equation for x_3, substitute this into the other equations, then solve for x_2 and so on.   A much better method for our later work is as follows.

Divide Equation 3 by 4:

$$16x_1 + 12x_2 + 10x_3 = 2000 \tag{1}$$

$$4x_2 + 2x_3 = 400 \tag{2}$$

$$x_3 = 100. \tag{4}$$

Subtract twice Equation 3 from Equation 2 and subtract ten times Equation 3 from Equation 1:

$$16x_1 + 12x_2 = 1000 \tag{5}$$

$$4x_2 = 200 \tag{6}$$

$$x_3 = 100. \tag{4}$$

Subtract three times Equation 2 from Equation 1:

$$16x_1 \qquad\qquad\qquad = 400 \qquad\qquad (7)$$

$$4x_2 \qquad = 200 \qquad\qquad (6)$$

$$x_3 = 100. \qquad\qquad (4)$$

Divide Equation 1 by 16 and Equation 2 by 4:

$$x_1 \qquad\qquad\qquad = \; 25 \qquad\qquad (8)$$

$$x_2 \qquad = \; 50 \qquad\qquad (9)$$

$$x_3 = 100. \; \blacksquare \qquad\qquad (4)$$

Row Reduction of Matrices

Now except for place holding, the x_1, x_2 and x_3 were not really used in this method of solving the equations. We see this by writing down the matrices that correspond to these sets of equations:

$$\begin{bmatrix} 16 & 12 & 10 \\ 0 & 4 & 2 \\ 0 & 0 & 4 \end{bmatrix}\begin{bmatrix} x_1 \\ x_2 \\ x_3 \end{bmatrix} = \begin{bmatrix} 2000 \\ 400 \\ 400 \end{bmatrix}$$

$$\begin{bmatrix} 16 & 12 & 10 \\ 0 & 4 & 2 \\ 0 & 0 & 1 \end{bmatrix}\begin{bmatrix} x_1 \\ x_2 \\ x_3 \end{bmatrix} = \begin{bmatrix} 2000 \\ 400 \\ 100 \end{bmatrix}$$

$$\begin{bmatrix} 16 & 12 & 0 \\ 0 & 4 & 0 \\ 0 & 0 & 1 \end{bmatrix}\begin{bmatrix} x_1 \\ x_2 \\ x_3 \end{bmatrix} = \begin{bmatrix} 1000 \\ 200 \\ 100 \end{bmatrix}$$

$$\begin{bmatrix} 16 & 0 & 0 \\ 0 & 4 & 0 \\ 0 & 0 & 1 \end{bmatrix}\begin{bmatrix} x_1 \\ x_2 \\ x_3 \end{bmatrix} = \begin{bmatrix} 400 \\ 200 \\ 100 \end{bmatrix}$$

$$\begin{bmatrix} 1 & 0 & 0 \\ 0 & 1 & 0 \\ 0 & 0 & 1 \end{bmatrix}\begin{bmatrix} x_1 \\ x_2 \\ x_3 \end{bmatrix} = \begin{bmatrix} 25 \\ 50 \\ 100 \end{bmatrix}.$$

The column vector of x_i's just "went along for the ride," so to speak, in this method of solving the equation. The operations we performed on the equations correspond to adding (or subtracting) a multiple of a row to (or from) another row and dividing a row by a number. In order to write less, we can forget the columns of x_i's and put the two matrices of numbers together with just a dividing line. We call the resulting matrix an *augmented* matrix. We then perform *row operations* which have easy symbolic descriptions and can rewrite the solution process as follows:

$$
\begin{bmatrix}
16 & 12 & 10 & 2000 \\
0 & 4 & 2 & 400 \\
0 & 0 & 4 & 400
\end{bmatrix}
\xrightarrow{R3 \div 4}
\begin{bmatrix}
16 & 12 & 10 & 2000 \\
0 & 4 & 2 & 400 \\
0 & 0 & 1 & 100
\end{bmatrix}
$$

$$
\xrightarrow[R2 - 2R3]{R1 - 10R3}
\begin{bmatrix}
16 & 12 & 0 & 1000 \\
0 & 4 & 0 & 200 \\
0 & 0 & 1 & 100
\end{bmatrix}
\xrightarrow{R1 - 3R2}
\begin{bmatrix}
16 & 0 & 0 & 400 \\
0 & 4 & 0 & 200 \\
0 & 0 & 1 & 100
\end{bmatrix}
$$

$$
\xrightarrow[R2 \div 4]{R1 \div 16}
\begin{bmatrix}
1 & 0 & 0 & 25 \\
0 & 1 & 0 & 50 \\
0 & 0 & 1 & 100
\end{bmatrix}.
$$

The solution to the system of equations is now the column vector to the right of the dividing line.

What is a Solution?

Clearly, we have the beginnings of a systematic method of solving systems of equations. Before going further, let us analyze what we *mean* by "solving a system of equations." The kinds of equations we will be discussing are linear equations in the variables $x_1, x_2, \ldots, x_n$; these are equations of the form

$$a_1x_1 + a_2x_2 + \cdots a_nx_n = c.$$

These are called linear equations in analogy with the equation

$$ax + by = c$$

which is the equation of a straight line in the plane. By a *solution* to *an equation* in the variables $x_1, x_2, \ldots, x_n$ we mean an assignment of numbers to the variables that makes the equation a true statement. Thus one solution to the equation

$$x_1 + 2x_2 - x_3 = 4$$

is $x_1 = 2, x_2 = 1, x_3 = 0$.   A second solution is $x_1 = 3, x_2 = 1, x_3 = 1$.   A system of equations is a list of equations.   A solution to a system of equations is an assignment of numbers to the variables that makes *each* equation a true statement. For example

$$x_1 + 2x_2 - x_3 = 4 \qquad\qquad (2.1)$$
$$x_1 - x_2 - 2x_3 = 0$$

is a system of equations.   The assignment $x_1 = 3, x_2 = 1, x_3 = 1$ is a solution to the system, but the assignment $x_1 = 2, x_2 = 1$ $x_3 = 0$ is not, because although

$$2 + 2 \cdot 1 - 0 = 4,$$

we have

$$2 - 1 - 2 \cdot 0 \neq 0.$$

Equivalent Systems of Equations

Two systems of equations are *equivalent* if they have exactly the same solutions. For example, the system of equations (2.1) is equivalent to the system of equations (2.2) below.

$$x_1 + 2x_2 - x_3 = 4 \qquad\qquad (2.2)$$
$$0 - 3x_2 - x_3 = -4.$$

Why is this?   We get the second equation in (2.2) by subtracting the first equation in (2.1) from the second equation, while the first equations in (2.1) and (2.2) are the same.   Therefore, any assignment of numbers that satisfies both equations in (2.1) will satisfy both equations in (2.2) below.

   But if we add the first equation in (2.2) to the second, the new equation we get is the second equation in (2.1).   This means that each solution to the equations in (2.2) will be a solution to both equations in (2.2).   Thus the equations in (2.2) have the same solutions as those in (2.1).   Therefore these two systems are equivalent.   There are many other ways to convert one system to another, equivalent, one.   Clearly, multiplying one equation on both sides by the same nonzero constants doesn't change its solutions.   Similarly, multiplying one equation in a system on both sides by the same nonzero number won't change the solutions of the equations, so it gives an equivalent system of equations.   Of course, interchanging two of the equations, for example, writing

$$x_1 - x_2 - 2x_3 = 0$$
$$x_1 + 2x_2 - x_3 = 4 \tag{2.3}$$

gives an equivalent system of equations.

The Three Basic Row Operations

Though we could make up many other ways of creating new equivalent systems of equations from old ones, the three methods just described (adding a multiple of one equation to another, multiplying equations by numbers, and interchanging equations) give us all the tools we need to develop a systematic way of discovering all the solutions to a system of equations.

Note that the system of linear equations in (2.1) can be written as the matrix equation

$$\begin{bmatrix} 1 & 2 & -1 \\ 1 & -1 & -2 \end{bmatrix} \begin{bmatrix} x_1 \\ x_2 \\ x_3 \end{bmatrix} = \begin{bmatrix} 4 \\ 0 \end{bmatrix}.$$

In fact, since each linear equation may be rewritten as an equation stating that a row vector of numbers times a column vector of variables is equal to some number, each system of linear equations can be written as a matrix equation. This equation will have the form of a matrix of numbers times a column vector of variables set equal to a column vector of numbers. Each of the three operations on equations we have described will correspond to an operation on the rows of the numerical matrix and column vector of numbers and leave the column vector of variables unchanged, just as in our analysis following Example 2.1. However, we will not always have the good fortune we had after Example 2.1 of being able to just read our solutions from the last column of an augmented matrix. In the example below, you may find it useful to write out explicitly the equations that correspond to each stage of the matrix operations.

Example 2.2. Find all solutions to the system of equations

$$x_1 + 2x_2 - x_3 = 4$$
$$x_1 - x_2 - 2x_3 = 0.$$

We proceed with an augmented matrix as in the method developed after Example 2.1.

$$\begin{bmatrix} 1 & 2 & -1 & | & 4 \\ 1 & -1 & -2 & | & 0 \end{bmatrix} \xrightarrow{R2 - R1} \begin{bmatrix} 1 & 2 & -1 & | & 4 \\ 0 & -3 & -1 & | & -4 \end{bmatrix}$$

$$\xrightarrow{R2 \div (3)} \begin{bmatrix} 1 & 2 & -1 & | & 4 \\ 0 & 1 & \frac{1}{3} & | & \frac{4}{3} \end{bmatrix} \xrightarrow{R1 - 2R2} \begin{bmatrix} 1 & 0 & -\frac{5}{3} & | & \frac{4}{3} \\ 0 & 1 & \frac{1}{3} & | & \frac{4}{3} \end{bmatrix}.$$

Now, however, it is not clear what the augmented matrix means or whether we have our solutions. The corresponding matrix equation is

$$\begin{bmatrix} 1 & 0 & -\frac{5}{3} \\ 0 & 1 & \frac{1}{3} \end{bmatrix} \begin{bmatrix} x_1 \\ x_2 \\ x_3 \end{bmatrix} = \begin{bmatrix} \frac{4}{3} \\ \frac{4}{3} \end{bmatrix}$$

or

$$x_1 - \frac{5}{3} x_3 = \frac{4}{3}$$

or

$$x_1 = \frac{4}{3} + \frac{5x_3}{3}$$

and

$$x_2 + \frac{1}{3} x_3 = \frac{4}{3}.$$

or

$$x_2 = \frac{4}{3} - \frac{x_3}{3}.$$

We have one equation which tells us how to determine x_1 from x_3 and one equation that tells us how to determine x_2 from x_3. These equations are equivalent to our original system of equations. However, for any number a, the assignment

$$x_1 = \frac{4}{3} + \frac{5a}{3} \qquad x_2 = \frac{4}{3} - \frac{a}{3} \qquad x_3 = a$$

is a solution to our system of equations. Thus we have infinitely many solutions of the form

$$\begin{bmatrix} x_1 \\ x_2 \\ x_3 \end{bmatrix} = \begin{bmatrix} \frac{4}{3} + \frac{5a}{3} \\ \frac{4}{3} - \frac{a}{3} \\ a \end{bmatrix}. \ \blacksquare$$

This representation of our solutions using an arbitrary number a is called a *parametric representation* of the solutions; the number a is called a *parameter*. The matrix

$$\begin{bmatrix} 1 & 0 & -\frac{5}{3} \\ 0 & 1 & \frac{1}{3} \end{bmatrix}$$

is said to be in "row–reduced echelon form". To describe what this means, it is convenient to introduce the idea of a *pivotal* entry in a matrix. An entry in the matrix will be called pivotal if it is the first nonzero entry in its row. Thus the two 1's in the matrix above are pivotal. In the matrix

$$\begin{bmatrix} 1 & 2 \\ 3 & 4 \\ 5 & 6 \end{bmatrix}$$

the entries 1, 3 and 5 are pivotal.

Row–reduced Form

We say a matrix is in *row–reduced form* if

(1) Each pivotal entry is 1.

(2) Each column has one pivotal entry and 0's as its other entries, or else has no pivotal entries.

Note that if a column has more than one pivotal element, then by performing row operations we may reduce the number of pivotal entries in that column to 1 without affecting any columns to the left of that column. Thus by moving from left to right, we can reduce a matrix to row–reduced form by performing row operations. We say that a matrix is in row–reduced *echelon* form if it is row–reduced and if

(3) Each pivotal entry is to the right of all pivotal entries in rows above it, and all rows of 0's entirely are below any nonzero rows.

Once a matrix is in row–reduced form, it can be put into row–reduced echelon form by interchanging rows. For example

$$\begin{bmatrix} 0 & 0 & 1 & 2 \\ 1 & 0 & 0 & 1 \end{bmatrix}$$

is not in echelon form, but is row–reduced. By interchanging row 1 and row 2, we get

$$\begin{bmatrix} 1 & 0 & 0 & 1 \\ 0 & 0 & 1 & 2 \end{bmatrix}$$

which is in echelon form.

A General Method of Solving Systems of Equations

Now we can write down a method of finding solutions to any system of equations. We call the variable x_j *pivotal* if there is a pivotal entry in column j. To solve a system of linear equations written in matrix form

$$AX = B$$

with

$$X = \begin{bmatrix} x_1 \\ x_2 \\ \cdot \\ \cdot \\ \cdot \\ x_n \end{bmatrix}$$

first write down the augmented matrix.

$$A|B.$$

Now perform row operations on the augmented matrix to put the part on the left–hand side of the vertical line in row–reduced echelon form. Finally, write down the system of equations that this new matrix corresponds to. One of two things will happen—either one (or more) equation will be a contradictory statement of the form $0 = b$ even though $b \neq 0$, or else the resulting equations will give a parametric representation of the solutions of the system in which the non-pivotal variables are the parameters and the pivotal variables are determined as functions of these parameters. Some examples should serve to make this abstract description of the process clear.

Example 2.3.   Solve the system of equations

$$\begin{aligned} x_1 + 3x_2 &= 4 \\ 2x_1 - x_2 &= 1 \\ -2x_1 + 4x_2 &= 2. \end{aligned}$$

This system of equations has the matrix representation

$$\begin{bmatrix} 1 & 3 \\ 2 & -1 \\ -2 & 4 \end{bmatrix} \begin{bmatrix} x_1 \\ x_2 \end{bmatrix} = \begin{bmatrix} 4 \\ 1 \\ 2 \end{bmatrix}.$$

The augmented matrix is

$$\begin{bmatrix} 1 & 3 & 4 \\ 2 & -1 & 1 \\ -2 & 4 & 2 \end{bmatrix}.$$

We reduce as follows:

$$\begin{bmatrix} 1 & 3 & 4 \\ 2 & -1 & 1 \\ -2 & 4 & 2 \end{bmatrix} \xrightarrow[R3 + 2R1]{R2 - 2R1} \begin{bmatrix} 1 & 3 & 4 \\ 0 & -7 & -7 \\ 0 & 10 & 10 \end{bmatrix} \xrightarrow[R3 \div 10]{R2 \div (-7)} \begin{bmatrix} 1 & 3 & 4 \\ 0 & 1 & 1 \\ 0 & 1 & 1 \end{bmatrix}$$

$$\xrightarrow[R3 - R2]{R1 - 3R2} \begin{bmatrix} 1 & 0 & 1 \\ 0 & 1 & 1 \\ 0 & 0 & 0 \end{bmatrix}.$$

Thus $x_1 = 1$, $x_2 = 1$ is the only solution to the system of equations. Note that despite the fact that we started with three equations in two unknowns, we still found a solution. ■

Inconsistent Systems of Equations

Example 2.4.   Find all solutions to the system of equations:

$$\begin{aligned} x_1 - 2x_2 + x_3 &= 4 \\ 2x_1 + x_2 - 3x_3 &= 5 \\ 3x_1 - x_2 - 2x_3 &= 10. \end{aligned}$$

We write

$$\begin{bmatrix} 1 & -2 & 1 & 4 \\ 2 & 1 & -3 & 5 \\ 3 & -1 & -2 & 10 \end{bmatrix} \xrightarrow[\begin{subarray}{c} R2 - 2R1 \\ R3 - 3R1 \end{subarray}]{} \begin{bmatrix} 1 & -2 & 1 & 4 \\ 0 & 5 & -5 & -3 \\ 0 & 5 & -5 & -2 \end{bmatrix}$$

$$\xrightarrow[\; R3 - R2 \;]{} \begin{bmatrix} 1 & -2 & 1 & 4 \\ 0 & 5 & -5 & -3 \\ 0 & 0 & 0 & 1 \end{bmatrix}.$$

The last row corresponds to the equation

$$0x_1 + 0x_2 + 0x_3 = 1,$$

or

$$0 = 1$$

so we know there is *no* solution to the system of equations despite the fact that we started with three equations in three unknowns. Note that because of our recognition that the last row corresponds to $0 = 1$, we didn't have to complete the row reduction process. This often happens when a set of equations is *inconsistent*, i.e., has no solutions. ■

Example 2.5.   Find all solutions to the system of equations:

$$\begin{aligned} x_1 + x_2 + x_3 + x_4 &= 8 \\ x_1 - 2x_2 + x_3 - 3x_4 &= -6 \\ 2x_1 - x_2 + 2x_3 - 2x_4 &= 2 \\ -x_1 + 5x_2 - x_3 + 7x_4 &= 20. \end{aligned}$$

The corresponding matrix equation is

$$\begin{bmatrix} 1 & 1 & 1 & 1 \\ 1 & -2 & 1 & -3 \\ 2 & -1 & 2 & -2 \\ -1 & 5 & -1 & 7 \end{bmatrix} \begin{bmatrix} x_1 \\ x_2 \\ x_3 \\ x_4 \end{bmatrix} = \begin{bmatrix} 8 \\ -6 \\ 2 \\ 20 \end{bmatrix}.$$

Beginning with the augmented matrix we write

$$\begin{bmatrix} 1 & 1 & 1 & 1 & | & 8 \\ 1 & -2 & 1 & -3 & | & -6 \\ 2 & -1 & 2 & -2 & | & 2 \\ -1 & 5 & -1 & 7 & | & 20 \end{bmatrix} \quad \begin{matrix} R2 - R1 \\ R3 - 2R1 \\ \overline{R4 + R1} \end{matrix} \longrightarrow \begin{bmatrix} 1 & 1 & 1 & 1 & | & 8 \\ 0 & -3 & 0 & -4 & | & -14 \\ 0 & -3 & 0 & -4 & | & -14 \\ 0 & 6 & 0 & 8 & | & 28 \end{bmatrix}$$

$$\begin{matrix} R3 - R2 \\ \overline{R4 + 2R2} \end{matrix} \longrightarrow \begin{bmatrix} 1 & 1 & 1 & 1 & | & 8 \\ 0 & -3 & 0 & -4 & | & -14 \\ 0 & 0 & 0 & 0 & | & 0 \\ 0 & 0 & 0 & 0 & | & 0 \end{bmatrix} \quad R2 \div (-3) \longrightarrow \begin{bmatrix} 1 & 1 & 1 & 1 & | & 8 \\ 0 & 1 & 0 & \frac{4}{3} & | & \frac{14}{3} \\ 0 & 0 & 0 & 0 & | & 0 \\ 0 & 0 & 0 & 0 & | & 0 \end{bmatrix}$$

$$\begin{matrix} R1 - R2 \end{matrix} \longrightarrow \begin{bmatrix} 1 & 0 & 1 & -\frac{1}{3} & | & \frac{10}{3} \\ 0 & 1 & 0 & \frac{4}{3} & | & \frac{14}{3} \\ 0 & 0 & 0 & 0 & | & 0 \\ 0 & 0 & 0 & 0 & | & 0 \end{bmatrix}.$$

The equations we get are

$$x_1 \quad\quad + x_3 - \frac{x_4}{3} = \frac{10}{3}$$

$$x_2 \quad\quad + \frac{4x_4}{3} = \frac{14}{3}.$$

Since x_1 and x_2 are the pivotal variables, we get a parametric representation of all the solutions by setting x_3 and x_4 equal to two arbitrary numbers r and s. Then the set of all solutions may be written in vector form as

$$\begin{bmatrix} x_1 \\ x_2 \\ x_3 \\ x_4 \end{bmatrix} = \begin{bmatrix} \frac{10}{3} - r + \frac{s}{3} \\ \frac{14}{3} - \frac{4s}{3} \\ r \\ s \end{bmatrix}. \quad \blacksquare$$

EXERCISES

1. Which of the following matrices is in row–reduced form?   Row–reduced echelon form?

(a) $\begin{bmatrix} 1 & 0 \\ 0 & 1 \end{bmatrix}$ (b) $\begin{bmatrix} 0 & 1 \\ 1 & 0 \end{bmatrix}$ (c) $\begin{bmatrix} 0 & 1 \\ 0 & 0 \end{bmatrix}$ (d) $\begin{bmatrix} 1 & 0 & 0 & 1 \\ 0 & 2 & 3 & 1 \end{bmatrix}$

(e) $\begin{bmatrix} 1 & 0 & 0 & 2 \\ 0 & 0 & 1 & 2 \\ 0 & 0 & 0 & 1 \end{bmatrix}$ (f) $\begin{bmatrix} 1 & 2 & 0 & 0 \\ 0 & 0 & 1 & 0 \\ 0 & 0 & 0 & 1 \end{bmatrix}$ (g) $\begin{bmatrix} 0 & 0 & 0 & 0 \\ 0 & 0 & 0 & 0 \\ 0 & 0 & 0 & 0 \end{bmatrix}.$

2. Solve the following systems of equations.   Use the method of row reduction of
an augmented matrix.

(a) $x_1 + 2x_2 = 3$
      $3x_1 - x_2 = 2$

(b) $-x_1 + 2x_2 = 5$
      $4x_1 - x_2 = 1$

(c) $x_1 + 2x_2 = 0$
      $x_1 - x_2 = 0$

(d) $ax + by = r$ (assume $ad \neq 0$
      $cx + dy = s$ and $ad - bc \neq 0$).

3. Solve the following systems of equations:

(a) $x_1 + x_2 - x_3 = 1$
      $2x_1 - x_2 + 2x_3 = 3$
      $3x_1 - 2x_2 + x_3 = 4$

(b) $2x_1 + x_2 - x_3 = 3$
      $x_1 - 3x_2 - x_3 = 2$
      $3x_1 - 2x_2 - x_3 = 2.$

4. Find all solutions to the following systems of equations:

(a) $3x_1 + 2x_2 - x_3 = 4$
      $x_1 + x_2 - x_3 = 0$
      $-x_1 + 2x_2 + x_3 = 6$

(b) $3x_1 + 2x_2 - x_3 = 4$
      $x_1 + x_2 - x_3 = 0$
      $2x_1 + x_2 = 4$

(c) $3x_1 + 2x_2 - x_3 = 4$
      $x_1 + x_2 - x_3 = 0$
      $2x_1 + x_2 = 5$

(d) $3x_1 + 2x_2 - x_3 = 4$
      $x_1 + x_2 - x_3 = 0$
      $-x_1 + 2x_2 + x_3 = 6$
      $x_1 + 3x_2 + x_3 = 10.$

5. Find all solutions to the systems of equations

(a) $x_1 + 2x_2 + x_3 + x_4 = 0$
      $x_1 + x_2 + x_4 = 0$
      $2x_1 + x_2 - x_3 = 0$

(b) $x_1 + 2x_2 + 3x_3 + 4x_4 = 0$
      $x_1 + 4x_2 + 9x_3 + 16x_4 = 0$
      $x_1 + 8x_2 + 27x_3 + 64x_4 = 0$

(c) $x_1 + 2x_2 - x_3 - x_4 = 0$
      $3x_1 + x_2 - x_3 - 2x_4 = 0$
      $2x_1 - x_2 - x_4 = 0$

(d) $3x_1 + 3x_2 + 3x_3 + 3x_4 = 0$
      $2x_1 + 4x_2 + 6x_3 + 8x_4 = 0$
      $2x_1 + x_2 - x_4 = 0.$

6. Milk is sold in quart and half–gallon boxes and gallon jugs. A grocer pays 50 cents for a quart of milk, 80 cents for a half–gallon and $1.40 for a gallon of milk.

(a) Write an equation which says the grocer buys x_1 quarts, x_2 half-gallons and x_3 gallons of milk and pays $100.

(b) A quart of milk takes up eight square inches of shelf space, a half–gallon 16 square inches and a gallon 36 square inches. Write an equation that says 2272 square inches of shelf space are available for and used by the milk.

(c) Find all possible combinations of milk that fulfill the conditions of part (a) and part (b).

(d) On the average, the grocer sells about the same number of half–gallons as gallons. Is there an order for milk that satisfies this condition as well?

(e) A salesperson tells the grocer to expect to sell on the average an equal number of quarts and half–gallons. Is there an order for milk satisfying conditions (a), (b), (d) and (e)? How about (a), (b) and (e)?

7. A company makes dark and light rye bread as well as white bread. Dark rye is made of $\frac{1}{2}$ white flour and $\frac{1}{2}$ rye flour, and light rye $\frac{1}{4}$ rye flour and $\frac{3}{4}$ white flour. The plant has a capacity of 1000 loaves a day and uses a pound of flour per loaf. On a given day, there are 200 pounds of rye flour and 800 pounds of white flour. The company makes five cents on a loaf of white bread, six cents on a loaf of light rye and eight cents on a loaf of dark rye. Can they design a production run that leads a profit of $60 and uses all the available flour? If so, how much of each bread should they make?

8. A cereal company is designing a new cereal with 10% protein. Assume wheat flour has 4% protein, oat flour 6% protein and soy flour 20% protein. Wheat flour costs 10 cents a pound, oat flour 15 cents a pound and soy flour 20 cents a pound. The company is willing to spend 15 cents for the pound of flour used in producing a box of cereal. What proportions of the ingredients should be used to make the cereal?

SECTION 3   MATRIX ALGEBRA AND INVERSE MATRICES

Addition of Matrices

We saw that addition operations on vectors arise naturally in order to keep track of data on sales. If our salesperson maintains a weekly table (matrix) whose entry in row i and column j is the amount Store i orders of item j, then a monthly summary table would have as its entry in row i and column j the sum of all the i, j entries of all the weekly matrices. Thus there is also a natural matrix addition operation; namely, to add matrices M and N to get a sum S, we let the i, j entry of S be the sum of the corresponding entries of M nd N. In symbols,

$$S_{ij} = M_{ij} + N_{ij}.$$

Clearly, this operation makes sense only when M and N have the same shape, i.e., the same number of rows of the same length. (When M has m rows of length n, we say M is an $m \times n$ matrix.)

Of course we expect rules like the commutative law $M + N = N + M$, the associative law $(M + N) + P = M + (N + P)$ and so on to hold because they hold for each entry. If 0 is a matrix consisting of 0's only and of the same "shape" as M, then $M + 0 = 0 + M = M$.

We might also want to multiply a matrix by a number for the same sort of reason we wanted to multiply vectors by numbers. The matrix rM, where r is a real number, is formed from the matrix M by multiplying each entry of M by r. In symbols, we can write

$$(rM)_{ij} = r(M_{ij}).$$

The matrix $-M$ is $-1M$; note that $-M + M = 0$, just as we expect. Of course, we expect that rules like $1M = M$ and $(r + s)M = rM + sM$ and $r(M + N) = rM + rN$ should hold because they hold for each entry.

Algebraic Laws for Matrix Operations

We now have the operations of addition, subtraction and multiplication for matrices, just as we have for numbers. So far, every law of arithmetic that we know of for numbers has an analogous law for matrices. Had we checked questions about these laws for matrix multiplication we would not have been so lucky. For example, for numbers a and b we know that $ab = ba$. However, if we let

$$A = \begin{bmatrix} 1 & 0 \\ 1 & 1 \end{bmatrix} \quad B = \begin{bmatrix} 2 & 0 \\ 0 & 1 \end{bmatrix},$$

then

$$AB = \begin{bmatrix} 1 & 0 \\ 1 & 1 \end{bmatrix}\begin{bmatrix} 2 & 0 \\ 0 & 1 \end{bmatrix} = \begin{bmatrix} 2 & 0 \\ 2 & 1 \end{bmatrix}$$

but

$$BA = \begin{bmatrix} 2 & 0 \\ 0 & 1 \end{bmatrix}\begin{bmatrix} 1 & 0 \\ 1 & 1 \end{bmatrix} = \begin{bmatrix} 2 & 0 \\ 1 & 1 \end{bmatrix}.$$

Thus the commutative law $AB = BA$ *does not hold* for matrix multiplication.

This example brings up two natural questions. First, how can we be sure that $B + A = A + B$ always holds when $AB = BA$ doesn't always hold? We said

$B + A = A + B$ holds because it holds for each entry. What does this mean? It means

$$(A + B)_{ij} = A_{ij} + B_{ij} = B_{ij} + A_{ij} = (B + A)_{ij}.$$

Thus, the i, j entry of $A + B$ equals the i, j entry of $B + A$ for each i and j; this is just what we mean when we say that two matrices are equal.

The second question that our example with $AB \neq BA$ brings up is "Do any of the laws of arithmetic hold for matrix multiplication?" Perhaps the most useful law is the associative law $(ab)c = a(bc)$ that we use either directly or intuitively when we rearrange parentheses in a product. This law *does* hold. To show this, we examine the i, j entry of $(AB)C$ and $A(BC)$.

Theorem 3.1.   If A is an m by r matrix, B an r by s matrix and C an S by n matrix, then the matrix products $(AB)C$ and $A(BC)$ are defined and equal.

Proof. AB is an m by s matrix so $(AB)C$ is defined and BC is an r by n matrix so $A(BC)$ is defined. Now the i, h entry of AB is $\sum\limits_{k=1}^{r} A_{ik}B_{kh}$, the product of row i of A with row h of B. Similarly,

$$[(AB)C]_{ij} = \sum_{h=1}^{s} (AB)_{hj}C_{hj} = \sum_{h=1}^{s} \left(\sum_{k=1}^{r} A_{ik}B_{kh} \right) C_{hj}$$
$$= \sum_{h=1}^{s} \sum_{k=1}^{r} A_{ik}B_{kh}C_{hj}.$$

Also

$$[A(BC)]_{ij} = \sum_{k=1}^{r} A_{ik}(BC)_{kj} = \sum_{k=1}^{r} A_{ik} \sum_{h=1}^{s} B_{kh}C_{hj}$$
$$= \sum_{k=1}^{r} \sum_{h=1}^{s} A_{ik}B_{kh}C_{hj}.$$

Both double sums add up exactly the same quantities, but in different orders. The order in which we add these quantities doesn't matter, so the two double sums have the same value. Thus $[(AB)C]_{ij} = [A(BC)]_{ij}$ so that $(AB)C = A(BC)$. ∎

Another useful rule of arithmetic is the rule we use when we convert the equation $2x = 2a$ into the equation $x = a$ by cancelling out the 2's. Note, however, that

$$\begin{bmatrix} 2 & 0 \\ 0 & 0 \end{bmatrix}\begin{bmatrix} a & 0 \\ 0 & a \end{bmatrix} = \begin{bmatrix} 2a & 0 \\ 0 & 0 \end{bmatrix} = \begin{bmatrix} 2 & 0 \\ 0 & 0 \end{bmatrix}\begin{bmatrix} a & 0 \\ 0 & 0 \end{bmatrix},$$

but

$$\begin{bmatrix} a & 0 \\ 0 & a \end{bmatrix} \neq \begin{bmatrix} a & 0 \\ 0 & 0 \end{bmatrix} \text{ if } a \neq 0.$$

Thus we *can't* always use the cancellation law. We explain the cancellation of 2 in the ordinary equation by multiplying both sides of the equation by $\frac{1}{2}$ and using the fact that $\frac{1}{2} \cdot 2 = 1$. In fact, for any number $b \neq 0$, there is a number $\frac{1}{b} = b^{-1}$ such that $b^{-1}b = bb^{-1} = 1$.

Notice that if

$$\begin{bmatrix} 2 & 0 \\ 0 & 2 \end{bmatrix} = B,$$

and if $\begin{bmatrix} x_1 & x_3 \\ x_2 & x_4 \end{bmatrix} = X$ and $\begin{bmatrix} a_1 & a_3 \\ a_2 & a_4 \end{bmatrix} = A,$

then whenever $BX = BA$ we have

$$\begin{bmatrix} 2 & 0 \\ 0 & 2 \end{bmatrix}\begin{bmatrix} x_1 & x_3 \\ x_2 & x_4 \end{bmatrix} = \begin{bmatrix} 2 & 0 \\ 0 & 2 \end{bmatrix}\begin{bmatrix} a_1 & a_3 \\ a_2 & a_4 \end{bmatrix},$$

so that by multiplication we get

$$\begin{bmatrix} \frac{1}{2} & 0 \\ 0 & \frac{1}{2} \end{bmatrix}\left(\begin{bmatrix} 2 & 0 \\ 0 & 2 \end{bmatrix}\begin{bmatrix} x_1 & x_3 \\ x_2 & x_4 \end{bmatrix}\right) = \left(\begin{bmatrix} \frac{1}{2} & 0 \\ 0 & \frac{1}{2} \end{bmatrix}\begin{bmatrix} 2 & 0 \\ 0 & 2 \end{bmatrix}\right)\begin{bmatrix} x_1 & x_3 \\ x_2 & x_4 \end{bmatrix}$$
$$= \begin{bmatrix} 1 & 0 \\ 0 & 1 \end{bmatrix}\begin{bmatrix} x_1 & x_2 \\ x_3 & x_4 \end{bmatrix} = \begin{bmatrix} x_1 & x_3 \\ x_2 & x_4 \end{bmatrix}$$

and similarly

$$\begin{bmatrix} \frac{1}{2} & 0 \\ 0 & \frac{1}{2} \end{bmatrix}\left(\begin{bmatrix} 2 & 0 \\ 0 & 2 \end{bmatrix}\begin{bmatrix} a_1 & a_3 \\ a_2 & a_4 \end{bmatrix}\right) = \left(\begin{bmatrix} \frac{1}{2} & 0 \\ 0 & \frac{1}{2} \end{bmatrix}\begin{bmatrix} 2 & 0 \\ 0 & 2 \end{bmatrix}\right)\begin{bmatrix} a_1 & a_3 \\ a_2 & a_4 \end{bmatrix} = \begin{bmatrix} a_1 & a_3 \\ a_2 & a_4 \end{bmatrix}.$$

Thus $X = A$.

The Identity Matrix

The matrix

$$\begin{bmatrix} 1 & 0 \\ 0 & 1 \end{bmatrix} = I_{2 \times 2} = I$$

is called the 2×2 *identity matrix*. The 3×3 identity matrix has 1's in the $(1, 1)$, $(2, 2)$, and $(3, 3)$ positions and 0's elsewhere. The *n* by *n identity matrix* is the $n \times n$ matrix with 1's as the diagonal entries from position $(1, 1)$ to (n, n) and zeros everywhere else. We use I to stand for the $n \times n$ identity matrix (ignoring the fact that for different values of n we are actually talking about different matrices.) Then clearly $IX = XI = X$.

Now we can see why it was possible to cancel out the matrix

$$B = \begin{bmatrix} 2 & 0 \\ 0 & 2 \end{bmatrix}$$

in the equations above; namely, there is a matrix B^{-1} such that $B^{-1}B = BB^{-1} = I$. We say B is invertible and the matrix

$$\begin{bmatrix} \frac{1}{2} & 0 \\ 0 & \frac{1}{2} \end{bmatrix}$$

is the inverse of B.

The Inverse of a Matrix

In general, we say a square matrix A is *invertible* if there is a matrix A^{-1} such that

$$AA^{-1} = A^{-1}A = I.$$

In this situation we say the matrix A^{-1} is the *inverse* to the matrix A. (Implicit in the use of the word "the" is that A^{-1} is completely determined by A.) We have seen that

$$\begin{bmatrix} 2 & 0 \\ 0 & 0 \end{bmatrix}$$

does not have an inverse but the matrix

$$\begin{bmatrix} 2 & 0 \\ 0 & 2 \end{bmatrix}$$

does have an inverse.

How do we determine whether a matrix has an inverse and what the inverse is if it exists?   An analysis of the 2×2 case will show us what happens in general.

The inverse matrix A^{-1} of a matrix A will be a solution to the matrix equation $AX = I$ (where X is a square matrix of unknowns).   For 2×2 matrices we have

$$\begin{bmatrix} a & b \\ c & d \end{bmatrix} \begin{bmatrix} x_1 & x_3 \\ x_2 & x_4 \end{bmatrix} = \begin{bmatrix} 1 & 0 \\ 0 & 1 \end{bmatrix}.$$

Thus

$$\begin{bmatrix} a & b \end{bmatrix} \begin{bmatrix} x_1 \\ x_2 \end{bmatrix} = 1$$

$$\begin{bmatrix} c & d \end{bmatrix} \begin{bmatrix} x_1 \\ x_2 \end{bmatrix} = 0$$

$$\begin{bmatrix} a & b \end{bmatrix} \begin{bmatrix} x_3 \\ x_4 \end{bmatrix} = 0$$

$$\begin{bmatrix} c & d \end{bmatrix} \begin{bmatrix} x_3 \\ x_4 \end{bmatrix} = 1.$$

This gives us the two matrix equations

$$\begin{bmatrix} a & b \\ c & d \end{bmatrix} \begin{bmatrix} x_1 \\ x_2 \end{bmatrix} = \begin{bmatrix} 1 \\ 0 \end{bmatrix}$$

and

$$\begin{bmatrix} a & b \\ c & d \end{bmatrix} \begin{bmatrix} x_3 \\ x_4 \end{bmatrix} = \begin{bmatrix} 0 \\ 1 \end{bmatrix}.$$

Thus if we could solve both systems of equations to get the four numbers x_1, x_2, x_3 and x_4, then we would have our inverse matrix, but if we couldn't solve both systems, then we would not have an inverse matrix. At first glance, this appears to call for two augmented matrices,

$$\begin{bmatrix} a & b & | & 1 \\ c & d & | & 0 \end{bmatrix} \quad \text{and} \quad \begin{bmatrix} a & b & | & 0 \\ c & d & | & 1 \end{bmatrix} \tag{3.1}$$

to be row reduced until the part to the left of the line is in row–reduced echelon form. However, we can perform these two row reductions at the same time and achieve the same result by starting with

$$\begin{bmatrix} a & b & | & 1 & 0 \\ c & d & | & 0 & 1 \end{bmatrix} \tag{3.2}$$

and row reducing until the part to the left of the line is in row–reduced echelon form. The first column to the right of the line in the reduced form of (3.2) will be the column to the right of the line in the reduced form of the first matrix in (3.1). The second column will be the column to the right of the line in the reduced form of the second matrix of (3.1).

Not only does this observation reduce the amount of computation needed; it turns out to give us a straightforward test for invertibility. Notice that after any row operation on (3.2), *each* row will have a nonzero entry somewhere to the right of the line. Further, if we continue performing row operations, we will continue to have at least one nonzero entry in each row to the right of the line. (Intuitively, this last remark probably seems clear. There are several proofs that it is correct; it also follows from later material in the chapter.)

Now once we have row reduced (3.2) so that the part on the left is in row–reduced echelon form, there can be no zero rows on the left if A is invertible. Why? Since each row on the right is nonzero, a zero row on the left would lead to the inconsistent equation $0 = e \neq 0$ when we write down the equation corresponding to the nonzero entry e in that row. Thus if A is invertible, we will have an identity matrix to the left of the line for our row–reduced echelon form. But then as we learned in the last section, x_1 will be the first entry in the first column to the right of the line, x_2 the second entry in the first column to the right of the line, x_3 the first entry in the second column to the right of the line and x_4 the second entry in the second column to the right of the line. Since x_1, x_2, x_3 and x_4 are the entries of A^{-1}, this means that A^{-1} will *be* the matrix to the right of the line. We summarize these observations as a theorem.

How to Find the Inverse Matrix

Theorem 3.2. A matrix has an inverse if and only if it can be row reduced to an identity. If A has an inverse, then A^{-1} may be found by row reducing the augmented matrix $[A|I]$ so that A is row–reduced to an identity; then the matrix to the right of the line in the row–reduced augmented matrix is A^{-1}.

Proof. Although our remarks above dealt with 2×2 matrices, they are equally valid for any square matrices. ∎

Example 3.1. Determine if either of the matrices below is invertible and find the inverse(s).

$$A = \begin{bmatrix} 1 & -1 \\ 2 & 3 \end{bmatrix} \quad B = \begin{bmatrix} 1 & -2 \\ -2 & 4 \end{bmatrix}$$

We form and reduce the augmented matrix

$$\left[\begin{array}{cc|cc} 1 & -1 & 1 & 0 \\ 2 & 3 & 0 & 1 \end{array}\right] \rightarrow \left[\begin{array}{cc|cc} 1 & -1 & 1 & 0 \\ 0 & 5 & -2 & 1 \end{array}\right] \rightarrow \left[\begin{array}{cc|cc} 1 & -1 & 1 & 0 \\ 0 & 1 & -\frac{2}{5} & \frac{1}{5} \end{array}\right] \rightarrow$$

$$\left[\begin{array}{cc|cc} 1 & 0 & \frac{3}{5} & \frac{1}{5} \\ 0 & 1 & -\frac{2}{5} & \frac{1}{5} \end{array}\right].$$

Thus, A is invertible and A^{-1} is the matrix.

$$\begin{bmatrix} \frac{3}{5} & \frac{1}{5} \\ -\frac{2}{5} & \frac{1}{5} \end{bmatrix}.$$

To check our work, we form the product AA^{-1}.

$$AA^{-1} = \begin{bmatrix} 1 & -1 \\ 2 & 3 \end{bmatrix}\begin{bmatrix} \frac{3}{5} & \frac{1}{5} \\ -\frac{2}{5} & \frac{1}{5} \end{bmatrix} = \begin{bmatrix} \frac{3}{5} + \frac{2}{5} & \frac{1}{5} - \frac{1}{5} \\ \frac{6}{5} - \frac{6}{5} & \frac{2}{5} + \frac{3}{5} \end{bmatrix} = \begin{bmatrix} 1 & 0 \\ 0 & 1 \end{bmatrix}.$$

For B we form and reduce the augmented matrix

$$\left[\begin{array}{cc|cc} 1 & -2 & 1 & 0 \\ -2 & 4 & 0 & 1 \end{array}\right] \rightarrow \left[\begin{array}{cc|cc} 1 & -2 & 1 & 0 \\ 2-2 & -4+4 & 2 & 1 \end{array}\right] \rightarrow \left[\begin{array}{cc|cc} 1 & -2 & 1 & 0 \\ 0 & 0 & 2 & 1 \end{array}\right].$$

Now we have a row of 0's on the left not matched by a row of 0's on the right. Therefore B has no inverse.

Inverses of Upper Triangular Matrices

Example 3.2.   Determine whether the matrix A below has an inverse and, if so, find its inverse.

$$A = \begin{bmatrix} 1 & 2 & 3 \\ 0 & 2 & 4 \\ 0 & 0 & 5 \end{bmatrix}.$$

We form and reduce the augmented matrix

$$[A|I] = \left[\begin{array}{ccc|ccc} 1 & 2 & 3 & 1 & 0 & 0 \\ 0 & 2 & 4 & 0 & 1 & 0 \\ 0 & 0 & 5 & 0 & 0 & 1 \end{array}\right] \rightarrow \left[\begin{array}{ccc|ccc} 1 & 2 & 3 & 1 & 0 & 0 \\ 0 & 1 & 2 & 0 & \frac{1}{2} & 0 \\ 0 & 0 & 1 & 0 & 0 & \frac{1}{5} \end{array}\right] \rightarrow$$

$$\left[\begin{array}{ccc|ccc} 1 & 2 & 0 & 1 & 0 & -\frac{3}{5} \\ 0 & 1 & 0 & 0 & \frac{1}{2} & -\frac{2}{5} \\ 0 & 0 & 1 & 0 & 0 & \frac{1}{5} \end{array}\right] \rightarrow \left[\begin{array}{ccc|ccc} 1 & 0 & 0 & 1 & -1 & \frac{1}{5} \\ 0 & 1 & 0 & 0 & \frac{1}{2} & -\frac{2}{5} \\ 0 & 0 & 1 & 0 & 0 & \frac{1}{5} \end{array}\right].$$

Thus A is invertible and

$$A^{-1} = \begin{bmatrix} 1 & -1 & \frac{1}{5} \\ 0 & \frac{1}{2} & -\frac{2}{5} \\ 0 & 0 & \frac{1}{5} \end{bmatrix}. \ \blacksquare$$

The matrix A in Example 3.2 is called *upper* triangular. An *upper triangular* matrix is an $n \times n$ matrix whose nonzero entries are all on or above the main diagonal positions $(1, 1)$, $(2, 2)$, . . . , (n, n).

By using the methods of the example, we can prove

Theorem 3.3.   If the diagonal entries of an upper triangular matrix are all nonzero, then the matrix has an inverse.

Proof.   The methods of Example 3.2 apply in general. $\blacksquare$

Inverse Matrices and Solutions of Equations

Remember that we began our study of inverse matrices by asking about cancellation of matrices in equations of the form

$$AX = AY.$$

Now suppose X is a column vector of n unknowns, B a column vector of n numbers and A an $n \times n$ invertible matrix. Then the matrix equation

$$AX = B$$

corresponds to the following system of n equations in n unknowns:

$$A_{11}x_1 + A_{12}x_2 + \cdots + A_{1n}x_n = b_1$$
$$A_{21}x_1 + A_{22}x_2 + \cdots \qquad\qquad = b_2$$
$$\vdots$$
$$A_{n1}x_1 + A_{n2}x_2 + \cdots + A_{nn}x_n = b_n.$$

Multiplying the matrix equation $AX = B$ by A^{-1} gives

$$A^{-1}(AX) = A^{-1}B$$
$$(A^{-1}A)X = A^{-1}B$$
$$IX = A^{-1}B$$
$$X = A^{-1}B.$$

$A^{-1}B$ is a column vector. The value of x_1 is the first entry of $A^{-1}B$; the value of x_i is the i–th entry of $A^{-1}B$. Thus when A is invertible, we have one and only one solution X to the matrix equation $AX = B$. This suggests one more result.

Theorem 3.4. The matrix equation $AX = B$, in which A is a square matrix and B a column matrix, has a unique solution (i.e., is satisfied by one and only one vector X) if and only if A has an inverse. The unique solution is $X = A^{-1}B$.

Proof. If A is invertible, it is clear that the equations $X = A^{-1}B$ and $AX = B$ are equivalent, so $X = A^{-1}B$ is the one and only solution to the equation. If the equation has one and only one solution, it may be found by row reducing $[A|B]$. If A could not be row reduced to I, then the number of pivotal variables would be less than the number of variables, so there would be infinitely many solutions or no solutions. Thus A must be row reducible to I, and by Theorem 3.2, A has an inverse. ∎

Example 3.3. Solve the system of equations

$$\begin{aligned} x_1 - x_2 &= 2 \\ 2x_1 + 3x_2 &= -1. \end{aligned}$$

The system of equations is equivalent to

$$\begin{bmatrix} 1 & -1 \\ 2 & 3 \end{bmatrix} \begin{bmatrix} x_1 \\ x_2 \end{bmatrix} = \begin{bmatrix} 2 \\ -1 \end{bmatrix}.$$

As shown in Example 3.1,

$$\begin{bmatrix} 1 & -1 \\ 2 & 3 \end{bmatrix}^{-1} = \begin{bmatrix} \frac{3}{5} & \frac{1}{5} \\ -\frac{2}{5} & \frac{1}{5} \end{bmatrix}.$$

so

$$\begin{bmatrix} x_1 \\ x_2 \end{bmatrix} = \begin{bmatrix} \frac{3}{5} & \frac{1}{5} \\ -\frac{2}{5} & \frac{1}{5} \end{bmatrix} \begin{bmatrix} 1 & -1 \\ 2 & 3 \end{bmatrix} \begin{bmatrix} x_1 \\ x_2 \end{bmatrix} = \begin{bmatrix} \frac{3}{5} & \frac{1}{5} \\ -\frac{2}{5} & \frac{1}{5} \end{bmatrix} \begin{bmatrix} 2 \\ -1 \end{bmatrix} = \begin{bmatrix} 1 \\ -1 \end{bmatrix} = \begin{bmatrix} x_1 \\ x_2 \end{bmatrix}. \quad ∎$$

EXERCISES

1. Using 2×2 matrices, show that $(A + B) + C = A + (B + C)$.

2. Applying the method we used for the commutative law, prove the associative law $A + (B + C) = (A + B) + C$ for *all* matrices.

3. Using 2×2 matrices, show that $A(B + C) = AB + AC$.

4. Applying methods like the one used to prove Theorem 3.1, prove that whenever the matrix product $A(B + C)$ is defined,

$$A(B + C) = AB + AC.$$

5. True or False. (Prove if true and give a counter–example if false.)

$$\begin{bmatrix} a & b \\ c & d \end{bmatrix}^{-1} = \begin{bmatrix} \dfrac{1}{a} & \dfrac{1}{b} \\ \dfrac{1}{c} & \dfrac{1}{d} \end{bmatrix}.$$

6. For what values of a and b does the 2×2 matrix

$$\begin{bmatrix} a & 0 \\ 0 & b \end{bmatrix}$$

commute (in products) with all 2×2 matrices?

7. Find an inverse for each of the matrices below.

(a) $\begin{bmatrix} 1 & 1 \\ 0 & 1 \end{bmatrix}$ (b) $\begin{bmatrix} 1 & -1 \\ 1 & 1 \end{bmatrix}$ (c) $\begin{bmatrix} 2 & -1 \\ 3 & 5 \end{bmatrix}$

(d) $\begin{bmatrix} 4 & 8 \\ 6 & 3 \end{bmatrix}$ (e) $\begin{bmatrix} 1 & 2 \\ 0 & 1 \end{bmatrix}$ (f) $\begin{bmatrix} 1 & 0 \\ 3 & 1 \end{bmatrix}$

8. Find the inverse of

$$\begin{bmatrix} 1 & a & b \\ 0 & 1 & c \\ 0 & 0 & 1 \end{bmatrix}.$$

9. Write out a proof of Theorem 3.3.

10. Test the following matrices to determine if they have inverses and, if so, find the inverses.

(a) $\begin{bmatrix} 2 & 4 & 12 \\ 0 & 4 & 8 \\ 0 & 0 & 6 \end{bmatrix}$ (b) $\begin{bmatrix} 1 & 0 & 1 \\ 1 & 1 & 0 \\ 0 & 1 & 1 \end{bmatrix}$ (c) $\begin{bmatrix} 1 & 0 & 1 \\ 1 & 1 & 0 \\ -1 & 2 & -3 \end{bmatrix}$

(d) $\begin{bmatrix} 1 & 1 & 1 \\ 0 & 1 & -1 \\ -1 & -2 & 0 \end{bmatrix}$ (e) $\begin{bmatrix} 1 & 2 & 3 \\ 1 & 1 & 0 \\ 0 & 1 & 1 \end{bmatrix}$ (f) $\begin{bmatrix} -2 & 0 & 4 \\ 3 & 0 & 2 \\ 1 & 1 & 1 \end{bmatrix}.$

11. Test the matrix below to determine whether it has an inverse and, if so, find it.

$$\begin{bmatrix} 1 & 1 & 1 & 1 \\ 1 & 2 & 3 & 4 \\ 1 & 4 & 9 & 16 \\ 1 & 8 & 27 & 64 \end{bmatrix}.$$

12. Find the solutions to the following systems of equations. (Hint: This should not involve a great deal of work.)

(a) $x - y = 3$
     $2x + 3y = 5$

(b) $x - y = 5$
     $2x + 3y = 6$

(c) $x - y = -1$
     $2x + 3y = 5$

(d) $x - y = 15$
     $2x + 3y = 10$

(e) $x - y = 10$
     $2x + 3y = -10$

(f) $x - y = 20$
     $-2x - 3y = 25.$

SECTION 4   DETERMINANTS

Why do We Need Determinants?

It is somewhat frustrating not to be able to just look at a matrix and determine at once whether or not it is invertible. After all, a number r has an inverse $\dfrac{1}{r} = r^{-1}$ if and only if $r \neq 0$. But the matrix.

$$\begin{bmatrix} 7 & 0 \\ 0 & 0 \end{bmatrix}$$

is nonzero and yet has no inverse. The matrix

$$\begin{bmatrix} 1 & -2 \\ -3 & 6 \end{bmatrix}$$

has all nonzero elements and yet it does not have an inverse. Remarkably, there is a single number we can determine from a square matrix that will be zero if and only if the matrix has no inverse. This number is called the determinant of the matrix. In the case of a 1×1 matrix, its determinant is just the number inside the matrix. We can discover the determinant of a 2×2 matrix by working out a formula for the inverse of a general 2×2 matrix.

Determinants of 2 by 2 Matrices

Example 4.1. Find the inverse of the matrix

$$\begin{bmatrix} a & b \\ c & d \end{bmatrix}$$

making any assumptions about a, b, c and d that are convenient. We write

$$\begin{bmatrix} a & b & 1 & 0 \\ c & d & 0 & 1 \end{bmatrix} \rightarrow \begin{bmatrix} 1 & \dfrac{b}{a} & \dfrac{1}{a} & 0 \\ c & d & 0 & 1 \end{bmatrix} \rightarrow \begin{bmatrix} 1 & \dfrac{b}{a} & \dfrac{1}{a} & 0 \\ 0 & d - \dfrac{cb}{a} & -\dfrac{c}{a} & 1 \end{bmatrix} \rightarrow$$

$$\begin{bmatrix} 1 & \dfrac{b}{a} & \dfrac{1}{a} & 0 \\ 0 & 1 & \dfrac{-c}{ad - bc} & \dfrac{a}{ad - bc} \end{bmatrix} \rightarrow \begin{bmatrix} 1 & 0 & \dfrac{1}{a} + \dfrac{bc/a}{ad - bc} & \dfrac{-b}{ad - bc} \\ 0 & 1 & \dfrac{-c}{ad - bc} & \dfrac{a}{ad - bc} \end{bmatrix}$$

$$= \begin{bmatrix} 1 & 0 & \dfrac{d}{ad - bc} & \dfrac{-b}{ad - bc} \\ 0 & 1 & \dfrac{-c}{ad - bc} & \dfrac{a}{ad - bc} \end{bmatrix}.$$

Now by assuming a and $ad - bc$ were not zero so we could divide by them, we found that

$$\begin{bmatrix} a & b \\ c & d \end{bmatrix}^{-1} = \begin{bmatrix} \dfrac{d}{ad - bc} & \dfrac{-b}{ad - bc} \\ \dfrac{-c}{ac - bc} & \dfrac{a}{ad - bc} \end{bmatrix}. \quad \blacksquare$$

In our example, we assumed that not one but two numbers were not 0 and from that found a formula for the inverse of our 2×2 matrix. Notice, though, that the only number that actually appears in a denominator in the final result is $ad - bc$. Thus perhaps there is another method of deriving the formula in which we need only assume that $ad - bc$ is nonzero. (Can you find such a method?) In fact

$$\begin{bmatrix} a & b \\ c & d \end{bmatrix} \begin{bmatrix} \dfrac{d}{ad - bc} & \dfrac{-b}{ad - bc} \\ \dfrac{-c}{ad - bc} & \dfrac{a}{ad - bc} \end{bmatrix} = \begin{bmatrix} \dfrac{ad - bc}{ad - bc} & \dfrac{-ab + ab}{ad - bc} \\ \dfrac{cd - cd}{ad - bc} & \dfrac{-bc + ad}{ad - bc} \end{bmatrix}$$

$$= \begin{bmatrix} 1 & 0 \\ 0 & 1 \end{bmatrix}.$$

This proves

Theorem 4.1.   The 2×2 matrix

$$\begin{bmatrix} a & b \\ c & d \end{bmatrix}$$

has an inverse if and only if $ad - bc \neq 0$.

Proof: The computation of matrix products above shows that if $ad - bc \neq 0$ the matrix has an inverse. If $ad - bc = 0$ then our matrix may be reduced to a matrix with a row of zeros and so has no inverse. ∎

Thus to determine whether or not our 2×2 matrix has an inverse, we need only check the single number $ad - bc$. Using the A_{ij} notation for the entries of a matrix, we can rewrite this to say that a 2×2 matrix A has an inverse if and only if $A_{11}A_{22} - A_{12}A_{21}$ is nonzero. We say that $A_{11}A_{22} - A_{12}A_{21}$ is the *determinant* of the 2×2 matrix A.

Properties of Determinants

Conceivably we could repeat the process above for 3×3 matrices and so on to find formulas for numbers we would want to call determinants. However, the computations would likely be unbearably messy (and boring as well). Instead, let us try to find a method of analyzing what properties this determinant function should have and see if we can use the properties to determine what the determinant is—and in fact to show that there is such a determinant that can be used to determine whether a matrix is invertible.

Now we know that a matrix is invertible if and only if it can be row reduced to an identity, so applying an elementary row operation to a matrix should not change its determinant from nonzero to zero or vice–versa. Thus one way of describing a determinant is describing what happens to it when we perform an elementary row operation on the matrix. By using the formula for 2×2 determinants, we can see what happens to determinants when we perform these operations on 2×2 matrices. From this perhaps we could say what should happen to a determinant of an $n \times n$ matrix when we perform an elementary row operation. The three elementary row operations are:

(1) Multiply a row by a real number $r \neq 0$.
(2) Add a multiple of row i to row j.
(3) Interchange row i and row j.

Let us see what each of these operations does to the determinant of a 2×2 matrix. We use det to stand for determinant. First, with operation 1,

$$\det \begin{bmatrix} ra & rb \\ c & d \end{bmatrix} = rad - rbc = r(ad - bc) = r\det \begin{bmatrix} a & b \\ c & d \end{bmatrix}$$

and

$$\det \begin{bmatrix} a & b \\ rc & rd \end{bmatrix} = ard - brc = r(ad - bc) = r\det \begin{bmatrix} a & b \\ c & d \end{bmatrix}.$$

Thus we should expect that if det stands for the determinant of an $n \times n$ matrix, then for any real number r,

$$\det \begin{bmatrix} \text{---}R_1\text{---} \\ \text{---}R_2\text{---} \\ \cdot \\ \cdot \\ \cdot \\ \text{---}rR_i\text{---} \\ \cdot \\ \cdot \\ \cdot \\ \text{---}R_n\text{---} \end{bmatrix} = r \det \begin{bmatrix} \text{---}R_1\text{---} \\ \text{---}R_2\text{---} \\ \cdot \\ \cdot \\ \cdot \\ \text{---}R_i\text{---} \\ \cdot \\ \cdot \\ \cdot \\ \text{---}R_n\text{---} \end{bmatrix}. \tag{4.1}$$

Here we are using the vertical list of the symbols R_i to stand for the matrix whose rows are the row vectors $R_1, R_2, \ldots, R_n$.

Now what should we expect to happen when we add a multiple of one row to another? Again, for 2×2 matrices

$$\det \begin{bmatrix} a & b \\ ra + c & rb + d \end{bmatrix} = a(rb + d) - b(ra + c)$$

$$= arb + ad - bra - bc = ad - bc = \det \begin{bmatrix} a & b \\ c & d \end{bmatrix}.$$

Surprisingly, this did not change the determinants. You can check that the same thing happens if we add r times the row vector $[c, d]$ to the first row. Thus we expect

$$
\det \begin{bmatrix} \text{—} R_1 \text{—} \\ \vdots \\ \text{—} R_i \text{—} \\ \vdots \\ \text{—} R_j + rR_i \text{—} \\ \vdots \\ \text{—} R_n \text{—} \end{bmatrix} = \det \begin{bmatrix} \text{—} R_1 \text{—} \\ \vdots \\ \text{—} R_i \text{—} \\ \vdots \\ \text{—} R_j \text{—} \\ \vdots \\ \text{—} R_n \text{—} \end{bmatrix}. \tag{4.2}
$$

What happens to the determinant of a 2×2 matrix if we interchange its rows?

$$
\det \begin{bmatrix} c & d \\ a & b \end{bmatrix} = cb - ad = -(ad - bc) = -\det \begin{bmatrix} a & b \\ c & d \end{bmatrix}.
$$

Thus we expect

$$
\det \begin{bmatrix} \text{—} R_1 \text{—} \\ \vdots \\ \text{—} R_j \text{—} \\ \vdots \\ \text{—} R_i \text{—} \\ \vdots \\ \text{—} R_n \text{—} \end{bmatrix} = -\det \begin{bmatrix} \text{—} R_1 \text{—} \\ \vdots \\ \text{—} R_i \text{—} \\ \vdots \\ \text{—} R_j \text{—} \\ \vdots \\ \text{—} R_n \text{—} \end{bmatrix}. \tag{4.3}
$$

In fact, rule 4.3 is a consequence of rules 4.1 and 4.2. (An outline of a proof is given in the exercises.)

What Does a Zero Row do to Determinants?

Now suppose we have a square matrix M and by row reduction, we get a matrix M' in which row i is a row of 0's. Then by multiplying row i by 0 we don't change M', so rule 4.1 tells us

$$\det M' = 0 \cdot \det M' = 0.$$

But since the elementary row operations on M only multiply $\det M$ by nonzero numbers, $\det M'$ is a nonzero multiple of $\det M$. Thus $\det M$ must be zero also. We have proved the statement below.

> *Theorem 4.2.*   If there is a function det defined on $n \times n$ matrices and satisfying rules 4.1, 4.2 and 4.3, then $\det M = 0$ if M can be row reduced to a matrix with a row of 0's.

Proof.   Given above. ∎

What if Row Reduction Doesn't Give a Row of Zeros?

Now if a square matrix M does not give a row of 0's when it is converted to a row–reduced echelon form matrix M', the matrix M' must be the $n \times n$ identity. By keeping track of the various numbers by which we multiply rows and the number of times we interchange rows as we convert M to I, we will get a list of numbers $r_1, r_2, \ldots, r_k$ such that

$$\det M = \pm r_1 r_2 \cdots r_k \det I.$$

Thus we also need to determine $\det I$. In the case of 2×2 matrices,

$$\det \begin{bmatrix} 1 & 0 \\ 0 & 1 \end{bmatrix} = 1 \cdot 1 - 0 \cdot 0 = 1,$$

and in the case of the 1×1 identity matrix $[1]$, $\det I = 1$, so a natural assumption is

$$\det I = 1. \tag{4.4}$$

Thus if there is a "determinant" function defined on $n \times n$ matrices satisfying rules 4.1–4.4, then we may compute the value of the determinant function by row reducing the matrix. Thus the only value the function can have is the one we get by a row reduction process. This is stated more formally below.

The Uniqueness Theorem

Theorem 4.3.   If there is a determinant function det defined on $n \times n$ matrices satisfying rules 4.1–4.4, then there is only one such function.

Note that Theorem 4.3 contains a big "if"—it doesn't tell us that there *is* a function satisfying rules 4.1–4.4. Rather it tells us that if we find two apparently different ways to define a determinant function on $n \times n$ matrices, then, in fact, these two methods will define the same function. By the way, it tells us that the function defined by

$$\det \begin{bmatrix} a & b \\ c & d \end{bmatrix} = ad - bc$$

is the *only* function on 2×2 matrices satisfying rules 4.1 to 4.4.

Now that we have observed (without proof yet) that a function satisfying rules 4.1 and 4.2 satisfies rule 4.3. To check that a function does satisfy all four rules, therefore, we need only check rules 4.1, 4.2 and 4.4.

Determinants of 3 by 3 Matrices

Example 4.2.   Show that the function defined by

$$\det \begin{bmatrix} a_1 & b_1 & c_1 \\ a_2 & b_2 & c_2 \\ a_3 & b_3 & c_3 \end{bmatrix} = a_1(b_2c_3 - c_2b_3) - a_2(b_1c_3 - c_1b_3) + a_3(b_1c_2 - b_2c_1)$$

satisfies rules 4.1, 4.2 and 4.4 and therefore is a determinant function defined on 3×3 matrices.

Since in an identity matrix a_2 and a_3 will be 0, rule 4.4 should be the easiest to check. In fact, in an identity matrix

$$a_1(b_2c_3 - c_2b_3) = 1(1 \cdot 1 - 0 \cdot 0) = 1.$$

Therefore, rule 4.4 is satisfied. Now suppose we multiply some row, say the first row, by r. Then

$$\det \begin{bmatrix} ra_1 & rb_1 & rc_1 \\ a_2 & b_2 & c_2 \\ a_3 & b_3 & c_3 \end{bmatrix} = ra_1(b_2c_3 - c_2b_3) - a_2(rb_1c_3 - rc_1b_3)$$

$$+ a_3(rb_1c_2 - rc_1b_2)$$
$$= ra_1(b_2c_3 - c_2b_3) - ra_2(b_1c_3 - c_1b_3)$$
$$+ ra_3(b_1c_2 - c_1b_2)$$
$$= r\det \begin{bmatrix} a_1 & b_1 & c_1 \\ a_2 & b_2 & c_2 \\ a_3 & b_3 & c_3 \end{bmatrix}.$$

Exactly the same kind of computation shows that multiplying row 2 or row 3 by r multiplies our function by r.

Now suppose we add r times the first row to the second row. Then we get

$$\det \begin{bmatrix} a_1 & b_1 & c_1 \\ a_2 + ra_1 & b_2 + rb_1 & c_2 + rc_1 \\ a_3 & b_3 & c_3 \end{bmatrix} = a_1(b_2c_3 + rb_1c_3 - b_3c_2 - rb_3c_1)$$

$$- (a_2 + ra_1)(b_1c_3 - c_1b_3) + a_3(b_1c_2 + rb_1c_1 - c_1b_2 - rc_1b_1)$$
$$= a_1(b_2c_3 - b_3c_2) + ra_1b_1c_3 - ra_1b_3c_1 - a_2(b_1c_3 - c_1b_3) - $$
$$ra_1b_1c_3 + ra_1c_1b_3 + a_3(b_1c_2 - c_1b_2) + rb_1c_1a_3 - ra_3c_1b_1$$
$$= a_1(b_2c_3 - b_3c_2) - a_2(b_1c_3 - c_1b_3) + a_3(b_1c_2 - c_1b_2)$$
$$= \det \begin{bmatrix} a_1 & b_1 & c_1 \\ a_2 & b_2 & c_2 \\ a_3 & b_3 & c_3 \end{bmatrix}.$$

The same kind of computation will show that adding r times any row to any other row will not change the value of det M. Thus we have found a function satisfying rules 4.1 to 4.4. Therefore we have the determinant function for 3×3 matrices. ■

Using the Rules to Find Determinants

Example 4.3. Compute the determinant of

$$\begin{bmatrix} 1 & 0 & 0 \\ 2 & 3 & 0 \\ 5 & 4 & 2 \end{bmatrix}.$$

By using our formula, we get

$$\det \begin{bmatrix} 1 & 0 & 0 \\ 2 & 3 & 0 \\ 5 & 4 & 2 \end{bmatrix} = 1(3 \cdot 2 - 0 \cdot 4) + 2(0 \cdot 2 - 0 \cdot 4) + 5(0 \cdot 0 - 0 \cdot 3)$$

$$= 3 \cdot 2 = 6.$$

Now we shall demonstrate a second method for computing this determinant. This method consists of row reducing the matrix and applying rules 4.1–4.4.

Observe that by subtracting row 1 twice from row 2, and then 5 times row 1 from row 3, and finally $\frac{4}{3}$ times row 2 from row 3, we get the matrix

$$\begin{bmatrix} 1 & 0 & 0 \\ 0 & 3 & 0 \\ 0 & 0 & 2 \end{bmatrix}.$$

By rule 2, this matrix has the same determinant as the original one. However, by rules 1 and 4,

$$\det \begin{bmatrix} 1 & 0 & 0 \\ 0 & 3 & 0 \\ 0 & 0 & 2 \end{bmatrix} = 3\det \begin{bmatrix} 1 & 0 & 0 \\ 0 & 1 & 0 \\ 0 & 0 & 2 \end{bmatrix} = 3 \cdot 2\det \begin{bmatrix} 1 & 0 & 0 \\ 0 & 1 & 0 \\ 0 & 0 & 1 \end{bmatrix} = 3 \cdot 2 \cdot 1 = 6.$$

For complicated matrices, this second method of finding determinants, namely row reducing and applying the rules, is easier to use. It also works for 4 × 4, 5 × 5 and bigger matrices. ∎

The formula in Example 4.2 is an example of a much more general formula that allows us to compute determinants of $n \times n$ matrices in terms of determinants of $n - 1$ by $n - 1$ matrices. Notice that each expression in parentheses in the formula is a determinant itself. Using the A_{ij} notation, we can rewrite the formula as:

$$\det \begin{bmatrix} A_{11} & A_{12} & A_{13} \\ A_{21} & A_{22} & A_{23} \\ A_{31} & A_{32} & A_{33} \end{bmatrix} = A_{11}\det \begin{bmatrix} A_{22} & A_{23} \\ A_{32} & A_{33} \end{bmatrix} - A_{21}\det \begin{bmatrix} A_{12} & A_{13} \\ A_{32} & A_{33} \end{bmatrix} \qquad (4.5)$$

$$+ A_{31}\det \begin{bmatrix} A_{12} & A_{13} \\ A_{22} & A_{23} \end{bmatrix} .$$

There is a much more compact way to write the formula. We use $A(i, j)$ to stand for the matrix obtained from A by deleting row i and column j. Then our formula becomes

$$\det A = A_{11}\det A(1, 1) - A_{21}\det A(2, 1) + A_{31}\det A(3, 1).$$

A General Formula for Determinants

This suggests the following formula for the determinants of an $n \times n$ matrix

$$\det A = A_{11}\det A(1, 1) - A_{21}\det A(2, 1) + \cdots + (-1)^{n+1}\det A(n, 1) \qquad (4.6)$$

$$= \sum_{i=1}^{n} (-1)^{i+1}A_{i1}\det A(i, 1).$$

This formula is in fact correct. We can use it to define 4×4 determinants using 3×3 determinants. Then we use it to define 5×5 determinants using 4×4 determinants and so on. However, to show that the formula can be used to define 4×4 determinants, we need an important property of 3×3 determinants which we haven't proved yet, namely the additivity property,

$$\det \begin{bmatrix} R_1 + R_1' \\ R_2 \\ R_3 \end{bmatrix} = \det \begin{bmatrix} R_1 \\ R_2 \\ R_3 \end{bmatrix} + \det \begin{bmatrix} R_1' \\ R_2 \\ R_3 \end{bmatrix}. \qquad (4.7)$$

The corresponding formula for 2×2 determinants is

$$\det \begin{bmatrix} R_1 + R_1' \\ R_2 \end{bmatrix} = \det \begin{bmatrix} R_1 \\ R_2 \end{bmatrix} + \det \begin{bmatrix} R_1' \\ R_2 \end{bmatrix}, \qquad (4.8)$$

or

$$\det \begin{bmatrix} A_{11} + A_{11}' & A_{12} + A_{12}' \\ A_{21} & A_{22} \end{bmatrix} = \det \begin{bmatrix} A_{11} & A_{12} \\ A_{21} & A_{22} \end{bmatrix} + \det \begin{bmatrix} A_{11}' & A_{12}' \\ A_{21} & A_{22} \end{bmatrix}.$$

This rule may be verified by just writing out a formula of the form $ad - bc$ for each determinant.

Now for a 3×3 determinant, we have

$$\det \begin{bmatrix} A_{11} + A'_{11} & A_{12} + A'_{12} & A_{13} + A'_{13} \\ A_{21} & A_{22} & A_{23} \\ A_{31} & A_{32} & A_{33} \end{bmatrix} = (A_{11} + A'_{11})\det \begin{bmatrix} A_{22} & A_{23} \\ A_{32} & A_{33} \end{bmatrix}$$

$$ - A_{21}\det \begin{bmatrix} A_{12} + A'_{12} & A_{13} + A'_{13} \\ A_{32} & A_{33} \end{bmatrix} + A_{31}\det \begin{bmatrix} A_{12} + A'_{12} & A_{13} + A'_{13} \\ A_{22} & A_{23} \end{bmatrix}$$

$$= (A_{11} + A'_{11})\det \begin{bmatrix} A_{22} & A_{23} \\ A_{32} & A_{33} \end{bmatrix} - A_{21}\det \begin{bmatrix} A_{12} & A_{13} \\ A_{32} & A_{33} \end{bmatrix} - A_{21}\det \begin{bmatrix} A'_{12} & A'_{13} \\ A_{32} & A_{33} \end{bmatrix}$$

$$+ A_{31}\det \begin{bmatrix} A_{11} & A_{13} \\ A_{22} & A_{23} \end{bmatrix} + A_{31}\det \begin{bmatrix} A'_{12} & A'_{13} \\ A_{22} & A_{23} \end{bmatrix},$$

where we have used Equation (4.8) twice in the last equality. We may rearrange the terms in this last sum as

$$A_{11}\det \begin{bmatrix} A_{22} & A_{23} \\ A_{32} & A_{33} \end{bmatrix} - A_{21}\det \begin{bmatrix} A_{12} & A_{13} \\ A_{32} & A_{33} \end{bmatrix} + A_{31}\det \begin{bmatrix} A_{11} & A_{13} \\ A_{22} & A_{23} \end{bmatrix}$$

$$+ A'_{11}\det \begin{bmatrix} A_{22} & A_{23} \\ A_{32} & A_{33} \end{bmatrix} - A_{21}\det \begin{bmatrix} A'_{12} & A'_{13} \\ A_{32} & A_{33} \end{bmatrix} + A_{31}\det \begin{bmatrix} A'_{12} & A'_{13} \\ A_{22} & A_{23} \end{bmatrix}$$

$$= \det \begin{bmatrix} A_{11} & A_{12} & A_{13} \\ A_{21} & A_{22} & A_{23} \\ A_{31} & A_{32} & A_{33} \end{bmatrix} + \det \begin{bmatrix} A'_{11} & A'_{12} & A'_{13} \\ A_{21} & A_{22} & A_{23} \\ A_{31} & A_{32} & A_{33} \end{bmatrix},$$

using Equation (4.5) twice. However this proves Equation (4.7).

The Addition Formulas for Determinants

Equations (4.7) and (4.8) are called *addition formulas* for determinants. In words, they say

> **Theorem 4.8.** If matrices A, A', and A'' are exactly the same in all positions but the positions in row i, and if row i of A is the sum of row i of A' and row i of A'', then $\det A = \det A' + \det A''$.

> *Proof.* This was proved for 3×3 matrices with $i = 1$ above. For $i = 2$ or 3, the same proof works. For $n \times n$ matrices, this result may be proved using mathematical induction in conjunction with the result of Theorem 4.5. ∎

Determinants of 4 by 4 Matrices.

Now we show how Equation (4.7) may be used to show that in the case $n = 4$, Equation (4.6) defines determinants of 4×4 matrices in terms of 3×3 determinants. Thus we want to show

$$\det A = A_{11}\det A(1,\ 1) - A_{21}\det A(2,\ 1) \qquad (4.9)$$
$$+ A_{31}\det A(3,\ 1) - A_{41}\det A(4,\ 1).$$

To prove the formula correct, we show that it satisfies rules 4.1, 4.2 (and, therefore, 4.3) and 4.4. Again 4.4 is the easiest to check, because if we use I_4 and I_3 to stand for 4×4 and 3×3 identity matrices, Equation (4.9) becomes the true statement

$$\det I_4 = 1 \cdot \det I_3.$$

Property (4.1) is not difficult to check. If we multiply row i of A by the real number r, we then multiply A_{i1}, by r and *one* row of each $A(k,\ 1)$ with $k \neq i$ by r. This multiplies each $\det A(k,\ 1)$ by r when $k \neq i$. Thus we multiply all terms on the right–hand side of Equation (4.9) by r.

Property 4.2 is the only slightly tricky property to check. We want to show that adding one row of A to another does not change the value of the right–hand side. To avoid notational problems, we assume $i = 1$ and $j = 2$. The proof for general values of i and j is essentially the same.

We use R_1, R_2, R_3 and R_4 to stand for the four rows of A. We use R_1', R_2', R_3' and R_4' to stand for the four row vectors of length 3 formed by deleting the first entry from each R_i. In this notation, Equation (4.9) becomes

$$\det \begin{bmatrix} R_1 \\ R_2 \\ R_3 \\ R_4 \end{bmatrix} = A_{11}\det \begin{bmatrix} R_2' \\ R_3' \\ R_4' \end{bmatrix} - A_{21}\det \begin{bmatrix} R_1' \\ R_3' \\ R_4' \end{bmatrix} + A_{31}\det \begin{bmatrix} R_1' \\ R_2' \\ R_4' \end{bmatrix} - A_{41}\det \begin{bmatrix} R_1' \\ R_2' \\ R_3' \end{bmatrix}. \qquad (4.10)$$

Thus we would have

$$\det \begin{bmatrix} R_1 \\ R_2 + R_1 \\ R_3 \\ R_4 \end{bmatrix} = A_{11}\det \begin{bmatrix} R_2' + R_1' \\ R_3' \\ R_4' \end{bmatrix} - (A_{21} + A_{11})\det \begin{bmatrix} R_1' \\ R_3' \\ R_4' \end{bmatrix}$$

$$+ A_{31}\det \begin{bmatrix} R_1' \\ R_2' + R_1' \\ R_4' \end{bmatrix} - A_{41}\det \begin{bmatrix} R_1' \\ R_2' + R_1' \\ R_3' \end{bmatrix}.$$

$$= A_{11}\det \begin{bmatrix} R_2' \\ R_3' \\ R_4' \end{bmatrix} + A_{11}\det \begin{bmatrix} R_1' \\ R_3' \\ R_4' \end{bmatrix} - A_{21}\det \begin{bmatrix} R_1' \\ R_3' \\ R_4' \end{bmatrix} - A_{11}\det \begin{bmatrix} R_1' \\ R_3' \\ R_4' \end{bmatrix}$$

$$+ A_{31}\det \begin{bmatrix} R_1' \\ R_2' \\ R_4' \end{bmatrix} - A_{31}\det \begin{bmatrix} R_1' \\ R_1' \\ R_4' \end{bmatrix} - A_{41}\det \begin{bmatrix} R_1' \\ R_2' \\ R_3' \end{bmatrix} - A_{41}\det \begin{bmatrix} R_1' \\ R_1' \\ R_3' \end{bmatrix}.$$

However, since the determinant of a matrix with two equal rows is zero (why?), after we cancel the two terms with equal absolute value but opposite sign, we get the right–hand side of Equation 4.10. Thus the function defined by Equation (4.10) satisfies rule 4.2 for determinants and so is a determinant function on 4×4 matrices.

Expanding by Minors for all Sizes

Perhaps the general pattern is becoming clear. To show that Equation (4.6), namely

$$\det A = \sum_{i=1}^{n} (-1)^{i+1}\det A(i, 1), \tag{4.6}$$

defines determinants for $n \times n$ matrices in terms of determinants of $n - 1$ by $n - 1$ matrices, we first show that determinants of $n - 1$ by $n - 1$ matrices satisfy the addition formula (Theorem 4.4). Then using Theorem 4.4 and rules 4.1–4.4 for determinants of $n - 1$ by $n - 1$ matrices, we show the Equation (4.6) defines a function on $n \times n$ matrices satisfying properties 4.1 to 4.4 (and Theorem 4.4). By the principle of mathematical induction, this proves that (4.1) defines determinants for matrices of all sizes.

We refer to Equation (4.6) as *expansion by minors on the first column*. With an appropriate choice of plus and minus signs, we may expand on any column: the proofs all carry over directly. Thus, we have

Theorem 4.5. There is a unique function satisfying rules 4.1–4.4 for determinants of $n \times n$ matrices. The values of this function for $n \times n$ matrices and $n - 1$ by $n - 1$ matrices are related by

$$\det A = \sum_{i=1}^{n} (-1)^{i+j}A_{ij}\det (A(i, j)),$$

where j is any fixed integer between 1 and n and $A(i, j)$ stands for the matrix obtained from A by deleting row i and column j.

Proof. Outlined above. ∎

(The special roles played by rows and columns above turn out not to be special at all—we may mix row and column reductions at will, or expand on rows by fixing i and summing over j in the formula of Theorem 4.5. This is a major result of the next section.)

EXERCISES

1. Find the determinants of the following matrices.

(a) $\begin{bmatrix} 1 & 4 & 5 \\ 0 & 2 & 6 \\ 0 & 0 & 3 \end{bmatrix}$
(b) $\begin{bmatrix} 1 & 1 & 1 \\ 1 & 2 & 3 \\ 1 & 4 & 9 \end{bmatrix}$
(c) $\begin{bmatrix} 1 & 0 & 1 \\ 0 & 1 & 0 \\ 0 & 1 & 1 \end{bmatrix}$

(d) $\begin{bmatrix} 1 & 2 \\ 3 & 4 \end{bmatrix}$
(e) $\begin{bmatrix} 1 & 3 & -1 \\ 2 & 2 & -1 \\ 3 & 1 & -1 \end{bmatrix}$
(f) $\begin{bmatrix} 2 & 4 & 6 \\ 0 & 0 & 8 \\ 0 & 0 & 5 \end{bmatrix}$

2. For arbitrary numbers a_1, a_2 and a_3, what is

$$\det \begin{bmatrix} a_1 & 0 & 0 \\ 0 & a_2 & 0 \\ 0 & 0 & a_3 \end{bmatrix} ?$$

3. The matrix $\text{diag}(a_1, a_2, \ldots, a_n)$ is the matrix with $A_{ii} = a_i$ and $A_{ij} = 0$ if $i \neq j$. What is its determinant?

4. Use elementary row operations to find the determinants of the following matrices.

(a) $\begin{bmatrix} 1 & 0 & 0 & 0 \\ 1 & 2 & 0 & 0 \\ 0 & 2 & 3 & 0 \\ 4 & 3 & 2 & 2 \end{bmatrix}$
(b) $\begin{bmatrix} 1 & 2 & -1 & -2 \\ 1 & 1 & 1 & 1 \\ 1 & 4 & 1 & 4 \\ 1 & 8 & -1 & -8 \end{bmatrix}$
(c) $\begin{bmatrix} 1 & -1 & 0 & 2 \\ 1 & 1 & 1 & 0 \\ -1 & 2 & 1 & 3 \\ -1 & 0 & 0 & 5 \end{bmatrix}$

5. We showed that $\det M = 0$ if M can be row reduced to a matrix with a row of 0's. Why does this mean that a square matrix which is not invertible has determinant equal to 0?

6. When we row reduce a matrix, we can multiply a row by a nonzero number, but never by zero. Why may we conclude that if M is invertible, its determinant is not zero?

7. What is the determinant of

$$\begin{bmatrix} 0 & 0 & 1 \\ 0 & 1 & a \\ 1 & b & c \end{bmatrix} ?$$

8. What is the determinant of

$$\begin{bmatrix} 0 & 0 & a_1 \\ 0 & a_2 & b_1 \\ a_3 & b_2 & b_3 \end{bmatrix} ?$$

9. If you perform the following sequence of row operations on a matrix, the result will be the same as a single elementary row operation. Which single operation is it? Add row i to row j. Subtract the new row j from row i. Multiply the new row i by -1. subtract the new row i from row j.

10. How can problem 9 be used to explain why rule 4.3 is a consequence of rules 4.1 and 4.2?

SECTION 5 ELEMENTARY MATRICES, INVERTIBILITY AND DETERMINANTS

Elementary Matrices

A matrix which we get from an identity matrix by performing an elementary row operation on it is called an *elementary matrix*. We use the symbol $E(i \leftrightarrow j)$ to stand for the matrix obtained from an identity matrix by interchanging rows i and j. $E(r \times i)$ stands for the matrix obtained from the identity by multiplying row i by r, and $E(j + r \times i)$ stands for the matrix obtained from the identity by adding r times row i to row j. Note, just as we don't specify the size of an identity matrix I, we don't specify the size of an elementary matrix except by the context. Some 3×3 examples are:

$$E(1 \leftrightarrow 2) = \begin{bmatrix} 0 & 1 & 0 \\ 1 & 0 & 0 \\ 0 & 0 & 1 \end{bmatrix}, E(r \times 2) = \begin{bmatrix} 1 & 0 & 0 \\ 0 & r & 0 \\ 0 & 0 & 1 \end{bmatrix}, E(3 + r \times 2) = \begin{bmatrix} 1 & 0 & 0 \\ 0 & 1 & 0 \\ 0 & r & 1 \end{bmatrix}.$$

Note that

$$E(1 \leftrightarrow 2)A = \begin{bmatrix} 0 & 1 & 0 \\ 1 & 0 & 0 \\ 0 & 0 & 1 \end{bmatrix} \begin{bmatrix} A_{11} & A_{12} & A_{13} \\ A_{21} & A_{22} & A_{23} \\ A_{31} & A_{32} & A_{33} \end{bmatrix} = \begin{bmatrix} A_{21} & A_{22} & A_{23} \\ A_{11} & A_{12} & A_{13} \\ A_{31} & A_{32} & A_{33} \end{bmatrix},$$

$$E(r \times 2)A = \begin{bmatrix} 1 & 0 & 0 \\ 0 & r & 0 \\ 0 & 0 & 1 \end{bmatrix} \begin{bmatrix} A_{11} & A_{12} & A_{13} \\ A_{21} & A_{22} & A_{23} \\ A_{31} & A_{32} & A_{33} \end{bmatrix} = \begin{bmatrix} A_{11} & A_{12} & A_{13} \\ rA_{21} & rA_{22} & rA_{23} \\ A_{31} & A_{32} & A_{33} \end{bmatrix}$$

and

$$E(3 + 2 \times r) = \begin{bmatrix} 1 & 0 & 0 \\ 0 & 1 & 0 \\ 0 & r & 1 \end{bmatrix} \begin{bmatrix} A_{11} & A_{12} & A_{13} \\ A_{21} & A_{22} & A_{23} \\ A_{31} & A_{32} & A_{33} \end{bmatrix}$$

$$= \begin{bmatrix} A_{11} & A_{12} & A_{13} \\ A_{21} & A_{22} & A_{23} \\ rA_{21} + A_{31} & rA_{22} + A_{32} & rA_{23} + A_{33} \end{bmatrix}.$$

This suggests the following theorem.

Elementary Matrices and Elementary Row Operations

Theorem 5.2.   Multiplying a matrix A on the left by an elementary matrix E yields the same result as performing on A the row operation performed on I to get E.

Proof.   Row i of EA is the matrix product of the i-th row of E with A. Thus if the i-th row of E is the j-th row of the identity matrix, then the i-th row of EA is the j-th row of A.   If the i-th row of E is r times the i-th row of the identity matrix, then the i-th row of EA is r times the i-th row of A. Finally, if the i-th row of E is the i-th row of I plus r times the j-th row of E, then the i-th row of EA is the i-th row of A plus r times the j-th row of A.   This describes all the rows of EA, and the only rows different from those of A correspond to the rows of E different from those of the identity.   ∎

Elementary Matrices and Determinants

The relationship between elementary matrices and determinants is given by the next two theorems.

Theorem 5.2.   The determinants of elementary matrices are, respectively,

$$\det E(i \leftrightarrow j) = -1$$
$$\det E(r \times i) = r$$
$$\det E(j + r \times i) = 1.$$

Proof.   Simply apply rules 4.1–4.4 of determinants.   ∎

Theorem 5.3.   If E is an elementary matrix, then

$$\det EA = \det E \det A.$$

Proof.   Apply Theorems 5.1, 5.2 and rules 4.1–4.4 of determinants.   ∎

Recall that we developed determinants as a way of testing a matrix for invertibility.   It turns out that invertible matrices are exactly those that can be factored as a product of elementary matrices.

Factoring Invertible Matrices into Elementary Matrices

Theorem 5.4.   A matrix is invertible if and only if it is a product of elementary matrices.

Proof.   First, note that $E(i \leftrightarrow j)^{-1} = E(i \leftrightarrow j)$, $E(r \times i)^{-1} = E(r^{-1} \times i)$ and $E(j + r \times i)^{-1} = E(j - r \times i)$.   These all follow from Theorem 5.1.   Next note that if the matrices A and B are invertible, then $B^{-1}A^{-1}$ is the inverse of AB (because $B^{-1}A^{-1}AB = ABB^{-1}A^{-1} = I$), so a product of invertible matrices is invertible.   Thus a product of elementary matrices is invertible.

Now assume A is invertible.   We have observed this means A can be row reduced to an identity, so by Theorem 5.1 we know there is a sequence of elementary matrices $E_1, E_2, \ldots, E_k$ such that

$$E_1 E_2 \cdots E_k A = I.$$

Then $A = E_k^{-1} E_{k-1}^{-1} \cdots E_2^{-1} E_1^{-1}$, and by our remarks at the beginning of the proof, the inverse of an elementary matrix is elementary.   Therefore, A is a product of elementary matrices.   ∎

Determinants and Invertibility

Theorem 5.5.   A square matrix is invertible if and only if its determinant is nonzero.

Proof.   If A is invertible, then by Theorems 5.3 and 5.4, its determinant is a product of nonzero numbers, and so is nonzero itself.

If A is not invertible then it may be row reduced to a matrix A' with a row of 0's, so that $A = E_1 E_2 \cdots E_k A'$ for some elementary matrices E_i.   Thus

$$\det A = \det E_1 \cdots \det E_k \det A' = 0.   ∎$$

The Product Theorem for Determinants

Theorem 5.6.   If A and B are square matrices of the same size, then $\det AB = \det A \det B$.

Proof. If A is invertible, then $AB = E_1E_2 \cdots E_kB$ and by Theorem 5.3, det $AB =$ det E_1det $E_2 \cdots$ det $E_k$det $B =$ det Adet B. If A is not invertible, then as in 5.5, there is a matrix A' with a row of 0's so that $A' = E_1E_2 \cdots E_kA$. Since A' has a row of 0's, $A'B$ has a row of 0's. But

$$AB = (E_k^{-1}E_{k-1}^{-1} \cdots E_1^{-1}E_1E_2 \cdots E_k)(AB)$$
$$= (E_k^{-1}E_{k-1}^{-1} \cdots E_1^{-1})A'B$$

and by Theorem 5.3,

$$\det AB = (\det E_k^{-1} \cdots \det E_1^{-1})\det (A'B) = 0 = \det A \det B. \ \blacksquare$$

Determinants and Transposes

Note that the transpose of an elementary matrix is an elementary matrix. In fact $E(r \times i)^t = E(r \times i)$ and $E(i \leftrightarrow j)^t = E(i \leftrightarrow j)$. It turns out that the transpose of a product of matrices is the product of the transposed matrices in the reverse order. That is

Theorem 5.7. $(AB)^t = B^tA^t$.

Proof.   The i, j entry of $(AB)^t$ is the j, i entry of AB. The i, j entry of B^tA^t is

$$\sum_{k=1}^{n} B_{ik}^t A_{kj}^t,$$

where n is the common dimension of A and B. But

$$\sum_{k=1}^{n} B_{ik}^t A_{kj}^t = \sum_{k=1}^{n} B_{ki}A_{jk} = \sum_{k=1}^{n} A_{jk}B_{ki} = (AB)_{ji}.$$

Thus the i, j entry of both $(AB)^t$ and B^tA^t is $(AB)_{ji}$. $\blacksquare$

By transposing matrices, we interchange the roles of rows and columns since A^t *is* the matrix whose rows are the columns of A. In particular, each elementary row operation on A corresponds to a similar operation on the columns of A^t. It is this relationship and the next theorem that together tell us we may compute a determinant by performing *both* row and column operations, multiplying together the scalars by which we multiply rows and columns and then multiplying the result by the minus ones that correspond to interchanging two rows or two columns.

Theorem 5.8.   det A^t = det A.

Proof.   If A is invertible it is a product of elementary matrices, so A^t is the product of the transposes of these elementary matrices in the reverse order.   But for an elementary matrix E, det E = det E^t.   Thus by Theorems 5.6 and 5.7, det A^t = det A for an invertible matrix.

However, if A^t is invertible, then it is a product of elementary matrices. Then by Theorem 5.7, A is a product of elementary matrices.   Thus if A is not invertible, then det A = 0, and since A^t cannot be invertible, det A^t = 0, so det A = det A^t.   ∎

The Elementary Column Operations and Determinants

Corollary 5.9.   The determinant function satisfies the three properties

$$\det[C_1, C_2, \ldots rC_i, \ldots, C_n] = r\det[C_1, C_2, \ldots, C_n] \quad (5.1)$$

$$\det[C_1 \ldots C_j \ldots C_i \ldots C_n] = -\det[C_1 \ldots C_i \ldots C_j \ldots C_n] \quad (5.2)$$
$$\det[C_1, C_2, \ldots C_i \ldots rC_i + C_j, \ldots C_n] =$$
$$\det[C_1, C_2, \ldots C_i \ldots C_j \ldots C_n]. \quad (5.3)$$

Example 5.1.   Find the determinant of the matrix

$$\begin{bmatrix} 1 & 4 & 2 & 3 \\ 2 & 4 & 1 & 5 \\ 3 & 0 & 6 & -6 \\ 4 & 0 & 5 & 1 \end{bmatrix}.$$

Using first (4.1) and then (5.1), we get

$$\det\begin{bmatrix} 1 & 4 & 2 & 3 \\ 0 & 4 & 1 & 5 \\ 3 & 0 & 6 & -6 \\ 4 & 0 & 5 & 1 \end{bmatrix} = 3\det\begin{bmatrix} 1 & 4 & 2 & 3 \\ 0 & 4 & 1 & 5 \\ 1 & 0 & 2 & -2 \\ 4 & 0 & 5 & 1 \end{bmatrix} = 12\det\begin{bmatrix} 1 & 1 & 2 & 3 \\ 0 & 1 & 1 & 5 \\ 1 & 0 & 2 & -2 \\ 4 & 0 & 5 & 1 \end{bmatrix}$$

$$= 12\det\begin{bmatrix} 0 & 1 & 2 & 3 \\ -1 & 1 & 1 & 5 \\ 1 & 0 & 2 & -2 \\ 4 & 0 & 5 & 1 \end{bmatrix} = 12\det\begin{bmatrix} 0 & 1 & 2 & 3 \\ -1 & 0 & -1 & 2 \\ 1 & 0 & 2 & -2 \\ 4 & 0 & 5 & 1 \end{bmatrix}$$

$$= 12\det \begin{bmatrix} 0 & 1 & 0 & 0 \\ -1 & 0 & -1 & 2 \\ 1 & 0 & 2 & -2 \\ 4 & 0 & 5 & 1 \end{bmatrix} = 12\det \begin{bmatrix} 0 & 1 & 0 & 0 \\ -1 & 0 & 0 & 0 \\ 1 & 0 & 1 & 0 \\ 4 & 0 & 1 & 9 \end{bmatrix}$$

$$= 108\det \begin{bmatrix} 0 & 1 & 0 & 0 \\ -1 & 0 & 0 & 0 \\ 0 & 0 & 1 & 0 \\ 4 & 0 & 1 & 1 \end{bmatrix} = 108\det \begin{bmatrix} 0 & 1 & 0 & 0 \\ -1 & 0 & 0 & 0 \\ 0 & 0 & 1 & 0 \\ 0 & 0 & 0 & 1 \end{bmatrix}$$

$$= 108\det \begin{bmatrix} 0 & 1 & 0 & 0 \\ 1 & 0 & 0 & 0 \\ 0 & 0 & 1 & 0 \\ 0 & 0 & 0 & 1 \end{bmatrix} = 108\det \begin{bmatrix} 1 & 0 & 0 & 0 \\ 0 & 1 & 0 & 0 \\ 0 & 0 & 1 & 0 \\ 0 & 0 & 0 & 1 \end{bmatrix} = 108 \ \blacksquare$$

A second consequence of Theorem 5.8 is that we may expand determinants along rows as well as columns.

Corollary 5.10. $\det A = \sum\limits_{j=1}^{n} (-1)^{i+j} A_{ij} \det A(i, j)$ for each i.

Proof.   Apply Theorem 5.8 and Theorem 4.5. ■

Example 5.2.   Compute

$$\det \begin{bmatrix} 2 & 0 & 3 & 4 \\ 0 & 3 & 1 & 0 \\ 1 & 5 & 0 & 2 \\ 1 & 1 & 1 & 1 \end{bmatrix}.$$

Expanding on row 2 we get

$$\det \begin{bmatrix} 2 & 0 & 3 & 4 \\ 0 & 3 & 1 & 0 \\ 1 & 5 & 0 & 2 \\ 1 & 1 & 1 & 1 \end{bmatrix} = 3\det \begin{bmatrix} 2 & 3 & 4 \\ 1 & 0 & 2 \\ 1 & 1 & 1 \end{bmatrix} - 1\det \begin{bmatrix} 2 & 0 & 4 \\ 1 & 5 & 2 \\ 1 & 1 & 1 \end{bmatrix}$$

$$= 3(-1)\det \begin{bmatrix} 3 & 4 \\ 1 & 1 \end{bmatrix} + 3(-1) \cdot 2\det \begin{bmatrix} 2 & 3 \\ 1 & 1 \end{bmatrix}$$

$$- 1 \cdot 2\det \begin{bmatrix} 5 & 2 \\ 1 & 1 \end{bmatrix} - 1 \cdot 4\det \begin{bmatrix} 1 & 5 \\ 1 & 1 \end{bmatrix}$$

$$= -3(-1) - 6(-1) - 2(3) - 4(-4)$$
$$= 3 + 6 - 6 + 16 = 19. \ \blacksquare$$

Often a clever mixture of elementary row and column operations and expansion by rows and/or columns will serve best for evaluating a determinant.

Example 5.3. Compute the determinant of Example 5.2 by a mixture of methods.

$$\det \begin{bmatrix} 2 & 0 & 3 & 4 \\ 0 & 3 & 1 & 0 \\ 1 & 5 & 0 & 2 \\ 1 & 1 & 1 & 1 \end{bmatrix} = \det \begin{bmatrix} 2 & 0 & 3 & 0 \\ 0 & 3 & 1 & 0 \\ 1 & 5 & 0 & 0 \\ 1 & 1 & 1 & -1 \end{bmatrix} = -1\det \begin{bmatrix} 2 & 0 & 3 \\ 0 & 3 & 1 \\ 1 & 5 & 0 \end{bmatrix}$$

$$= -1\left(2\det \begin{bmatrix} 3 & 1 \\ 5 & 0 \end{bmatrix} + 3\det \begin{bmatrix} 0 & 3 \\ 1 & 5 \end{bmatrix} \right)$$

$$= -2(-5) - 3(-3) = 19. \quad \blacksquare$$

EXERCISES

1. Suggest two elementary matrices whose product is

$$\begin{bmatrix} 2 & 0 \\ 0 & 3 \end{bmatrix}$$

and multiply them together to check your suggestion.

2. Suggest three elementary matrices whose product is

$$\begin{bmatrix} 2 & 6 \\ 0 & 3 \end{bmatrix}$$

and multiply them together to check your suggestion.

3. In Exercises 1 and 2, analyze what happens if you change the order of multiplication. (Note: There are six possible orders of multiplication in Exercise 2.)

4. Suggest six elementary matrices whose product is

$$\begin{bmatrix} 1 & 2 & 4 \\ 0 & 4 & 6 \\ 0 & 0 & -2 \end{bmatrix}$$

and multiply them together to check your suggestion.

5. (a) On the basis of Exercises 1, 2 and 4, how many elementary matrices will be needed (at most) to factor a 3×3 upper–triangular matrix with nonzero diagonal entries into elementary matrices?

(b) What is the maximum number of matrices needed to factor a 3 × 3 upper–triangular matrix with 1's on the main diagonal into elementary matrices?

6. Write down explicitly the elementary 4 × 4 matrices given symbolically below.

   (a) $E(1, 3)$ (b) $E(4 + 1 \times 2)$ (c) $E(-3 \times 2)$
   (d) $E(4, 2)$ (e) $E(3 - 3 \times 2)$ (f) $E(5 \times 3)$

7. Write down explicitly the matrices which are the *transposes* of the matrices of Exercise 6. What way does this suggest of describing the transpose of an elementary matrix in terms of column operations?

8. Let A be a matrix of integers.
   (a) Explain why det A is an integer.
   (b) Explain why there is an integer a such that det $AA^t = a^2$.

9. Explain why det AB = det BA even if $AB \neq BA$.

10. Explain why the determinant of an upper–triangular matrix is the product of its diagonal entries.

11. Compute the following determinants. (Hint: Appropriate row and column operations simplify the computations.)

(a)
$$\det \begin{bmatrix} 2 & 4 & 0 & 2 \\ 2 & 4 & 0 & 4 \\ 0 & 2 & 6 & 0 \\ 4 & -4 & -12 & -4 \end{bmatrix}$$

(b)
$$\det \begin{bmatrix} 1 & 0 & 1 & 2 \\ 1 & 1 & 2 & 3 \\ -2 & -2 & -3 & -5 \\ 2 & 2 & 3 & 1 \end{bmatrix}$$

(c)
$$\det \begin{bmatrix} 4 & 0 & 3 & 2 \\ 8 & 4 & -2 & 2 \\ 0 & 6 & 3 & 0 \\ 4 & 0 & -3 & 0 \end{bmatrix}$$

12. Compute the following determinants.

(a)
$$\det \begin{bmatrix} 1 & 8 & 3 & 1 \\ 4 & 4 & 0 & 2 \\ 0 & 1 & 0 & 0 \\ 6 & 6 & 1 & 2 \end{bmatrix}$$

(b)
$$\det \begin{bmatrix} 0 & 1 & 1 & 0 \\ 1 & 0 & 1 & 1 \\ 1 & 1 & 0 & 1 \\ 1 & 0 & 1 & 1 \end{bmatrix}$$

(c)
$$\det \begin{bmatrix} 4 & 0 & 2 & 0 \\ -2 & 4 & -2 & 6 \\ 0 & 2 & 0 & 2 \\ 4 & 0 & -2 & -2 \end{bmatrix}$$

13. Which of the following matrices is invertible?

(a)
$$\begin{bmatrix} 1 & 0 & 1 \\ 0 & 1 & 1 \\ 1 & 1 & 1 \end{bmatrix}$$

(b)
$$\begin{bmatrix} 1 & 0 & 1 \\ 0 & 1 & 1 \\ 1 & 1 & 0 \end{bmatrix}$$

(c)
$$\begin{bmatrix} 3 & 2 & 1 & -1 \\ -1 & 2 & 1 & 1 \\ -2 & 1 & 4 & -2 \\ 4 & 3 & -2 & 2 \end{bmatrix}$$

(d)
$$\begin{bmatrix} 1 & 0 & 2 & 4 \\ 0 & -1 & 3 & 1 \\ -3 & 2 & 1 & 0 \\ 4 & 1 & 0 & -1 \end{bmatrix}$$

14. Which of the following systems has one and only one solution?

(a)
$$x_1 \qquad\; + x_3 = 3$$
$$x_2 + x_3 = 4$$
$$x_1 + x_2 - x_3 = 5$$

(b)
$$x_1 \qquad\; + x_3 = 3$$
$$x_2 + x_3 = 4$$
$$-x_1 + x_2 \qquad\; = 5$$

(c)
$$3x_1 + 2x_2 + \; x_3 - \; x_4 = 11$$
$$-x_1 + 2x_2 + \; x_3 + \; x_4 = 6$$
$$-2x_1 + \; x_2 + 4x_3 - 2x_4 = -7$$
$$4x_1 + 3x_2 - 2x_3 + 2x_4 = 8$$

(d)
$$x_1 \qquad\qquad + 2x_3 + x_4 = 6$$
$$- \; x_2 - 3x_3 + x_4 = 5$$
$$-3x_1 + 2x_2 + \; x_3 \qquad\; = 1$$
$$4x_1 + \; x_2 \qquad\quad - x_4 = -7$$

Index